GEO-METRICS IIIm

GEO-METRICS IIIm

AS BASED UPON
HARMONIZATION OF NATIONAL AND
INTERNATIONAL STANDARDS PRACTICES

The Metric Application of Geometric Tolerancing Techniques

LOWELL W. FOSTER

ADDISON-WESLEY PUBLISHING COMPANY, INC.

Reading, Massachusetts   Menlo Park, California   New York
Don Mills, Ontario   Wokingham, England   Amsterdam   Bonn
Sydney   Singapore   Tokyo   Madrid   San Juan
Paris   Seoul   Milan   Mexico City   Taipei

The publisher offers discounts on this book when ordered in quantity for
special sales. For more information please contact:

> Corporate & Professional Publishing Group
> Addison-Wesley Publishing Company
> One Jacob Way
> Reading, Massachusetts   01867

Library of Congress Cataloging-in-Publication Data

Foster, Lowell W.
    Geo-metrics IIIm : the metric application of geometric dimensioning
and tolerancing techniques, as based upon harmonization of national and
international standards practices /Lowell W. Foster
            p. cm
        Includes index.
        ISBN 0-201-63343-4
        1. Drawing-room practice.   2. Tolerance (Engineering)   I. Title.
T352.F6915   1994
604.2'4--dc20   93-5660
                                                            CIP

ISBN 0-201-63343-4

Text printed on recycled and acid-free paper.
2 3 4 5 6 7 8 9 10 -CRW- 969594
Second Printing, April 1994

CONTENTS

Preface

This text has been prepared to promote worldwide use of geometric dimensioning and tolerancing and to up-grade knowledge on the subject. Its title *Geometrics IIIm* attempts to represent and encourage the giant stride being taken by United States industry in its transition toward greater use of national and international standards on this subject. This text also aspires to be of value to all users worldwide as a catalyst for better understanding and to improve the state of the art. This text is a revised and expanded version of its predecessor *Geometrics II.*

Another major objective of this text is to promote and encourage use of the principles involved as an effective engineering, production, and quality-control language or tool which can provide economic and technical advantages.

To further the foregoing objectives, this text is therefore presented in a format suitable for use as a teaching mechanism, both for the classroom and in-industry situations, as well as for providing a daily ready reference for engineering or production work. *Geometrics IIIm* is dedicated to the promoting of standardization of engineering drawing techniques in support of the American National Standard "Dimensioning and Tolerancing," ANSI Y14.5M–1982 and any subsequent issues that follow the date of publication of the text. This text is also dedicated to furthering worldwide efforts through the International Standards Organization (ISO) and its standards and programs (e.g., ISO/TCIO/SC5 "Geometrical Dimensioning and Tolerancing"). To this end, this text attempts to extend the principles of the current standards both national and international, into a harmonization of practices that reflect the imminent international state-of-the-art. See the Introduction of this text for further details on the harmonization of concepts and standards on this subject.

The data contained in this book must be considered advisory and are to be used at the discretion of the reader. As a new feature of this book, "author advisory" comments are inserted at various places in the text. These offer "rules of thumb," helpful hints, suggestions, etc., as based upon the author's experience and assist the reader in determining options and making choices. All such comments adhere within the spirit and sense of the governing standards while refining the content of those standards.

The author wishes to express his gratitude to his many colleagues and professional friends in this field of endeavor around the world who also have made contributions to the state of the art. He also wishes to give credit to the reference documents listed and to those responsible for them.

Finally, the author wishes to express particular gratitude to his wife, Marion, daughter, Janette, and son, John, for their assistance, patience, and understanding for enduring the often untimely sacrifices of family together time. Without their encouragement and support, this work could not have been completed.

Minneapolis, Minnesota
August 1993

L.W.F.

ACKNOWLEDGMENTS

Credit is gratefully given to the following persons and to the documents used as reference material:

1. American National Standard ANSI Y14.5M–1982. "Dimensioning and Tolerancing." American Society of Mechanical Engineers, New York City, and American National Standards Institute, New York City.

2. American National Standard ANSI Y14.5–1973. "Dimensioning and Tolerancing for Engineering Drawings." American Society of Mechanical Engineers, New York City, and American National Standards Institute, New York City.

3. American National Standard USASI Y14.5–1966. "Dimensioning and Tolerancing for Engineering Drawings." American Society of Mechanical Engineers, New York City, and American National Standards Institute, New York City.

4. International Standards Organization Standards, ISO/R1101–1969, *Tolerances of Form and of Position, Part I, Generalities, Symbols, Indications on Drawings:* ISO/DIS2691, 1972. *Technical Drawings—Tolerances of Form and Position, Part II,* ISO/R1660–1971, *Technical Drawings—Tolerances of Form and Position, Part III, Dimensioning and Tolerances of Profiles.*

5. SAE Aerospace–Automotive Drawing Standards, Sections A 6, 7, and 8, September 1963.

6. Military Standard 8C, 16 October 1963.

7. "True Position Dimensioning–Rev. 1962," Bibeau & Sweet, Scintilla Division, Bendix Aviation Corporation, Sidney, New York.

8. "Process Standard 9900011," Sandia Corporation, Albuquerque, New Mexico.

9. "Concepts of the True Position Dimensioning System," Sandia Corporation, Albuquerque, New Mexico.

10. Ordnance Corps, "Standard for Dimensioning and Tolerancing ORD 30–1–7."

11. "Synopsis of MIL–STD–8B," J.V. LaPointe, Honeywell, Inc., Minneapolis, Minnesota.

12. "Fundamentals of Position Tolerance," John V. Liggett, Massey Ferguson, Inc., Detroit, Michigan.

13. Military Standard 8B, 16 November 1959.

14. "Clarification of Problem Areas of MIL–STD–8B Dimensioning and Tolerancing"—code ident 10001 OD15292, Department of the Navy, Bureau of Naval Weapons, 19 June 1961.

15. "A Treatise on Geometric and Positional Dimensioning and Tolerancing," Lowell W. Foster, Honeywell, Inc., Minneapolis, Minnesota. First printing December 1963; revised edition November 1964.

16. "A Treatise on Geometric Dimensioning and Tolerancing," Lowell W. Foster, Honeywell, Inc., Minneapolis, Minnesota. First printing, January 1966; revised edition, first printing, September 1966; second printing, March 1967; third printing, July 1967; fourth printing, February 1968; fifth printing, July 1968.

17. "Geometric Dimensioning and Tolerancing," Lowell W. Foster, Honeywell, Inc., Minneapolis, Minnesota. Society of Automotive Engineers, New York City. SAE Paper 680488.

18. *Geometric Dimensioning and Tolerancing,* A Working Guide, Lowell W. Foster, Minneapolis, Minnesota. Published by Addison-Wesley Publishing Company, 1970.

19. *Geometric Dimensioning and Tolerancing,* A Working Guide with 1973 Addendums, Lowell W. Foster, Minneapolis, Minnesota. Published by Addison-Wesley Publishing Company, 1974.

20. "True Position Tolerancing—Before and After the Fact," Lowell W. Foster, Honeywell, Inc., Minneapolis, Minnesota. American Society of Tool and Manufacturing Engineers, Dearborn, Michigan. ASTME Paper 1068-409.

21. *Geo-Metrics: The Metric Application of Geometric Tolerancing,* Lowell W. Foster, Minneapolis, Minnesota. Published by Addison-Wesley Publishing Company, 1974.

22. *Geo-Metrics: The Metric Application of Geometric Tolerancing,* with Addendum as Based upon ANSI Y14.5M–1982 Practices, Lowell W. Foster, Minneapolis, Minnesota. Published by Addison-Wesley Publishing Company, 1982.

23. *Geo-Metrics II,* with Addendum as Based upon ANSI Y14.5M–1982 Practices, Lowell W. Foster, Minneapolis, Minnesota. Published by Addison-Wesley Publishing Company, 1979.

24. *Geo-Metrics II,* with Addendum as Based upon ANSI Y14.5M–1982 Practices, Lowell W. Foster, Minneapolis, Minnesota. Published by Addison-Wesley Publishing Company, 1983, 1979.

25. The Lowell W. Foster Geometric Dimensioning and Tolerancing Module, "Geo-Calc" and User's Guide, used with Texas Instruments' Compact Computer (CC40). Foster/Marker/Rennix; Lowell W. Foster Associates, Inc., Minneapolis, Minnesota.

26. *Geometrics, The Metric Application of Geometric Tolerancing* as based upon ANSI Y14.5M–1982 practices. Lowell W. Foster, Minneapolis, Minnesota. Published by Addison-Wesley Publishing Company, Revised Edition 1986.

27. *Geo-Metrics II, The Application of Geometric Tolerancing Techniques Using Customary Inch System* as based on ANSI Y14.5M–1982 practices. Lowell W. Foster, Minneapolis, Minnesota. Published by Addison-Wesley Publishing Company, Revised 1986 Edition.

28. *ISO 1101 Technical Drawings, Geometric Tolerancing,* International Organization for Standardization, Geneva, Switzerland.

29. *ISO 5458, Positional Tolerancing,* International Organization for Standardization, Geneva, Switzerland.

30. *ISO 5459, Datums and Datum Systems,* International Organization for Standardization, Geneva, Switzerland.

31. *ISO 2692, Maximum Material Principle,* International Organization for Standardization, Geneva, Switzerland.

32. *ISO 3040, Cones,* International Organization for Standardization, Geneva, Switzerland.

33. *ISO 1660, Profiles,* International Organization for Standardization, Geneva, Switzerland.

34. *ISO 129,* International Organization for Standardization, Geneva, Switzerland.

35. *ISO 406, Linear and Angular Dimensions,* International Organization for Standardization, Geneva, Switzerland.

36. *ISO 2692: 1988 DAM1. Least Material Principle,* International Organization for Standardization, Geneva, Switzerland.

37. *ISO 8015, Fundamental Tolerancing,* International Organization for Standardization, Geneva, Switzerland.

38. *ISO 7083, Symbols Proportions,* International Organization for Standardization, Geneva, Switzerland.

39. *ISO 10578, Projected Tolerance Zone,* International Organization for Standardization, Geneva, Switzerland.

40. *ISO 10579, Non-rigid Parts,* International Organization for Standardization, Geneva, Switzerland.

41. *ISO 1000, SI Units,* International Organization for Standardization, Geneva, Switzerland.

42. *ISO 286, Limits and Fits,* International Organization for Standardization, Geneva, Switzerland.

43. *ISOTR5460, Technical Drawings, Verification Principles,* International Organization for Standardization, Geneva, Switzerland.

44. *ISO 2768-2 General Geometrical Tolerances,* International Organization for Standardization, Geneva, Switzerland.

45. *ISO 2768-1 Tolerances for Linear and Angular Dimensions,* International Organization for Standardization, Geneva, Switzerland.

46. *ISO 9000, 9001, 9002, 9003, 9004, Quality Management and Quality Systems Series,* International Organization for Standardization, Geneva, Switzerland.

INTRODUCTION

BRIEF OVERVIEW OF BASIS AND STATUS OF GD&T

Geometric dimensioning and tolerancing (GD&T) is a means of specifying engineering design and drawing requirements with respect to actual "function" and "relationship" of part features. Furthermore, it is a technique which, properly applied, ensures the most economical and effective production of these features. Thus geometric dimensioning and tolerancing can be considered both an engineering design drawing language and a functional production and inspection technique. Uniform understanding and interpretation among design, production, and inspection groups are the major objectives of the system. This text discusses the subject step by step, focusing on practical application. Before presenting these details, however, we wish to provide the reader with a brief overview of the basis and status of geometric dimensioning and tolerancing.

The authoritative document governing the use of geometric dimensioning and tolerancing in the United States is ANSI/ASME Y14.5M, "Dimensioning and Tolerancing" (latest revision now in preparation). This standard evolved out of a consolidation of earlier standards, ANSI Y14.5M–1982, ANSI Y14.5–1973, USASI Y14.5–1966, ASA Y14.5–1957, SAE Automotive Aerospace Drawing Standards (Sections A6, 7, and 8–September 1963) and MIL–STD-8C, October 1963. This consolidation was accomplished over many years by committee actions representing military, industrial, and educational interests in developing national and international standards. The work of these committees has had and continues to have three prime objectives:

1. to provide a single standard for practices in the United States and harmonize as much as possible USA practices with International Standards Organization (ISO) standards.

2. to update existing practices in keeping with technological advances and extend the principles into new areas of application.

3. to establish a single basis and "voice" for the United States in the interest of international trade, in keeping with the United States' desire to be more active, gain greater influence, and pursue a more extensive exchange of ideas with other nations in the area of international standards development.

The historical evolution of geometric dimensioning and tolerancing in the United States is an interesting story which, however, is not discussed in this text. It suffices to say that the early introduction of functional gaging, giving rise to the possibility of new techniques, along with the growing need for more specifically and economically stated engineering design requirements, has caused its growth. Advancing product sophistication and complexity, computerization, rapid industrial expansion, diversification, etc., have created an environment in which more exacting engineering drawing communication is not only desirable but mandatory for competitive and effective operation.

Updated and expanded practices have been initiated in the latest Y14.5 standard. Further expansion will no doubt occur as growth in the area continues. In the process of extending into new areas, this expansion is confronted by the challenge of ensuring progress without upsetting stability. Rapid advances in this subject, although desirable, must be tempered by the ability to make the transition with no loss of continuity or understanding.

United States coordination and compatibility with international dimensioning and tolerancing practices have been extended significantly in the latest Y14.5M standard.

The symbology and influence from ISO (International Standards Organization) and earlier ABCA (America, Britain, Canada, Australia) documents, activities, and committees have found their way into the latest Y14.5 standard. This influence continues in the ongoing development and use of geometric tolerancing in the United States and as the USA plays a major role in developing, coordinating, and adopting international standards.

Many of the United States industrial concerns and the military have overseas affiliations and markets. Thus, the increasing need for understanding and for more uniform practices throughout the world is evident as Total Quality Management (TQM), World Class Engineering, ISO 9000 quality standards and USA competitiveness require urgent consideration.

This text presents the subject of dimensioning and tolerancing in order of complexity of the details, and attempts to clarify and promote the use of Y14.5. It also emphasizes the importance of the ongoing effort to expand the principles and to more closely incorporate international practices.

THE NEW GD&T—A PERVASIVE TECHNOLOGY

Much has been written about GD&T, Geometric Dimensioning and Tolerancing. The annals of literature on the subject would fill many volumes and still brim over. The author alone has written papers on the subject more times than can be remembered. Many other authors and standards writers of various persuasions and objectives have also contributed over the years; no doubt, all with noble intentions in a desire to be noted, published, and to play a role in the parade of industrial progress. All such efforts are to be applauded for their contributions and many for enriching the sphere of knowledge in significant measure.

It is not without some hesitation that again an effort is made in this introduction to promote the subject. An apology is in order, if the reader deems it appropriate, if this appears as another "sales pitch" GD&T missile. Such an effort sometimes tends to present the material in the sunshine of creation as if introducing GD&T as a new technology and new idea for the first time. The subject has been, of course, propounded long before. Yet by wearing another hat, it must also be remembered that what might be "warmed-over words" to one person may be a dawn of awakening to another. Admittedly, the string must go on with the basic raw materials of GD&T presented sufficiently and clearly in the standards of authority (presently ANSI/ASME Y14.5M) and in supporting papers, texts, and the many necessary other training resources also required.

As in almost any subject, an aura of stagnation can appear to predominate if the same time-worn old GD&T story, as valid as it may be, is continually put forth in editorials, published texts, and papers. We do *not* need to rediscover GD&T again. What is needed is to find ways and means to better implement it and make it the core of our engineering design and manufacturing culture and its integrated communication network. First, there is a need for a new accounting of the subject in a more dynamic light. This is both as the dependable well-adapted and long-established engineering communication tool that it is, but also as the catalyst and mechanism for a real breakthrough of new technology. It could be identified as "The New GD&T—A Pervasive Technology."

Over many years of observing, working with, and promoting GD&T, tremendous advances have been made. The subject has progressed from one of early nominal use by those primarily in military work, using the MIL–STD–8 series of standards, to a modern engineering tool absolutely essential to the technical and economic well-being of the company or organization. ANSI/ASME Y14.5M, ISO/1101, ISO standards, and standards of other countries, have made GD&T the technical drawing language of the world. It has bridged local and international barriers of communication and provided that tool to "say what we mean and mean what we say" better than ever before. Its advantages can best be realized from the experiences in one's own facility, dealing with sub-contractors, multi-divisional coordination, military contracts, and reaching out the arm of communication anywhere in the world. There yet remain some minor differences between the latest ANSI/ASME Y14.5M (USA) standard and the latest ISO (International) standards. However, it can be roughly estimated that there is about 90 to 95% compatibility between these standards now for all practical purposes, and work continually goes on to improve that compatibility. The graphic language of symbols and numbers can "speak for itself" and surmount any language barrier around the world. Later in this introduction the harmonizing of USA and ISO standards is discussed.

The term "pervasive technology" is not new to industrial parlance. It has been used as a term of reference in a number of ways. Used here it implies that geometric dimensioning and tolerancing, GD&T, can now be considered a new and pervasive technology in two distinct, albeit related, ways. One, the standards and the contained universal GD&T techniques and language have "arrived"; it is no longer considered an optional approach. Its presence, influence and use is mandated to fill the void of communication which has persisted in many areas for years. It is a new day and age for GD&T and in many ways it is coming to the rescue; rescue from the doldrums of poor quality, complacency, archaism and resistance to the changing world; a life-ring to the future.

A Renaissance of GD&T Use

In every sense there is occurring a renaissance of use and recognition of GD&T. All should take advantage of the new frontiers of use of the subject. One cannot afford to be left behind. Those that do not have at least some speaking acquaintance with these standards may be left incommunicado.

This renaissance of new life and recognition is not a circumstantial quirk of fate. It has been a continual evolutionary trend and development, based upon hard-earned experience, economics, state of the art, and a desperate need for improved technical communication. Continually, more attention has been drawn to these standards as an essential authoritative base of credibility as we face the advantages and realities of the world in which we now live. The second reason, and probably the predominant one for the "new" GD&T age, is the advent and growth of the computer with all it offers and demands. The computer has revolutionized our world. The use and development of this technology has advanced to the point where the limits of communication and innovation have virtually been obliterated. The standards and the computer have provided the ideal marriage; one begets the other.

Standard ANSI Y14.5M (GD&T). The Catalyst and Base for the New Technology

Active and numerous projects are underway which are exploring and developing new frontiers relative to computer application and research in mechanical engineering. This is in addition to the hundreds of CAD, CAM, CIM, CADD, CNC, CAI, etc., program applications already in place and functioning as a part of everyday operations. What has been discovered and is being increasingly recognized, is that for nearly all such programs, the fundamental "standard" or language must be used if these efforts are to have a broad base of credibility, authority and be usable and coherent to others. ANSI Y14.5M (GD&T) *is the standard and language* which provides that base.

Further developments are underway in the USA for example ASME B89.1.12 committee which deal with coordinate measuring machines (CMM's) to better implement computerized inspection operations as based upon the concepts of the Y14.5 standard. ASME, industry, government, and academia are also cooperating on developmental and standards projects which deal with "Mathematical Definition of Dimensioning and Tolerancing Principles" (the ASME

Y14.5.1 committee), "Certification of GD&T Professionals" (the ASME Y14.5.2 committee) and national and internationally coordinated projects dealing with such subjects as "Mechanical Tolerancing." The basic standards are required as the essential key ingredients to give such research and development the common bond for consolidation and departure to new areas of analysis and technological enlightenment. There are also many additional computer explorations underway that would seem to need that same "home-base" anchor or ground zero origin.

The Technology Gap

While the renaissance of GD&T provides the necessary communication link in the computer revolution and the future appears revitalized, there is a "catch-22" to threaten the successes envisioned. Experience tends to convince that wherever there is "good news" there probably will also be a down side or price to pay somewhere. In this case, no search for the negatives is needed. Personal experience in daily contact with a cross section of academia, government, industry and the international arena will adequately convey the impression that there is a serious growing technology gap on this subject. Unfortunately, many are not even aware of the standard, do not understand it, misuse it, don't want to use it, fear it, etc.

There is a dilemma here; a direction is already charted and in place to address this technology gap, but all too many are not yet on course. Standard ANSI/ASME Y14.5M is the course and direction provided; on a "silver platter" just waiting to be used. All we have to do is use it. Many have spent a good deal of an industrial life sharing in the experience and development of the standard and its forerunners. Now is the best time in history to "bite the bullet", face facts, and take advantage of the technology we ourselves have created. History also reveals that where supposed disadvantages seem to prevail, there must also be advantages therein which can evolve out of the turmoil.

Implementing the New GD&T—A Pervasive Technology

The foregoing commentary can really be summed up by saying it another way. Our own technology is out-pacing us. Many cannot or will not make that effort to "learn something new" or reach for that higher plane of communication. It can also be added that many possibly have not had a chance or may not even be aware of the "bridge" needed to survive in the world of engineering communication today. As stated before, continual progress is being made but tends to be like walking in sand. Two steps forward with one step backward. The technology gap continually widens.

The down side of this situation then is that there is an engineering standard technology well designed and adapted to the present and future; but, its understanding and use is left wanting by a monumental share of the industrial populace. As the technology expands and the industrial population increases, the technology gap increases commensurately. The chasm between those knowledgeable and conversant with the standard and the GD&T state-of-the-art are being further separated from the pack. At the present pace there will never be enough highly-qualified persons to fill the need. There must be devised some way to package the technology where at least some of the needed intelligence can be supplied in simplified form. If somehow the implementation of the product design-to-manufacture cycle could be facilitated through "GD&T smart" aids, this could be accomplished. These simplified "tools" could bring the technology to a more nominal-user level. This approach could also be implemented in manufacturing and ensure correct and efficient inspection follow-through. The technology gap is then dramatically narrowed. The need-to-know depth becomes more identifiable and manageable.

The foregoing obviously describes computer programs and the present new technology already in place at many levels. This again introduces the pervasive technology theme. The Y14.5 standard (GD&T) provides the mechanism to fill the void and provides the authoritative language and technique for specific communication. The catch is: How well is the standard understood and used in design and in the computer programs developed? Those accepting the design-model data in integrated computer application place great "trust" in its correctness and build on or integrate the data directly. If the data is not clearly representative of the requirements, done well, and in the language of the standard, communication again is headed for failure. Worse yet, its shortcomings are transmitted to other operations.

The above scenario is put on-track and becomes a positive pervasive technology when the product designer and programmers responsible for creating the computer output are qualified in the language and dynamics of the standard. Then truly the bridge is constructed across the technology gap and each side is brought closer with limits and responsibilities better defined. There are high hopes and achievable goals then set down. Progress, of course, is already well down this road in many of the activities before mentioned.

The key to successful application of GD&T lies in education, training, and proper use of the standard and its contained techniques. It is the bridge needed to span the fate of further slipping into the chaos of our own doing. As before mentioned, certification requirements are being considered as a means of ensuring that those who play a key role in initially applying GD&T are qualified and where such input provides a dependable base for all involved. Emphasizing the bright side, the renaissance of GD&T use and prominence, provides the tools, advantage, and the language for both manual and automated communication in defining and building products. GD&T is playing a timely role as a partner to the computer. It is almost as if it was planned that way from the beginning; and who is to say that it was not ordained to happen.

Automated Intelligence—GD&T Spoken Here

Relative to the state of things as cited in the preceding commentary, what is being done to build the bridge over the technology gap? Many things; some have already been mentioned. Any observer or participant in the changing industrial scene, will detect many notable efforts now in place or underway to couple the computer to daily tasks. When GD&T is married to the computer great things can happen. With the type of quantum leaps forward, represented by the creative and innovative breakthroughs now being realized, the technology gap and crisis is addressed in great measure. The new GD&T, as the required engineering language, and the pervasive avenue provided by the computer programs, opens the heretofore locked doors of communication. Now everyone can, and everyone must, participate in the exciting new GD&T and the age of pervasive technology. This text modestly tries to play a role in this effort by providing the daily "ready reference" tool and the teaching mechanism for classroom use or self-study.

Harmonizing ANSI Y14.5 and ISO standards

The ultimate objective of facilitating world-wide technical design standards, using the language of geometric dimensioning and tolerancing (GD&T), has been, and is, a lofty goal to which many have aspired for decades. As the world "gets smaller," the need for such standards becomes more proven as ways and means are sought to better communicate product requirements on a global scale. Thus, the dedicated efforts by many over the years have come around full circle and now appear to be able to help fulfill far-reaching communication needs.

Geometric dimensioning and tolerancing, as provided in ANSI/ASME Y14.5M and comparable ISO standards, has now evolved to be the major tool to achieve such global communication. Its role in the industrial world has matured to a point where these techniques are called upon as a mandatory avenue for defining a product for most universal meaning, effective production, and economic advantage.

Emphasis on quality

There is a major emphasis today also upon all aspects of "quality" while referencing TQM, World Class Engineering, SPC, ISO 9000, etc. These are the "buzzwords" that are commonly heard and used as indicators of desirable goals and representative of progress for any given organization. GD&T supplies a fundamental basis and ground zero for most of these allied disciplines; quality seems to start with a well-defined product.

It is not a new idea, but a more commonly conceived one in recent years, that a company or organization decides to also "go to ISO standards." Apparently, it is the trend; sometimes in necessity as a multinational mandate or as a move toward the "big picture" emphasis in the organization's future. It seems that the "quality" and international emphases often become integrated into a common objective. On the surface, this move to ISO appears as a significant, yet readily achievable one, by simply shifting to a higher level standard usage. This is not so easy, but, it is possible to an extent, as described below. This goal has also been the ultimate objective as the long range plan for the national (USA) GD&T standards developers for at least forty-five years (possibly longer). However, to "what extent" does, and can, the aspiration to adopt and use ISO standards go?

What are ISO standards?

Since ISO standards are developed via an international forum, compromises and differences prevail. The finalization and approval of ISO standards, as well as their original initiation and progressive formulation, are based upon the very principles of cooperation and compromise by the countries involved. Therefore, ISO standards achieve the highest level of stature in the spirit of international cooperation and the common good. It follows then that each country attempts, as the obvious, to influence and input into the standards development process, their own preferences, philosophy and practices as much as possible. In this way their own national standards and habits are preserved as a higher order standard to which compliance is nicely accommodated. Volumes could be written on the details of the USA and its representatives' involvement and contribution to ISO/GD&T standards development over the years. Suffice it to say, it has been a very significant and successful endeavor reaching back about thirty years.

ISO; is it "them" or "us"?

A major misunderstanding seems to prevail when there is consideration of adopting ISO standards on GD&T. That is, that such an action seems to require adoption of "their" standards which could involve ideas totally alien to us. This is not so; it is "our" standard just as much, if not more, in some cases, than any other country could claim. Of course, the spirit of ISO standards is really a joint "ownership" by all signatories of the final standard. The point being made is, that the USA has been active as member "Experts" and/or "Conveners" (Chair) on all GD&T-related standards that have been developed in the ISO arena since the beginning of the evolution of ISO standards on this subject.

When the USA adopts ISO standards, it's "us" & "them"!

When ISO standards are adopted on any GD&T subject, it can be said that the USA has had previous input, contribution, voting influence, and privilege on that standard. The author alone has been convener (chairman) of six of the committees that developed the later listed ISO standards and, in addition, served on *all* of the committees which developed these standards. USA delegates have convened or served on all Working Groups (WG's). The WG's are ISO committees which do the preliminary spade work to develop the ISO technical standards from inception. The ISO/GD&T standards have USA fingerprints all over them. That does not mean, of course, that we (the USA) were totally successful in winning every point. We have learned from, and gained great respect for, our ISO colleagues from other countries and their contribution, as well, to the ISO. Together, the ISO standards efforts have borne fruit. The USA influence has been significant. When there is an adoption of an ISO/GD&T standard, it constitutes an adoption of something of our own doing; perhaps not totally satisfactory, but a product of cooperative effort in which USA influence was included. Our "batting average" in promoting our proposals to ultimately become ISO standards is a "ball park" .750 in the opinion of the author.

ISO standards versus ANSI/ASME Y14.5M

In adopting GD&T ISO standards it will be discovered that about fifteen to twenty ISO standards are directly or indirectly involved to approximate the coverage of USA ANSI/ASME Y14.5M. In the ISO standards development agenda it is, "one subject, one standard." This philosophy and history alone could deserve volumes of coverage. Yet, it can be said in brief, that this method evolves as the only reasonable way when being exposed to the lengthy travel, involved time consumed, cost outlays, national pride, language differences, due process required, etc. One subject at a time surfaces as the only practical and achievable approach. Obviously, related subjects must be discussed simultaneously (such as position tolerance and datums), but the detailed coverage on each ends up in separate documents of minimum size and very basic coverage.

ISO standards necessary for GD&T coverage

The following documents must be considered when adopting ISO/GD&T standards:

1. ISO/1101– Technical Drawings Geometical tolerancing
2. ISO/5458– Technical Drawings Positional tolerancing
3. ISO/5459– Technical Drawings Datums and Datum Systems
4. ISO/2692– Technical Drawings Maximum material principle
5. ISO/3040– Technical Drawings Cones
6. ISO/1660– Technical Drawings Profiles
7. ISO/129– Technical Drawings General principles
8. ISO/406– Technical Drawings Linear and angular dimensions
9. ISO/10578 Technical Drawings Projected tolerance zones
10. ISO/2692:1988/DAM 1 Technical Drawings Least material principle

11.	ISO/8015	Technical Drawings Fundamental tolerance principle
12.	ISO/7083	Technical Drawings Symbols proportions
13.	ISO/10579	Technical Drawings Non-rigid parts

Additional ISO standards involved:

1. ISO/1000 – SI Units
2. ISO/286 – Limits & Fits
3. ISO/TR5460 Technical Drawings–Verification principles
4. ISO/2768–2 General geometrical tolerances
5. ISO/1302 – Surface Texture
6. ISO/2768–1 Tolerances for linear and angular dimensions
7. Other peripheral standards on screw threads, gears, drills, welding, etc., may also be required for coverage beyond Y14.5 for product design.

USA contribution to ISO standard concepts

Some examples of USA-originated concepts, principles or standard methodology which are now contained within ISO standards:

The datum reference frame (three plane concept)

Datum precedence

Datum targets, point, lines, areas

Cylindricity

Projected Tolerance zone

Total runout

Composite positional tolerancing

Least material condition

Non-rigid part specification

Multiple/compound datums

Conicity using profile tolerance

Spherical diameter symbol

Formulas for positional tolerancing

USA delegates attendance at ISO meetings

Since 1967, USA delegates have attended ISO/TC10/SC5 plenary or working group meetings in:

Moscow	Stockholm	Vilnius	Carmel, CA
London	Ottawa	Zurich	Beijing
Berlin	New York	Prague	Paris
Cologne	The Hague	Gothenburg	
Oslo	Orlando	Copenhagen	

Cities such as New York, London, Paris, Berlin, Zurich, Ottawa have been common repeat locations.

ANSI/ASME Y14.5M and ISO compatibility

As of this date, in the opinion of the author, the ANSI Y14.5 standard and ISO principles are 90–95% in agreement. This is based upon major concepts and methodology, and not a myriad of minor details of preference where some discrepancies exist and probably will continue to exist. Of the differences that did exist, the major concerns were the datum feature symbol and methodology, the RFS symbol (US used only), and projected tolerance zone placement within the feature control frame. Other differences of some note are: Concentricity applied on MMC basis (USA allows only RFS), ISO permits position tolerance application to non-size feature (USA uses profile), and ISO uses flatness tolerance on a coplanarity application (USA uses profile). There are some other differences of minor note which may continue to exist.

Current considerations, in the new revision to USA ANSI/ASME Y14.5 subcommittee activities, have addressed all of these foregoing matters. Some have been completely resolved. The ISO datum feature symbol and methodology will be the USA standard. The RFS symbol will be removed and will imply the RFS principle (by default) with no indicator (RFS symbol may be used by those who desire to retain temporarily the redundant modifier), the projected tolerance zone symbol may be placed within the feature control frame.

Since ISO 1101, ISO 5458 and ISO 2692 are work items for possible future revision within ISO/TC10/SC5, some of the remaining USA/ISO differences have been already discussed in meetings and via correspondence. Resolution of these matters favorable to USA preferences, are possible in the future. This could eliminate many more variations between the USA Y14.5 standard and the ISO family of GD&T standards.

By completion of the new revision to Y14.5M, there is a very close harmonization between Y14.5 and ISO standards up to the 95% range (in the opinion of the author) in those matters of major consequence.

Unpredictables and new concepts

Another phase of the effort to harmonize concepts between ISO and new USA GD&T technology, brings new challenges.

The USA standards occasionally reach out beyond the ISO work agenda with newer concepts and thus some more sophisticated differences may result for a period of time. The impact of those differences can be minimized, however, by the now established, improved communication atmosphere between the USA and ISO participants. That is, the newer concepts are now being discussed somewhat earlier due to improved modern communication networks (FAX, computer, phone). Such concepts as the "axis of the actual mating envelope" using position tolerance RFS; the "centerplane of the actual mating envelope" using position tolerance RFS; coaxial use of position tolerance RFS; coaxial use of position tolerance MMC; modified definitions to accommodate the "mathematizing" of GD&T; "tangent plane" application and symbol; added symbols for countersink, counterbore, depth, arc length, composite position tolerancing, composite profile tolerancing, Resultant Condition, Inner and Outer Boundary, Coplanarity; and other refinements may cause the standards to struggle some to stay abreast (i.e., Y14.5 & ISO). However, the USA will be working these matters jointly through

US/Technical Advisory Groups (TAG's) and ISO/TC10/SC5, SC1, etc. The USA plays a major role in both USA and ISO interests; they are both on our work agenda. The USA is now the Secretariat of the ISO/TC10/SC5 international committee (Secretariat = Chairmanship) as a relatively new development beginning in late 1992.

Third angle projection versus First angle projection

This matter is not directly a GD&T concern but has, nevertheless, a major impact on international engineering drawings. No permanent resolution of this matter seems visible; we have "agreed to disagree." So long as the symbolic logo to designate one or the other is found on the drawing, the problem is minimized and the two systems can coexist comfortably; at least for the foreseeable future.

Future meetings of ISO/TC10/SC5

Future meetings, to further advance standardization of GD&T, both nationally and internationally, and its compatibility with electronic (computer) application, will be scheduled as necessary.

Conclusion

The future seems very bright for the use of GD&T as a major factor in the years ahead. Its use as a technical language for global communication, enabling quality programs to thrive and improve international trade, economics, and understanding seems ensured. Its use will be a necessity and a rewarding culmination of the years of dedication to its purpose. Its continuance, maintenance, and continued success in the future lies in our collective hands.

Following in this text is our best effort to assist this important endeavor. That is, to present GD&T in a digestible manner and with the flexibility necessary to accommodate varying needs, interests, and background to pursue the desired educational goals on this important subject.

WHY USE GEOMETRIC DIMENSIONING AND TOLERANCING?

Why is it that we should be so interested in this subject?

FIRST AND FOREMOST ITS USE SAVES MONEY!

It saves money directly by providing for maximum producibility of the part through maximum production tolerances. It provides "bonus" or extra tolerances in many cases.

It ensures that design dimensional and tolerance requirements, as they relate to actual function, are specifically stated and thus carried out.

It adapts to, and assists, computerization techniques in design and manufacture.

It ensures interchangeability of mating parts at assembly.

It provides uniformity and convenience in drawing delineation and interpretation, thereby reducing controversy and guesswork.

Aside from these primary reasons there are others of a more general nature:

The intricacies of today's sophisticated engineering design demand new and better ways of accurately and reliably communicating requirements. Old methods simply no longer suffice.

Diversity of product line and manufacture makes considerably more stringent demands of the completeness, uniformity, and clarity of drawings.

It is increasingly becoming the "spoken word" throughout industry, the military, and, internationally, on engineering drawing documentation. Every engineer or technician involved in originating or reading a drawing should have a working knowledge of this new state of the art.

WHAT IS GEOMETRIC DIMENSIONING AND TOLERANCING?

In particular, it is a means of dimensioning and tolerancing a drawing with respect to the actual function or relationship of part features which can be most economically produced. *Function* and *relationship* are the key words.

In general, it is a system of building blocks for good drawing practice which provides the means of stating necessary dimensional or tolerance requirements on the drawing not otherwise covered by implication or standard interpretation.

WHEN SHOULD GEOMETRIC DIMENSIONING AND TOLERANCING BE USED?

When part features are critical to function or interchangeability;
when functional gaging techniques are desirable;

when datum references are desirable to ensure consistency between design, manufacturing and verification operations;

when computerization techniques in design and manufacture are desirable;

when standard interpretation or tolerance is not already implied.

GEOMETRIC CHARACTERISTICS AND SYMBOLS

The geometric characteristics and symbols that are used as the building blocks for geometric dimensioning and tolerancing are:

▱	Flatness
—	Straightness
○	Circularity (Roundness)
⌭	Cylindricity
⊥	Perpendicularity (Squareness)
∠	Angularity
//	Parallelism
⌒	Profile of a Surface
⌒	Profile of a Line
↗	Circular Runout
↗↗	Total Runout
⊕	Position
◎	Concentricity
＝	Symmetry

USING SYMBOLS

The general use of symbols instead of notes on a drawing provides a number of advantages. The illustrations below incorporate the geometric characteristic symbols with datum and feature control symbols. Some of the advantages of symbols over notes are

1.   The symbol has uniform meaning. A note can be stated inconsistently, with a possibility of misunderstanding.

2.   Symbols are compact, quickly drawn, and can be placed on the drawing where the control applies; symbols adapt readily to computer applications.

Notes require much more time and space, tend to be scattered on the drawing, often appear as footnotes which separate the note from the feature to which it applies.

3.   Symbols are the international language and surmount individual language barriers.

Notes may require translation if the drawing is used in another country.

4.   Symbols can be applied with drafting templates or computer techniques and retain better legibility in various forms of copy reproduction.

5.   Geometric tolerancing symbols follow the established precedent of other well-known symbol systems, e.g., electrical and electronic, welding, surface texture, etc.

USING SYMBOLS

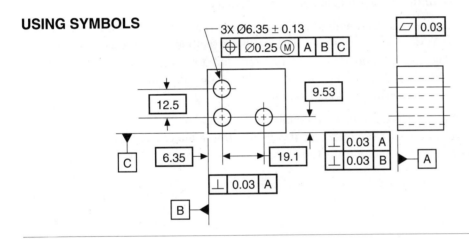

USING NOTES

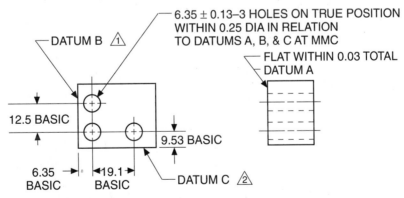

△2   SURFACE C PERPENDICULAR TO DATUM A WITHIN 0.03
     & WITH DATUM B WITHIN 0.03

△1   SURFACE B PERPENDICULAR TO DATUM A WITHIN 0.03

MAXIMUM MATERIAL CONDITION PRINCIPLE
SYMBOL Ⓜ ABBREVIATION (MMC)

One of the fundamental and most important principles of geometric dimensioning and tolerancing is MAXIMUM MATERIAL CONDITION. A thorough understanding of its meaning is therefore essential.

Note in the figure below that the "maximum material condition" size of the $6.25 \, ^{+0.1}_{0}$ diameter hole is 6.25 or its *low* limit size. The hole at its low limit obviously retains more material than if it were at its *high* limit or larger size; thus the term "maximum material condition" defines the *low* limit when it applies to a hole or similar feature.

Note similarly that the $6.15 \, ^{0}_{-0.1}$ diameter pin is at its "maximum material condition" size when it is at its *high* limit of size of 6.15. In this instance it is more readily seen that more material exists in the pin when it is at its maximum permissible size. However, the same principle exists in both hole and pin MMC situations. Relating mating part features in this manner ensures their functional relationships, and as will be seen later in the text, establishes the criteria for determining necessary form, orientation, and position tolerances.

The symbol for "maximum material condition," the M enclosed in a circle, and the occasionally used abbreviation MMC are shown above. The symbolic method is to be used with feature control frames only. The abbreviation MMC may be used with note callouts but not with symbolic representations. We shall discuss later the application of the "maximum material condition" principle and illustrate it with practical examples.

Generally, the use of the "maximum material condition" principle permits greater possible tolerance as part feature sizes vary from their calculated "maximum material condition" limits. It also ensures interchangeability and permits functional gaging techniques. It is one of the

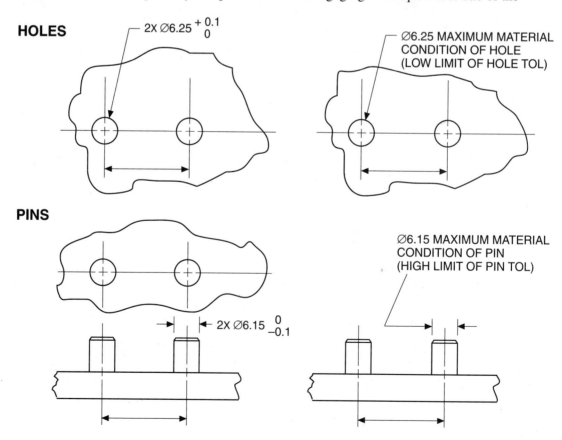

HOLES

2X Ø6.25 $^{+0.1}_{0}$

Ø6.25 MAXIMUM MATERIAL CONDITION OF HOLE (LOW LIMIT OF HOLE TOL)

PINS

2X Ø6.15 $^{0}_{-0.1}$

Ø6.15 MAXIMUM MATERIAL CONDITION OF PIN (HIGH LIMIT OF PIN TOL)

fundamental principles upon which the system of geometric dimensioning and tolerancing is based. Below is the definition of maximum material condition and the usual prerequisites for application. We shall later expand the use of the principle by means of examples.

Definition. The condition in which a feature of size contains the maximum amount of material within the stated limits of size: for example, minimum hole diameter and maximum shaft diameter.

The "maximum material condition" principle is normally valid only when both of the following conditions exist:

1. Two or more features are interrelated with respect to location or orientation (e.g., a hole and an edge or surface, two holes, etc.). At least *one* of the related features is to be a feature of size.

2. The feature (or features) to which the MMC principle is to apply must be a feature of size (e.g., a hole, slot, pin, etc.) with an axis or center plane.

"Maximum material condition" has two connotations. One as seen in the preceding figure; the ⌀6.25 MAXIMUM MATERIAL CONDITION HOLE further means that there is a Perfect Form at MMC boundary at ⌀6.25 and the pin likewise at ⌀6.15. (See Rule 1). The second connotation is where the material condition symbol Ⓜ is used in a feature control frame to apply the principle of MMC to the geometric tolerance specified to locate the pins and holes. See numerous figures in this text.

"Maximum material condition" might also be considered as a "new" term for an "old" situation, such as the familiar terms "worst condition," "critical size," etc., used in the past for relating mating part features.

Where the maximum material condition principle is not appropriate, the "regardless of feature size" principle or "least material principle" may be applied. (See below.)

REGARDLESS OF FEATURE SIZE
ABBREVIATION (RFS)

Definition. The term used to indicate that a geometric tolerance or datum reference applies at any increment of size of the feature within its size tolerance.

"Regardless of feature size" is another principle of geometric dimensioning and tolerancing which must be well understood. Unlike maximum material condition, the "regardless of feature size" principle permits *no* additional positional, form or orientation tolerance, no matter to which size the related features are produced. It is really the independent form of dimensioning and tolerancing which has always been used prior to the introduction of the MMC principle.

The abbreviation for "regardless of feature size" is RFS. We shall later clarify the principle by means of examples.

The RFS principle is valid only when applied to features of size (for example, a hole, slot, pin, etc., with an axis or center plane). The size connotation cannot be applied to a feature which does not have "size."

NOTE The symbol Ⓢ was previously used with positional tolerance.

LEAST MATERIAL CONDITION PRINCIPLE
SYMBOL Ⓛ ABBREVIATION (LMC)

Definition. The condition in which a feature of size contains the least amount of material within the stated limits of size: for example, maximum hole diameter and minimum shaft diameter.

The least material condition principle may be desirable as an alternative to MMC or RFS in certain design considerations. See also "Least Material Condition," pages 150 and 152.

Note that the actual local sizes of the hole at ⌀6.35 and the pin at ⌀6.05 of the figures on page 14 are also their least material condition sizes.

BASIC AND DATUM

The terms BASIC and DATUM are most important. Proper application of the principles implied by these terms greatly contributes to effective geometric dimensioning and tolerancing.

BASIC

Definition. A numerical value used to describe the theoretically exact size, profile, orientation, or location of a feature or datum target. It is the basis from which permissible variations are established by tolerances in feature control frames or on other dimensions or notes.

Use of a BASIC dimension, which is a theoretical and exact value, requires also a *tolerance* stating the permissible variation from this exact value (most often relative to a position, angularity or profile requirement). A BASIC dimension states only *half* the requirement. To complete it, a tolerance must be associated with the features involved in the BASIC dimension.

In the past, a BASIC dimension was identified on the drawing by the word BASIC (or the accepted abbreviation BSC) adjacent to, or below, the dimension, or by a general note on the drawing. The symbolic method that follows on page 17 is the recommended method in keeping with latest standards and international practice.

EXAMPLE

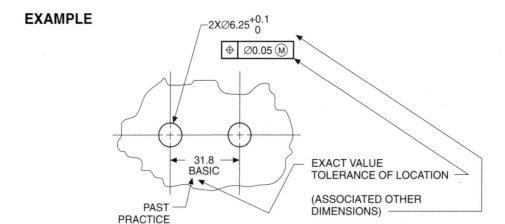

Some companies use a "naked" or untoleranced dimension instead of BASIC. The same meaning as BASIC is invoked by adding a drawing footnote, title block notation, or company standards.

The term TP (for True Position) derived from British standards has also been used in the past. It has the same meaning as BASIC.

BASIC dimensions are also used in other applications such as tapers for which tolerances must be derived from associated size dimensions.

The use of BASIC dimensions on datum targets assumes standard tooling or gagemaker's tolerances (see DATUM section for more detail). BASIC dimensions used to indicate a limited area or portion of a surface where a tolerance (e.g., Runout) applies, assumes standard inspection set-up precision.

SYMBOLIC METHOD OF STATING A BASIC OR THEORETICAL EXACT VALUE—RECOMMENDED

The preferred method of stating an "exact" value replacing BASIC, BSC, TP, etc., is the international (ISO) method recommended by ANSI Y14.5. According to this method, the exact value is enclosed in a frame or box (see example below).

EXAMPLE

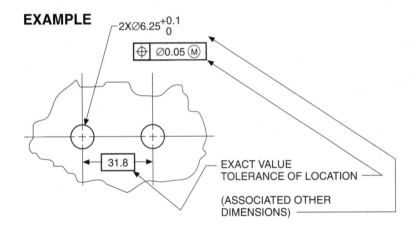

The symbolic method for "exact" values may be used with symbolic *or* notated geometric tolerancing. Because of the need to standardize U.S. practices and encourage compatibility with international practices, using the symbolic or boxed "exact" value is strongly recommended.

DATUMS AND DATUM FEATURE SYMBOL

Definitions. A theoretically exact point, axis, or plane derived from the true geometric counterpart of a specified datum feature. A datum is the origin from which the location or geometric characteristics of features of a part are established.

Datum surfaces and datum features are actual part surfaces or features used to establish datums. They include all the surface or feature inaccuracies.

DATUM FEATURE SYMBOL

To identify a feature as a datum, the following datum feature symbol is used:

(Datum feature triangle may be filled or open. Leader may be appropriately directed to a feature.)

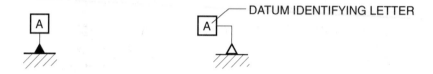

The datum feature symbol consists of a capital letter enclosed in a square frame, a leader line extending from the frame to the concerned feature and terminating with a triangle.

Each datum requiring identification is assigned a different reference letter. Do not use letters I, O, Q. If the single letter alphabet is exhausted, double letters may be used, i.e., AA, AB, etc.

Where datum feature symbol is repeated to identify the same feature in other locations of a drawing, it need not be identified as reference.

PLACEMENT OF THE DATUM FEATURE SYMBOL
APPLICATION TO PLANE SURFACES

The datum feature symbol is applied to the concerned feature surface outline, extension line, dimension line or feature control frame as follows:

1.   Placed on the outline of a feature surface or an extension line of the feature outline (but clearly separated from the dimension line when the datum feature is represented by the extension line or feature surface itself.)

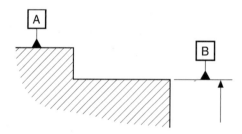

PLACEMENT OF THE DATUM FEATURE SYMBOL
APPLICATION TO SIZE FEATURES

2.   Placed on an extension of the dimension line of a size feature when the datum is the axis or median center plane. If there is insufficient space for the two arrows, one of them may be replaced by the datum feature triangle.

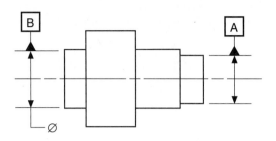
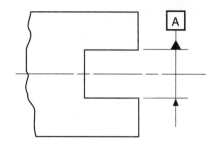

3.   Placed on the outline of a cylindrical feature surface or an extension line of the feature outline, separated from the size dimension, when the datum is the axis. For CAD systems, the triangle may be tangent to the feature.

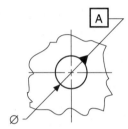
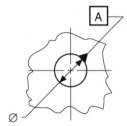
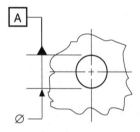

4.   Placed below or above and attached to the feature control frame when the feature, or group of features, controlled is the datum axis or datum centerplane.

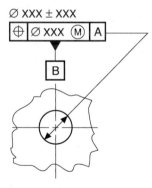

5.   Placed on the planes established by datum targets on complex or irregular datum features, equalizing datums, etc., as an option for clarification (see pages 263–265), or to re-identify previously established datum axes or planes, as reference, on repeated or multi-sheet drawing

6.   Placed on a dimension leader line to the feature size dimension where no geometrical tolerance and feature control frame is used.

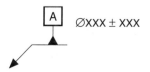

FEATURE AND FEATURE CONTROL FRAME

FEATURE

The general term applied to a physical portion of a part, such as a surface, hole or slot.

Features are specific component portions of a part and may include one or more surfaces such as holes, faces, screw threads, profiles, or slots. Features may be "individual" or "related."

FEATURE CONTROL FRAME

The feature control frame consists of a box containing the geometric characteristic symbol, datum references, tolerance, and the material condition symbol (e.g., for MMC) if applicable.

The example below shows this feature control frame as used on a part drawing.

EXAMPLE

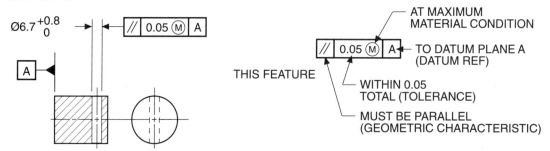

PLACEMENT OF THE FEATURE CONTROL FRAME

The feature control frame is associated with the feature(s) being tolerated by one of the following methods:

1.   Attaching a side, end, or corner of the symbol box to an EXTENSION LINE or leader from the feature (used on most form tolerances). See Fig. 1.

2.   Attaching a side or end of the symbol box to the DIMENSION LINE or EXTENSION LINE pertaining to the feature when it is cylindrical. See Fig. 2.

3.   Placing the symbol box below or closely adjacent to the dimension or note pertaining to that feature. See Fig. 3.

4.   Running a leader line from the symbol to the feature. See Fig. 4.

FIGURE 1

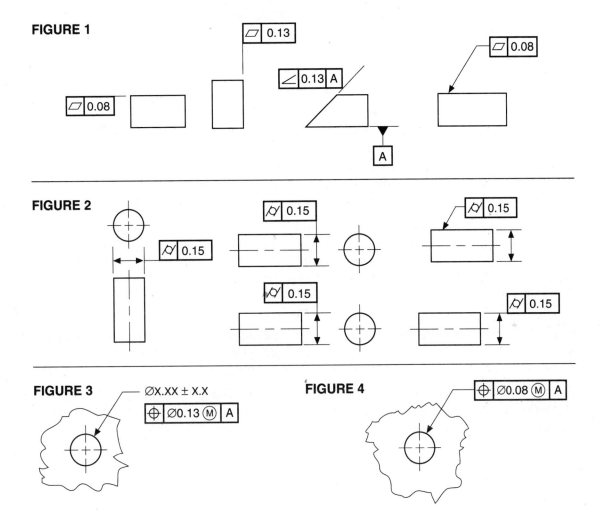

FIGURE 2

FIGURE 3

FIGURE 4

COMBINED FEATURE CONTROL FRAME AND DATUM FEATURE SYMBOL

When a feature serves as a datum and is also controlled by a geometric tolerance, the feature control frame and the datum feature symbol may be combined as shown.

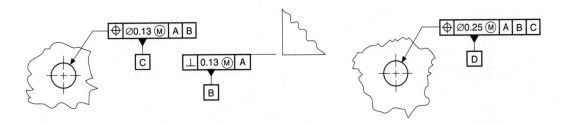

REFERENCE TO DATUM

When an orientation, profile, runout, or location tolerance must be related to a datum, this relationship is stated by placing the datum reference letter following the geometric characteristic symbol and the tolerance.

The illustrations on page 23 show additional examples of the feature control frames with reference to datums.

Figure 1 is a typical feature control frame using a single datum reference. The symbol reads "This feature shall be within a 0.03 tolerance zone perpendicular to datum A."

Figure 2 shows a feature control frame with *two* datums. The symbol reads "This feature shall be located at true position within 0.05 diameter at maximum material condition with respect to both datums A and B."

Note that vertical lines are used to separate the characteristic symbol, the feature tolerance, and the datum references. These vertical lines are used on all feature control frames to ensure clarity. One reason for this is illustrated in Fig. 3, in which the maximum material condition symbol is used. The vertical lines clearly show that MMC condition symbols apply only to those datums or tolerances with which they appear in the subdivision of the symbol box.

Figure 4 illustrates primary, secondary, and tertiary datums showing the order of precedence. When the order of precedence of datums is significant to function, datum references should be classified as primary, secondary, and tertiary. The datum precedence is shown by placing each datum reference letter in the proper order. The first datum letter (left to right) is considered the primary datum, the second letter secondary, and the third letter tertiary. Thus the datum reference letters will not necessarily be in alphabetical order. See section on DATUMS for further explanation.

Figure 5 illustrates a feature control frame in which multiple datum features are used simultaneously to establish a single datum reference (equal precedence of datum features) e.g., to establish a common datum axis. See section on DATUMS and RUNOUT for further details.

Figure 6 illustrates a possibly questionable use of a datum reference. Note that datum A applies at MMC, whereas the feature controlled applies at RFS. This means that the datum reference is subject to variation and cannot serve as a fixed reference for any RFS relationship. Although there may be exceptions under special circumstances, generally, wherever MMC is used on any datum, the feature controlled should also be controlled at MMC.

DIAMETER SYMBOL ⌀

The symbol used to designate a diameter (DIA) or cylindrical feature or tolerance zone is shown above. The symbol (⌀) precedes the specified tolerance. See Figs. 2, 3, 4, and 6 on the following page. The symbol is used elsewhere on the drawing to indicate that the feature is circular or cylindrical. The symbol (⌀) precedes the stated nominal size value of the concerned feature. See figures on page 27 and throughout this text.

FIGURE 1

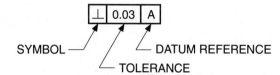

⊥ | 0.03 | A
SYMBOL
DATUM REFERENCE
TOLERANCE

FIGURE 2

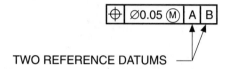

⌖ | ∅0.05 Ⓜ | A | B
TWO REFERENCE DATUMS

FIGURE 3

MMC APPLICABLE TO DATUM
⊥ | ∅ 0.05 Ⓜ | A Ⓜ
MMC APPLICABLE TO TOLERANCE
DATUM REFERENCE

FIGURE 4

ORDER OF PRECEDENCE OF DATUM

PRIMARY DATUM REFERENCE
SECONDARY DATUM REFERENCE
⌖ | ∅0.25 Ⓜ | A | B | C
TERTIARY DATUM REFERENCE

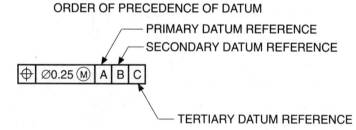

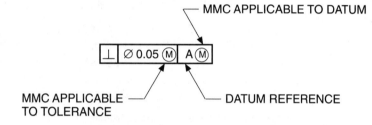

FIGURE 5

MULTIPLE DATUM FEATURES USED SIMULTANEOUSLY TO ESTABLISH A SINGLE DATUM REFERENCE, FOR EXAMPLE, TO ESTABLISH DATUM AXIS.
↗ | 0.08 | A – B

FIGURE 6

⊥ | ∅0.05 | A Ⓜ
FEATURE CONTROL RFS
DATUM MMC

NOT NORMALLY PRACTICAL

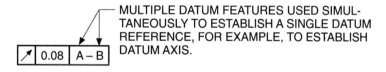

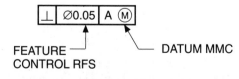

GEOMETRIC CHARACTERISTICS—FORM, ORIENTATION, PROFILE, RUNOUT, AND LOCATIONAL TOLERANCE—OTHER SYMBOLS AND TERMS

Illustrated at right are the geometric characteristics and symbols which are the basis for the language of geometric dimensioning and tolerancing.

FIVE TYPES OF GEOMETRIC CHARACTERISTICS

Expanding on preceding text explanation, it is seen that the geometric characteristics are of five types (see "5-TYPES" column at right):

1. FORM tolerance—A form tolerance states how far an actual surface or feature is permitted to vary from the desired form implied by the drawing.

2. ORIENTATION tolerance—An orientation tolerance states how far an actual surface or feature is permitted to vary relative to a datum or datums.

3. PROFILE tolerance—A profile tolerance states how far an actual surface or feature is permitted to vary from the desired form on the drawing and/or vary relative to a datum or datums.

4. RUNOUT tolerance—A runout tolerance states how far an actual surface or feature is permitted to vary from the desired form implied by the drawing during full (360°) rotation of the part on a datum axis.

5. LOCATION tolerance—A location tolerance states how far an actual size feature is permitted to vary from the perfect location implied by the drawing as related to a datum, or datums, or other features.

KINDS OF FEATURES TO WHICH A GEOMETRIC CHARACTERISTIC IS APPLICABLE

The geometric characteristics are also divisible into three "kinds" of features to which a particular characteristic is applicable. (See "KIND OF FEATURE" column at right.)

1. *INDIVIDUAL* feature—A single surface, element, or size feature which relates to a perfect geometric counterpart of itself as the desired form; no datum is proper nor used. (characteristics ▱, —, ○, ⌖).

2. *RELATED* feature—A single surface or element feature which relates to a datum, or datums, in form and orientation. (characteristics ⊥, ∠, ∥). A size feature (e.g., hole, slot, pin, shaft) which relates to a datum, or datums, in form, attitude (orientation), runout, or location. (characteristics ⊥, ∠, ∥, ↗, ↗↗, ⊕, ○, ≡).

3. *INDIVIDUAL* or *RELATED* feature—A single surface or element feature whose perfect geometric profile is described which may, or may not, relate to a datum, or datums, (characteristics ⌒, ⌓).

OTHER SYMBOLS AND TERMS

For review, other symbols and characteristics are shown at right. The Projected Tolerance Zone, Least Material Condition, and Datum Target symbols are explained in later text.

GEOMETRIC CHARACTERISTICS, SYMBOLS, AND TERMS

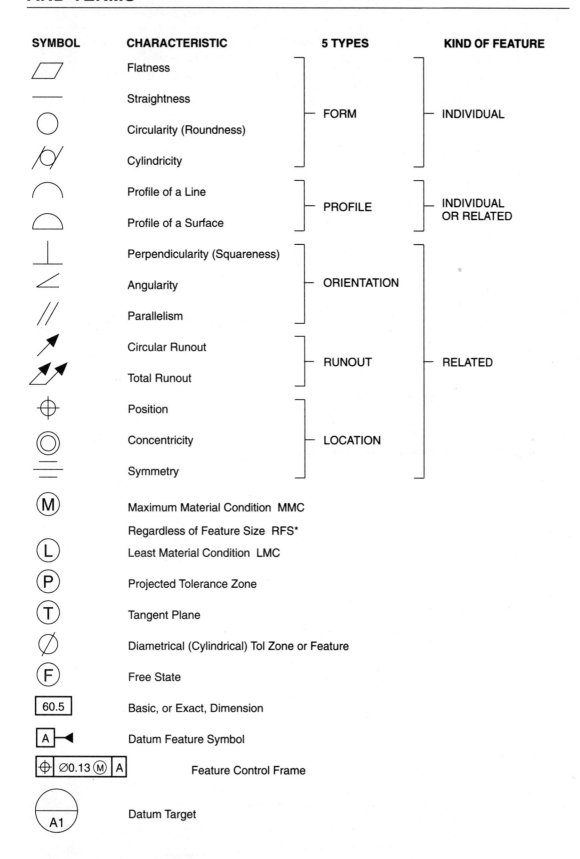

SYMBOL	CHARACTERISTIC	5 TYPES	KIND OF FEATURE
▱	Flatness		
—	Straightness	FORM	INDIVIDUAL
○	Circularity (Roundness)		
⌭	Cylindricity		
⌒	Profile of a Line	PROFILE	INDIVIDUAL OR RELATED
⌓	Profile of a Surface		
⊥	Perpendicularity (Squareness)		
∠	Angularity	ORIENTATION	
//	Parallelism		RELATED
↗	Circular Runout	RUNOUT	
↗↗	Total Runout		
⊕	Position		
◎	Concentricity	LOCATION	
≡	Symmetry		

SYMBOL	TERM
Ⓜ	Maximum Material Condition MMC
	Regardless of Feature Size RFS*
Ⓛ	Least Material Condition LMC
Ⓟ	Projected Tolerance Zone
Ⓣ	Tangent Plane
⌀	Diametrical (Cylindrical) Tol Zone or Feature
Ⓕ	Free State
60.5	Basic, or Exact, Dimension
A◀	Datum Feature Symbol
⊕ ⌀0.13 Ⓜ A	Feature Control Frame
A1	Datum Target

* (RFS implied unless otherwise specified under Material Condition Rule , See Rule 2 and Alternate Rule 2A.)

25

GENERAL RULES

Like any discipline, geometric dimensioning and tolerancing is based on certain fundamental rules. Some of these follow from standard interpretation of the various characteristics, some govern specification, and some are general rules applying across the entire system.

The various rules appropriate to given geometric characteristics and related nomenclature will be discussed later. The general rules are described below and on succeeding pages.

APPLICABILITY OF GENERAL RULES

ANSI Y14.5M contains four general rules. ANSI Y14.5M must be referenced whenever these rules are to be applied.

INDIVIDUAL FEATURES OF SIZE

LIMITS OF SIZE

Unless otherwise specified the limits of size of a feature prescribe the extent within which variations of geometric form, as well as size, are allowed. This control applies solely to individual features of size.

RULE 1—LIMITS OF SIZE RULE

Individual Feature of Size

Where only a tolerance of size is specified, the limits of size of an individual feature prescribe the extent to which variations in its geometric form as well as size are allowed.

Variations of Size

The actual local size of an individual feature at any cross-section shall be within the specified tolerance of size.

Variations of Form (Envelope Principle)

The form of an individual feature is controlled by its limits of size to the extent prescribed in the following paragraph and illustration:

a.   The surface, or surfaces, of a feature shall not extend beyond a boundary (envelope) of perfect form at MMC. This boundary is the true geometric form represented by the drawing. No variation in form is permitted if the feature is produced at its MMC limit of size.

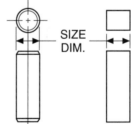

INDIVIDUAL SIZE FEATURES

(continued)

(Rule 1 continued)

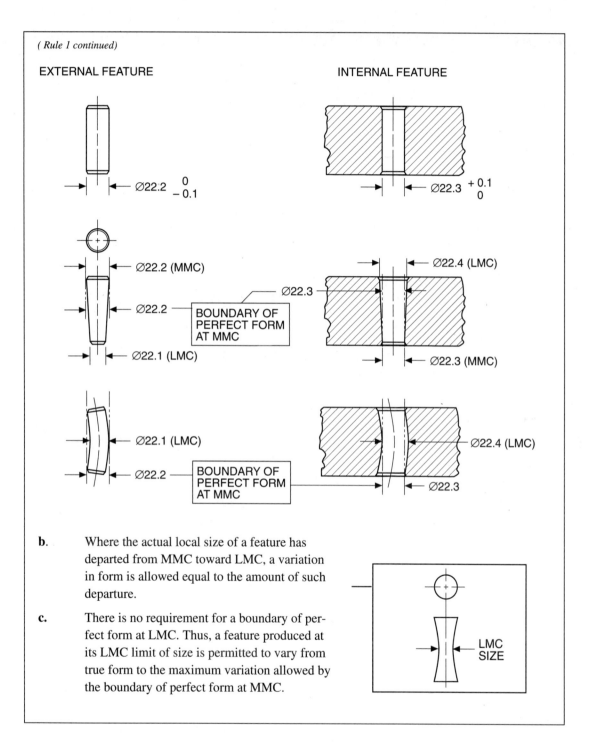

EXTERNAL FEATURE   INTERNAL FEATURE

Ø22.2 $^{0}_{-0.1}$ Ø22.3 $^{+0.1}_{0}$

Ø22.2 (MMC)   Ø22.4 (LMC)

Ø22.2 — Ø22.3

BOUNDARY OF
PERFECT FORM
AT MMC

Ø22.1 (LMC)   Ø22.3 (MMC)

Ø22.1 (LMC)   Ø22.4 (LMC)

Ø22.2 — BOUNDARY OF
PERFECT FORM
AT MMC   Ø22.3

LMC
SIZE

b. Where the actual local size of a feature has departed from MMC toward LMC, a variation in form is allowed equal to the amount of such departure.

c. There is no requirement for a boundary of perfect form at LMC. Thus, a feature produced at its LMC limit of size is permitted to vary from true form to the maximum variation allowed by the boundary of perfect form at MMC.

WHEN PERFECT FORM AT MMC (RULE 1) DOES *NOT* APPLY:

The control of geometric form prescribed by limits of size does not apply to the following:

a.   Stock such as bars, sheets, tubing, structural shapes, and other items produced to established industry or government standards that prescribe limits for straightness, flatness, and other geometric characteristics. Unless geometric tolerances are specified on the drawing of a part made from these items, standards for the items govern the surfaces that remain in the "as furnished" condition on the finished part.

b.   Parts subject to free state variation in the unrestrained condition.

PERFECT FORM AT MMC *NOT* REQUIRED
(RULE 1 REMOVED)

Where it is desirable to permit a surface, or surfaces, of a feature to exceed the boundary of perfect form at MMC, a note such as PERFECT FORM AT MMC NOT REQUIRED is specified exempting the pertinent size dimension from the provision of RULE 1.

RELATIONSHIP BETWEEN INDIVIDUAL FEATURES

The limits of size *do not* control the runout, orientation, or location relationship *between* individual features.

Features shown perpendicular, coaxial, or symmetrical or otherwise geometrically related to each other must be controlled for runout, orientation, or location to avoid incomplete drawing requirements. These controls may be specified by the use of appropriate geometric tolerances.

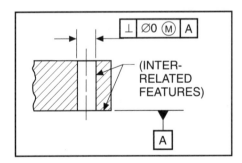

If it is necessary to establish a boundary of perfect form at MMC to control the relationship between features, the following methods are used:

a.   Specify a zero tolerance of orientation at MMC, including a datum reference (at MMC if applicable), to control perpendicularity, parallelism, or angularity of the feature.

b.   Specify a zero positional tolerance at MMC, including a datum reference to control locational, coaxial, or symmetrical features.

c.   Relate the dimensions to a datum reference frame, or by a local or general note indicating datum precedence.

d.   Indicate this control for the features involved by a note such as PERFECT ORIENTATION (OR COAXIALITY OR SYMMETRY) AT MMC REQUIRED FOR RELATED FEATURES.

APPLICABILITY OF RFS, MMC, AND LMC

Applicability of RFS, MMC, or LMC is *limited to features subject to variations in size.* They may be datum features or other features whose axes or center planes are controlled by geometric tolerances. In such cases the following practices apply:

RULE 2—MATERIAL CONDITION RULE

For *all* applicable geometric tolerances, RFS applies with respect to the individual tolerance, datum reference, or both, where no modifying symbol is specified. MMC, Ⓜ, or LMC, Ⓛ, must be specified on the drawing where it is required. Such as:

(MAY BE SPECIFIED ON SIZE FEATURES AS APPROPRIATE)

NOTE The below characteristics and controls are *always* applicable at RFS and due to the nature of the control cannot be applied at MMC or LMC.

PITCH DIAMETER RULE

Each tolerance of orientation or position and datum reference specified for a screw thread applies to the axis of the thread derived from the pitch cylinder. Where an exception to this practice is necessary, the specific feature of the screw thread (such as MAJOR ⌀ or MINOR ⌀) shall be stated beneath the feature control frame or beneath the datum feature symbol, as applicable.

MAJOR ⌀   MAJOR ⌀

Each tolerance of orientation or location and datum reference specified for gears, splines, etc., must designate the specific feature of the gear, spline, etc., to which it applies (such as PITCH ⌀, PD, MAJOR ⌀, or MINOR ⌀). This information is stated beneath the feature control frame or beneath the datum feature symbol.

PD   PD

DATUM/VIRTUAL CONDITION RULE

A virtual condition exists for a datum feature of size where its axis or center plane is controlled by a geometric tolerance. In such cases, the datum feature applies at its virtual condition even though it is referenced in a feature control frame at MMC or LMC.

NOTE   Alternate practice (Rule 2A). For a tolerance of position RFS may be specified on the drawing with respect to the individual tolerance, datum reference, or both, as applicable. (Past Practices.) (Not compatible with ISO practices.)

APPLICABILITY OF MMC, RFS, OR LMC

	Characteristic	Applicability to feature	Applicability to datum reference
⟋⟋	Flatness	Not applicable	
—	Straightness	MMC Ⓜ or RFS applicable if tolerance applies to axis or center plane of a feature with size; not applicable if considered feature is a single plane surface	No datum reference
◯	Circularity	Not applicable	
⌀	Cylindricity		
⌒ ⌒	Profile of a surface Profile of a line	Not applicable	RFS on datum reference applicable if datum feature has size and has an axis or center plane; not applicable if datum feature is a single plane surface. MMC Ⓜ not applicable†
⟂ // ∠	Perpendicularity Parallelism Angularity	MMC Ⓜ, RFS, or LMC Ⓛ applicable if tolerance applies to axis or center plane of a feature with size; not applicable if considered feature is a single plane surface	MMC Ⓜ, RFS, or LMC Ⓛ on datum reference applicable if datum feature has size and has an axis or center plane; not applicable if datum feature is a single plane surface
⊕	Position	MMC Ⓜ, RFS, or LMC Ⓛ applicable if tolerance applies to axis or center plane of a feature with size; not applicable if considered feature is a single plane surface	MMC Ⓜ, RFS, or LMC Ⓛ on datum reference applicable if datum feature has size and has an axis or center plane; not applicable if datum feature is a single plane surface
↗ ↗↗ ◎ ≡	Circular Runout Total Runout Concentricity Symmetry	Only RFS applicable	Only RFS applicable

† May have exceptions under special conditions. (See PROFILE section)

SHAPE OF TOLERANCE ZONE FOR FORM, ORIENTATION, PROFILE, RUNOUT, AND LOCATIONAL TOLERANCES

Where the specified tolerance value represents the diameter of a cylindrical zone, the diameter symbol $\varnothing$ shall be included in the feature control frame.

Where the tolerance zone is other than a diameter, the tolerance value represents the distance between two parallel straight lines or planes or the distance between two uniform boundaries.

EXAMPLES

TOLERANCE ZONE SHAPE IS: Where the diameter (cylindrical) symbol $\varnothing$ is specified, the tolerance is a diameter (or cylindrical) shape. Where no $\varnothing$ is specified, the tolerance zone is between two parallel lines or planes in the direction of the dimension arrows. The tolerance indicated is the TOTAL tolerance permitted.

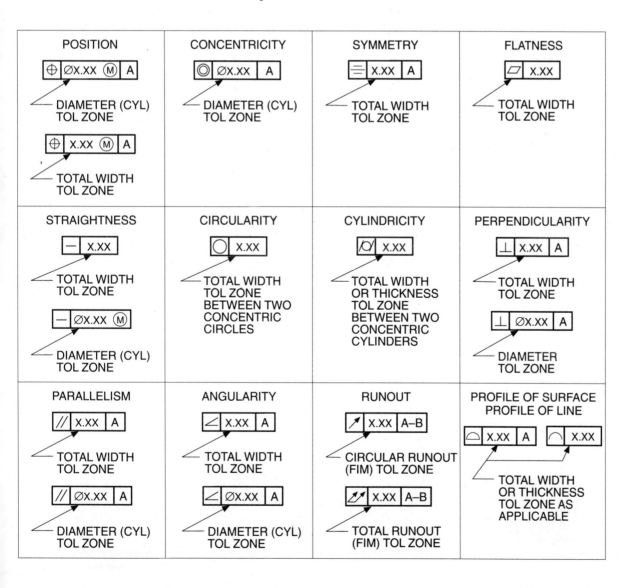

VIRTUAL CONDITION (MMC)

It is necessary to understand Virtual Condition as it applies to features. The definition below and the examples below and throughout the text will clarify its meaning.

Definition. Virtual Condition. A constant boundary generated by the collective effects of a size feature's specified MMC or LMC material condition and the geometric tolerance for that material condition.

WHERE MMC IS SPECIFIED:

The virtual condition of a feature is the extreme boundary of that feature which represents the "worst case" for, typically, such concerns as a clearance or fit possibility relative to a mating part or situation.

- FOR PIN: In the case of an external feature such as a pin or shaft, the virtual condition is determined by: MMC + TOL = VC; e.g., Pin MMC + stated orientation or position tolerance = Pin virtual condition.

Virtual condition of a pin or shaft is always a "constant value" and can also be referred to as the "outer boundary (locus)" in worst case analysis calculations.

EXAMPLE   **MEANING**

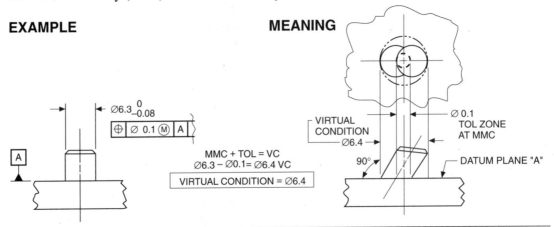

- FOR HOLE: In the case of an internal feature such as a hole, the virtual condition is determined by: MMC – TOL = VC; e.g., Hole MMC – stated orientation or position tolerance = Hole virtual condition.

Virtual condition of a hole is always a "constant value" and can also be referred to as the "inner boundary (locus)" in worst case analysis calculations.

EXAMPLE   **MEANING**

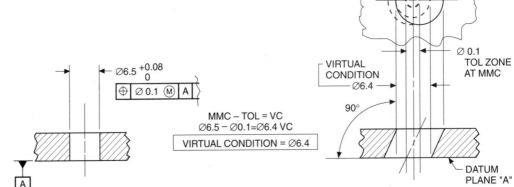

VIRTUAL CONDITION (LMC)

Definition. Virtual Condition. A constant boundary generated by the collective effects of a size feature's specified MMC or LMC material condition and the geometric tolerance for that material condition.

WHERE LMC IS SPECIFIED:

The virtual condition of a feature is the extreme boundary of that feature which represents the "worst case" for such concerns as cross-sectional mass, strength, alignment, wall thickness, compensating effects, interference, etc., relative to a mating part or situation.

- FOR PIN: In the case of an external feature such as a pin or shaft, the virtual condition is determined by: LMC – TOL = VC; e.g., Pin LMC – stated orientation position tolerance = Pin virtual condition.

Virtual condition of a pin or shaft is always a "constant value" and can also be referred to as the "inner boundary (locus)" in worst case analysis calculations.

EXAMPLE   **MEANING**

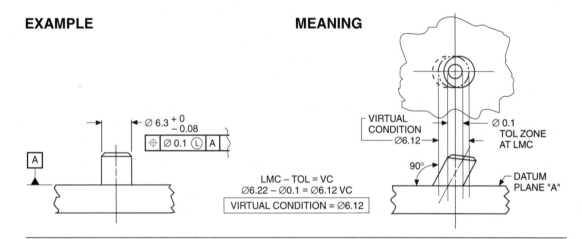

- FOR HOLE: In the case of an internal feature, such as a hole, the virtual condition is determined by: LMC + TOL = VC; e.g., Hole LMC + stated orientation or position tolerance = Hole virtual condition.

Virtual condition of a hole or shaft is always a "constant value" and can also be referred to as the "outer boundary (locus)" in worst case analysis calculations.

EXAMPLE   **MEANING**

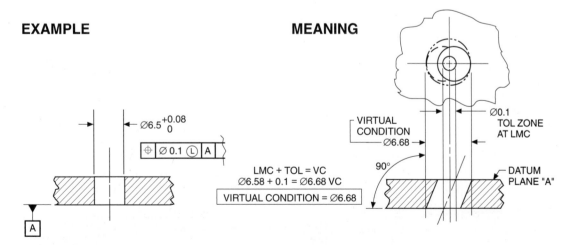

RESULTANT CONDITION — (MMC)

It is necessary to understand Resultant Condition as it applies to features. The definition below and the examples shown will clarify its meaning.

Definition. Resultant Condition.

The variable boundary generated by the collective effects of a size feature's specified MMC or LMC material condition, the geometric tolerance for that material condition, the size tolerance, and the additional geometric tolerance derived from the feature's departure from its specified MMC or LMC material condition.

WHERE MMC IS SPECIFIED:

The resultant condition of a feature is the extreme boundary which represents the "worst case" of that feature (in the opposite direction from its virtual condition).

- FOR PIN: In the case of an external feature such as a pin or shaft, the resultant condition is determined by: AMES – TOL – BTOL = RC; e.g., Pin Actual Mating Envelope Size (AMES) – stated position or orientation Tolerance (TOL) – Bonus Tolerance (BTOL) = Resultant Condition (RC).

  Resultant condition of a pin is a variable "worst case value" and can also be referred to as the "inner boundary (locus)" in worst case analysis calculations.

AMES	– TOL –	BTOL	= RC
Ø 6.3 MMC	Ø 0.1	· 0	Ø 6.2
Ø 6.28	Ø 0.1	Ø 0.02	Ø 6.16
Ø 6.26	Ø 0.1	Ø 0.04	Ø 6.12
Ø 6.24	Ø 0.1	Ø 0.06	Ø 6.08
Ø 6.22 LMC	Ø 0.1	Ø 0.08	Ø 6.04

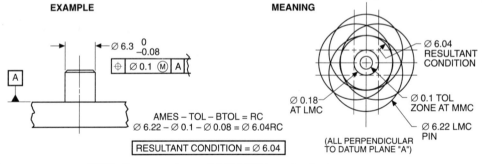

EXAMPLE   **MEANING**

Ø 6.3 $^{0}_{-0.08}$

⊕ | Ø 0.1 Ⓜ | A

A

AMES – TOL – BTOL = RC
Ø 6.22 – Ø 0.1 – Ø 0.08 = Ø 6.04RC

RESULTANT CONDITION = Ø 6.04

Ø 6.04 RESULTANT CONDITION

Ø 0.18 AT LMC

Ø 0.1 TOL ZONE AT MMC

Ø 6.22 LMC PIN

(ALL PERPENDICULAR TO DATUM PLANE "A")

(NOTE: Above example uses Ø 6.22 (LMC) for explanation.)

- FOR HOLE: In the case of an internal feature such as a hole, the resultant condition is determined by: AMES + TOL + BTOL = RC; e.g., Hole Actual Mating Envelope Size (AMES) + stated position or orientation Tolerance (TOL) + Bonus Tolerance (BTOL) = Resultant Condition (RC).

  Resultant Condition of a hole is a variable "worst case value" and can also be referred to as the "outer boundary (locus)" in worst case analysis calculations.

AMES	+ TOL +	BTOL	= RC
Ø 6.5 MMC	Ø 0.1	0	Ø 6.6
Ø 6.52	Ø 0.1	Ø0.02	Ø 6.64
Ø 6.54	Ø 0.1	Ø0.04	Ø 6.68
Ø 6.56	Ø 0.1	Ø0.06	Ø 6.72
Ø 6.58 LMC	Ø 0.1	Ø0.08	Ø 6.76

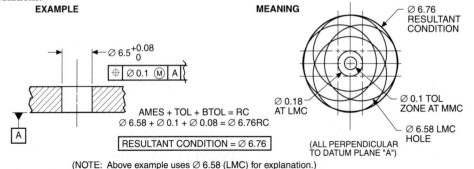

EXAMPLE   **MEANING**

Ø 6.5 $^{+0.08}_{0}$

⊕ | Ø 0.1 Ⓜ | A

A

AMES + TOL + BTOL = RC
Ø 6.58 + Ø 0.1 + Ø 0.08 = Ø 6.76RC

RESULTANT CONDITION = Ø 6.76

Ø 6.76 RESULTANT CONDITION

Ø 0.18 AT LMC

Ø 0.1 TOL ZONE AT MMC

Ø 6.58 LMC HOLE

(ALL PERPENDICULAR TO DATUM PLANE "A")

(NOTE: Above example uses Ø 6.58 (LMC) for explanation.)

RESULTANT CONDITION — (LMC)

It is necessary to understand Resultant Condition as it applies to features. The definition below and the examples shown will clarify its meaning.

Definition. Resultant Condition.

The variable boundary generated by the collective effects of a size features specified MMC or LMC material condition, the geometric tolerance for that material condition, the size tolerance, and the additional geometric tolerance derived from the feature's departure from its specified material condition.

WHERE LMC IS SPECIFIED:

The resultant condition of a feature is the extreme boundary which represents the "worst case" of that feature (in the opposite direction from its virtual condition).

- **FOR PIN:**   In the case of an external feature such as a pin or shaft, the resultant condition is determined by: AMES + TOL + BTOL = RC; e.g., Pin Actual Mating Envelope Size (AMES) + stated position or orientation Tolerance (TOL) + Bonus Tolerance (BTOL) = Resultant Condition (RC).

AMES	+ TOL	+ BTOL	= RC
Ø6.22 LMC	Ø 0.1	0	Ø6.32
Ø6.24	Ø 0.1	Ø0.02	Ø6.36
Ø6.26	Ø 0.1	Ø0.04	Ø6.4
Ø6.28	Ø 0.1	Ø0.06	Ø6.44
Ø6.3 MMC	Ø 0.1	Ø0.08	Ø6.48

Resultant condition of a pin is a variable "worst case value" and can also be referred to as the "outer boundary (locus)" in worst case analysis calculations.

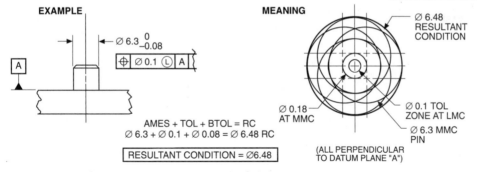

EXAMPLE

MEANING

$\emptyset\ 6.3\ ^{0}_{-0.08}$

⊕ Ø 0.1 Ⓛ A

A

Ø 6.48 RESULTANT CONDITION

Ø 0.18 AT MMC

Ø 0.1 TOL ZONE AT LMC

Ø 6.3 MMC PIN

AMES + TOL + BTOL = RC
Ø 6.3 + Ø 0.1 + Ø 0.08 = Ø 6.48 RC

RESULTANT CONDITION = Ø6.48

(ALL PERPENDICULAR TO DATUM PLANE "A")

(NOTE: Above example uses Ø 6.3 (MMC) for explanation.)

- **FOR HOLE:**   In the case of an internal feature such as a hole, the resultant condition is determined by: AMES – TOL – BTOL = RC; e.g., Hole Actual Mating Envelope Size (AMES) – stated position or orientation Tolerance (TOL) – Bonus Tolerance (BTOL) = Resultant Condition (RC).

AMES	– TOL	– BTOL	= RC
Ø6.58 LMC	Ø 0.1	0	Ø6.48
Ø6.56	Ø 0.1	Ø0.02	Ø6.44
Ø6.54	Ø 0.1	Ø0.04	Ø6.4
Ø6.52	Ø 0.1	Ø0.06	Ø6.36
Ø6.5 MMC	Ø 0.1	Ø0.08	Ø6.32

Resultant Condition of a hole is a variable "worst case value" and can also be referred to as the "inner boundary (locus)" in worst case analysis calculations.

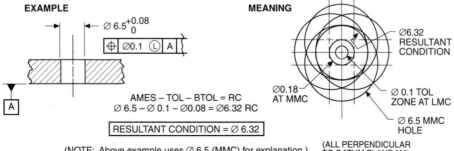

EXAMPLE

MEANING

$\emptyset\ 6.5\ ^{+0.08}_{0}$

⊕ Ø0.1 Ⓛ A

A

Ø6.32 RESULTANT CONDITION

Ø0.18 AT MMC

Ø 0.1 TOL ZONE AT LMC

Ø 6.5 MMC HOLE

AMES – TOL – BTOL = RC
Ø 6.5 – Ø 0.1 – Ø0.08 = Ø6.32 RC

RESULTANT CONDITION = Ø 6.32

(NOTE: Above example uses Ø 6.5 (MMC) for explanation.)

(ALL PERPENDICULAR TO DATUM PLANE "A")

TOLERANCES OF FORM, ORIENTATION, PROFILE, AND RUNOUT

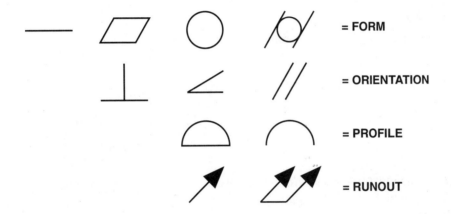

= FORM

= ORIENTATION

= PROFILE

= RUNOUT

Tolerances of form, orientation, profile and runout state how far actual surfaces or features are permitted to vary from those implied by the drawing. Expressions of form and orientation tolerances refer to flatness, straightness, circularity, cylindricity, parallelism, perpendicularity and angularity. Profile of a surface, profile of a line, circular runout and total runout tolerances are unique variations and combinations of form, orientation and sometimes location and are considered as separate types of characteristics.

Form, orientation, profile, or runout tolerances should be specified for all features critical to the design requirements:

a. where established workshop practices cannot be relied upon to provide the required accuracy;

b. where documents establishing suitable standards of workmanship cannot be prescribed;

c. where tolerances of size and location do not provide the necessary control.

The various tolerances of form, orientation, profile and runout often have an effect on one another; that is, parallelism could include flatness or straightness, and runout could include circularity, straightness, or cylindricity.

The following series of form, orientation, profile, and runout tolerance examples address each of these characteristics and their individual purpose and meaning, in order to explain the basic principles.

TOLERANCES OF FORM, ORIENTATION, AND PROFILE

Tolerances of form, orientation, and profile provide methods by which to control part geometry where size or location does not adequately do so. Such tolerances state how far an actual surface or feature is permitted to vary from the desired geometric shape implied by the drawing.

KINDS OF FEATURES TO WHICH A FORM, ORIENTATION, OR PROFILE TOLERANCE IS APPLICABLE

To correctly apply form, orientation, or profile tolerances, an understanding of the kind of features, INDIVIDUAL OR RELATED, upon which each characteristic can be used is required. Such tolerances can be applied as follows:

INDIVIDUAL feature—A *single surface, element,* or *size* feature which relates to a perfect geometric counterpart of itself as the desired form; no datum is proper or used.

Form Characteristics which can be applied:

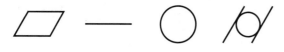

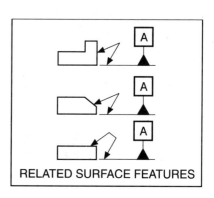

INDIVIDUAL FEATURES

RELATED surface feature—A *single surface* or *element* feature which relates to a datum, or datums, in orientation.*

Orientation Characteristics which can be applied:

RELATED SURFACE FEATURES

---

*Those tolerances involving related features and datums are sometimes also referred to as attitude tolerances.

RELATED size feature—A single *size* feature which relates to a datum, or datums, in form and orientation.*

Orientation Characteristics which can be applied:

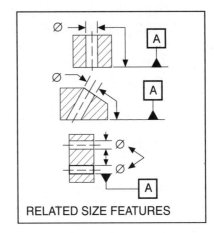

RELATED SIZE FEATURES

INDIVIDUAL or *RELATED* surface or size feature—A *single surface, element* or *size* feature whose perfect geometric profile is described, which may or may not relate to a datum, or datums, in form, orientation, and profile.*

Profile Characteristics which can be applied:

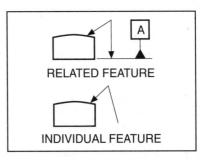

RELATED FEATURE

INDIVIDUAL FEATURE

TOLERANCES OF FORM—INDIVIDUAL FEATURES—NO DATUM

Tolerances of form used on individual features where no datum is proper nor used involve the characteristics below:

(FLATNESS)   (STRAIGHTNESS)   (CIRCULARITY)   (CYLINDRICITY)

These characteristics are used to describe form tolerances of *single surface, element,* or *size* features and relate to a perfect geometric counterpart of itself.

See following pages for details of application.

---

*Those tolerances involving related features and datums are sometimes also referred to as attitude tolerances.

FLATNESS ▱

Definition. Flatness is the condition of a surface having all elements in one plane.

FLATNESS TOLERANCE

Flatness tolerance specifies a tolerance zone confined by two parallel planes within which the surface must lie.

FLATNESS TOLERANCE APPLICATION

The example below shows how a flatness symbol is applied.

The symbol is interpreted to read "This surface shall be flat within 0.05 total tolerance zone over entire surface." Note that the 0.05 tolerance zone is *total* variation. To be acceptable, the entire actual surface must fall within the parallel plane extremities of the 0.05 tolerance zone. The 0.05 flatness tolerance is based upon the design requirement which is derived from the precision determined necessary for mating part interface, seal-off, appearance, etc.

A flatness tolerance is a form control of all elements of a surface as it compares to a simulated perfect geometric counterpart of itself. The perfect geometric counterpart of a flat surface is a plane. The tolerance zone is established as a width or thickness zone relative to this plane as established from the actual part surface.

EXAMPLE

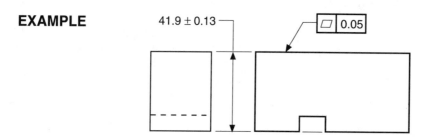

MEANING

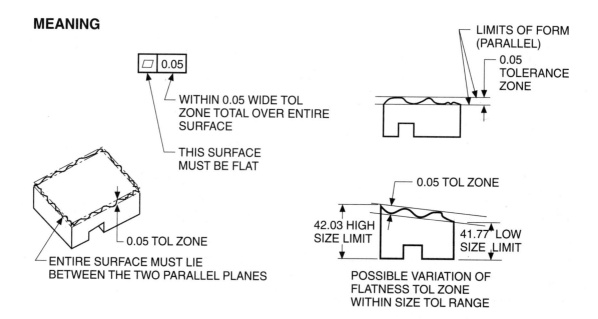

Note that the extremities or high points of the surface determine one limit or plane of the tolerance zone, with the other limit or plane being established 0.05 (the specified tolerance) parallel to it.

Since flatness tolerancing control is essentially a relationship of a feature to itself, no datum references are required or proper.

Also, note that since flatness is a form tolerance controlling surface elements only, it is not applicable to RFS or MMC considerations.

In the absence of a flatness tolerance specification, the size tolerance and method of manufacture of a part will exercise some control over its flatness. However, when a flatness tolerance is specified, as applicable to a single surface, the flatness tolerance zone must be contained *within* the size tolerance limits.* It cannot be additive to the size tolerance. Therefore, a flatness tolerance should normally be *less than* the part size tolerance.

Where necessary the terms "MUST NOT BE CONCAVE" or "MUST NOT BE CONVEX" may be added beneath the feature control frame.

Author advisory — as a "rule-of-thumb," as based upon the norms of production and probabilities, it may be well to consider that a design calculated flatness tolerance requirement on a stable rigid part, should be equal to, or less than, one-half of the overall size tolerance (i.e. 0.26 on illustrated part) for justification as a specified flatness tolerance. The 0.05 tolerance illustrated being less than 0.13 (1/2 of 0.26 size tolerance) is an example of this logic.

STRAIGHTNESS —

Definition. Straightness is a condition where an element of a surface or an axis is a straight line.

STRAIGHTNESS TOLERANCE

A straightness tolerance specifies a tolerance zone within which the considered element or derived median line must lie.

STRAIGHTNESS TOLERANCE APPLICATION

A straightness tolerance is applied in the view where the elements to be controlled are represented by a straight line.

STRAIGHTNESS TOLERANCE—SURFACE ELEMENT CONTROL

Straightness tolerance is typically used as a form control of individual surface elements such as those on cylindrical or conical surfaces. Since surfaces of this kind are made up of an infinite number of longitudinal elements, a straightness requirement applies to the entire surface as controlled in single line elements in the direction specified.

The example following illustrates straightness control of individual longitudinal surface elements on a cylindrical part. Note that the symbol is directed to the feature surface (or extension line) and not to the dimension lines. The straightness tolerance must be less than the size tolerance.

---

*Under special circumstances, a note specifically exempting the pertinent size dimension, i.e., PERFECT FORM AT MMC NOT REQUIRED, may be specified. In such cases the flatness may be greater than the size tolerance.

Author advisory—as a "rule-of-thumb," as based upon the norms of production and probabilities, it may be well to consider that a design calculated surface element straightness tolerance requirement on a stable rigid part, should be equal to, or less than, one half of the overall diameter size tolerance (i.e. 0.16 on illustrated part) for justification as a specified straightness tolerance.

All actual local size (circular elements) of the surface must be within the specified size tolerance and the boundary of perfect form at MMC.* Also, each longitudinal element of the surface must lie in a tolerance zone defined by two parallel lines spaced apart by the amount of the prescribed tolerance where the two lines and the nominal axis of the part share a common plane.

NOTE Since surface element control is specified, the tolerance zone applies uniformly whether the part is of a bowed, waisted, or barreled shape.

EXAMPLE

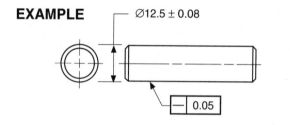

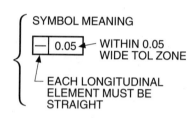

MEANING

THE FEATURE MUST BE WITHIN THE SPECIFIED TOL OF SIZE AND THE BOUNDARY OF PERFECT FORM AT MMC (12.58). EACH LONGITUDINAL ELEMENT OF THE SURFACE MUST LIE BETWEEN TWO PARALLEL LINES (0.05 APART) WHERE THE TWO LINES AND THE NOMINAL AXIS OF THE PART SHARE A COMMON PLANE.

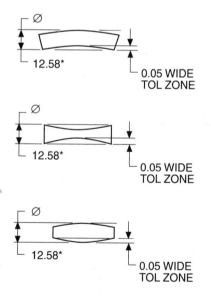

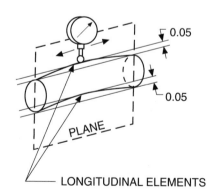

*BOUNDARY OF PERFECT FORM AT MMC.

Note the absence of any datum reference when straightness tolerancing is used. Straightness is a form control of a single element, or an axis, as it relates to a perfect geometric counterpart of itself. Therefore, datum references are neither required nor proper.

---

*Under special circumstances a note specifically exempting the pertinent size dimensioning, i.e., PERFECT FORM AT MMC NOT REQUIRED, may be specified. In such cases, the straightness tolerance may be greater than the size tolerance.

STRAIGHTNESS TOLERANCE APPLIED TO FLAT SURFACE

Straightness tolerancing may be applied to a flat surface to provide element surface control in a specific direction as a refinement of size tolerance or other form tolerance such as flatness. If so used, the straightness tolerance must be less than the size tolerance or the refined tolerance.

The example below illustrates straightness tolerance of surface elements as a refinement of size tolerance. Note that the tolerance is applied in the view in which the elements to be controlled appear as a straight line.

The individual straightness elements must be within both the size* tolerance and the straightness tolerance zone of 0.08, whereas element to element in the other view, variation within the size tolerance may occur.

EXAMPLE

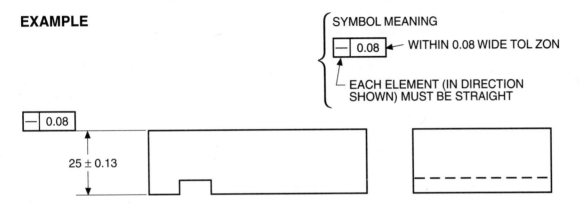

MEANING

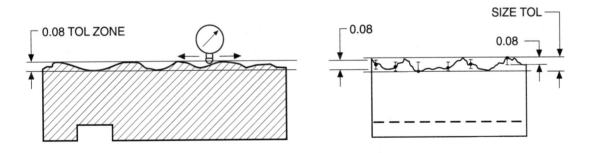

---

*Under special circumstances, a note specifically exempting the pertinent size dimensioning, i.e., PERFECT FORM AT MMC NOT REQUIRED, may be specified. In such cases, the straightness tolerance may be greater than the size tolerance.

STRAIGHTNESS TOLERANCE RFS AND MMC TO AXIS

Where function of a size feature permits a collective result of *size* and *form* variation and the perfect form at MMC boundary may be exceeded, the RFS or MMC principles may be used.

In this instance, where the appropriate symbology and specifications are stated, the part is not confined to the perfect form at MMC boundary. All actual local sizes (cross-sectional elements) of the surface are to remain within the specified size tolerance, but the total part surface may exceed the perfect form at MMC boundary to the extent of the straightness tolerance.

This principle may be applied to individual size features such as pins, shafts, bars, etc., where the longitudinal elements are to be specified with a straightness tolerance independent of, or in addition to, the size tolerance as a design reality or manufacturing necessity. In such a case, a new outer boundary or virtual condition is developed which represents the collective effect of the size and form error that must be considered in determining the fit, mating feature relationship clearance or between parts.

STRAIGHTNESS—RFS

Where a cylindrical feature is to be controlled on an RFS basis, as below, the feature control symbol must be located with the size dimension or attached to the dimension line, and the diameter symbol must precede the straightness tolerance.

EXAMPLE

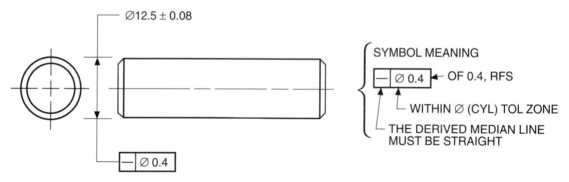

MEANING

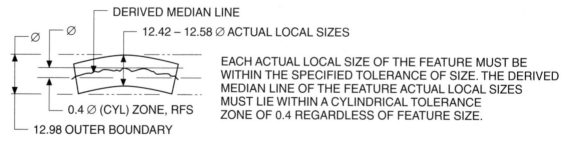

Author advisory—In such cases, the straightness tolerance need *not* be greater than the size tolerance; however, it is a common situation. Where the part length may approach *ten times* the part diameter or more, maintaining the perfect form at MMC boundary may be impractical. Therefore, this method could be a necessary alternative in the design considerations.

STRAIGHTNESS—MMC—VIRTUAL CONDITION

Where a cylindrical feature has a functional relationship with another feature, such as a pin or shaft and a hole, the control of straightness on an MMC basis may be desirable. If the pin or shaft, for example, is to fit into a hole of a given diameter, the collective effect of the pin size and its straightness error must be considered in relationship to the hole size minimum, i.e., their virtual conditions must be considered relative to one another.

On the part below, the size must be maintained at all actual local sizes (cross-sectional elements) within the stated limits. Likewise its straightness, using the axis as the criteria, must be within tolerance, but only when the part is at MMC. Therefore, the part develops (or is based upon) a virtual condition of 12.98, which represents the extreme condition the part can have and yet perform its function, or fit the mating part.

By stating the requirements on an MMC basis, the allowable straightness tolerance may increase an amount equal to the actual local size departure from MMC. The feature control symbol must be located with the size dimension, or attached to the dimension line; the diameter symbol must precede the straightness tolerance; and the MMC symbol must be inserted following the tolerance. In this manner maximum tolerance is achieved, part fit is guaranteed, and functional gaging techniques may be used.

EXAMPLE

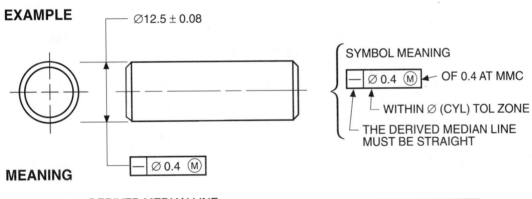

⌀12.5 ± 0.08

SYMBOL MEANING

— ⌀ 0.4 Ⓜ ◄— OF 0.4 AT MMC

└— WITHIN ⌀ (CYL) TOL ZONE

└— THE DERIVED MEDIAN LINE MUST BE STRAIGHT

MEANING

— ⌀ 0.4 Ⓜ

DERIVED MEDIAN LINE

FEATURE ACTUAL LOCAL SIZES, SIZE ⌀

⌀

⌀

⌀ (CYL) TOL ZONE

12.98 VIRTUAL CONDITION

ACTUAL LOCAL SIZE ⌀	⌀ TOL ZONE IS
12.58 MMC	0.40
12.56	0.42
12.54	0.44
12.52	0.46
12.50	0.48
12.48	0.50
12.46	0.52
12.44	0.54
12.42 LMC	0.56

EACH ACTUAL LOCAL SIZE OF THE FEATURE MUST BE WITHIN THE SPECIFIED TOLERANCE OF SIZE. THE DERIVED MEDIAN LINE OF THE FEATURE MUST LIE WITHIN A CYLINDRICAL TOLERANCE ZONE OF 0.4 AT MMC. AS THE ACTUAL LOCAL SIZES OF THE FEATURE DEPART FROM MMC, AN INCREASE IN THE STRAIGHTNESS TOLERANCE IS ALLOWED WHICH IS EQUAL TO THE AMOUNT OF SUCH DEPARTURE.

Where straightness tolerance on an MMC basis is specified, functional gaging techniques may be used. The below gage and conditions demonstrate how these principles can be applied to the preceding part.

GAGE   CONDITIONS

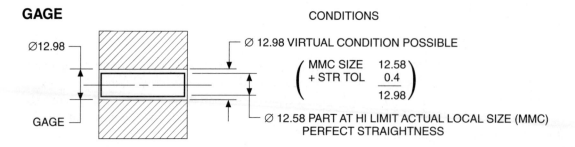

Ø12.98

GAGE

— Ø 12.98 VIRTUAL CONDITION POSSIBLE

$$\left(\begin{array}{cc} \text{MMC SIZE} & 12.58 \\ + \text{STR TOL} & \underline{0.4} \\ & 12.98 \end{array}\right)$$

— Ø 12.58 PART AT HI LIMIT ACTUAL LOCAL SIZE (MMC)
  PERFECT STRAIGHTNESS

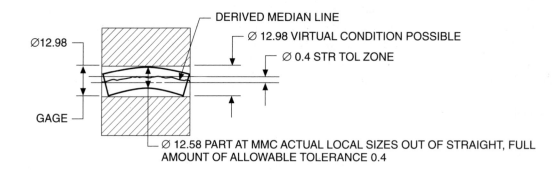

— DERIVED MEDIAN LINE
— Ø 12.98 VIRTUAL CONDITION POSSIBLE
— Ø 0.4 STR TOL ZONE

Ø12.98

GAGE

— Ø 12.58 PART AT MMC ACTUAL LOCAL SIZES OUT OF STRAIGHT, FULL
  AMOUNT OF ALLOWABLE TOLERANCE 0.4

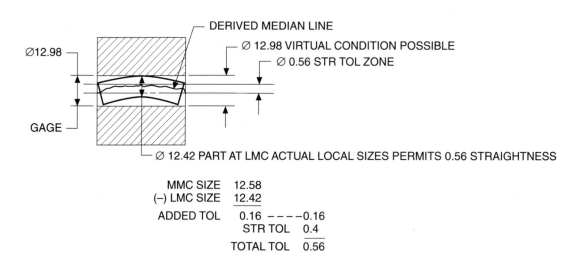

— DERIVED MEDIAN LINE
— Ø 12.98 VIRTUAL CONDITION POSSIBLE
— Ø 0.56 STR TOL ZONE

Ø12.98

GAGE

— Ø 12.42 PART AT LMC ACTUAL LOCAL SIZES PERMITS 0.56 STRAIGHTNESS

MMC SIZE	12.58
(–) LMC SIZE	12.42
ADDED TOL	0.16 – – – –0.16
STR TOL	0.4
TOTAL TOL	0.56

From the above it is seen that straightness applied on an MMC basis provides control of mating part conditions to facilitate design and provides maximum production tolerance. Functional gaging techniques are permissible. Note that the above gage simulates both the extreme permissible condition of the part as well as the condition of the mating part.

Author Advisory—As a practical matter it is not realistic to try to determine *different* resulting straightness tolerance values at each actual local size. It is, therefore, necessary to base the results of the varying sizes on a collective basis. As the individual actual local sizes of the feature depart from maximum material condition (MMC), they develop an actual mating envelope. This actual mating envelope may be of any varying size to the maximum value which would then reach "virtual condition" size. Thus, the straightness tolerance of such a part is gradually increased as the centers of the actual local sizes derive the median line resulting from the feature size and form deviations (i.e., the shaft gets smaller and more bow is permitted). This simultaneously develops a gradually increasing actual mating envelope size as the collective result of the various involved surface elements. Thus, the specific value of the amount of the increased tolerance would be difficult to state since that amount would be equal the collective result of these conditions on any one part. The illustrated functional gage shows this principle and demonstrates the validity of the functional gage method with the "virtual condition" gage hole size representing the "worst case" condition.

STRAIGHTNESS ON UNIT LENGTH BASIS

Where required, straightness may be applied on a unit length basis. This method is occasionally used to facilitate special design requirements and to prevent abrupt surface variations within a relatively short length of the feature. To prevent extreme variations of bow over the total length of the part, the amount of permissible straightness variation allowable on the total length of the part should be specified. If the unit variations are permitted to continue along the length of the part with no maximum tolerance indicated, an unsatisfactory part could result.

The example below illustrates a part with unit straightness specified on an RFS basis (MMC principles could also be used if desired.)

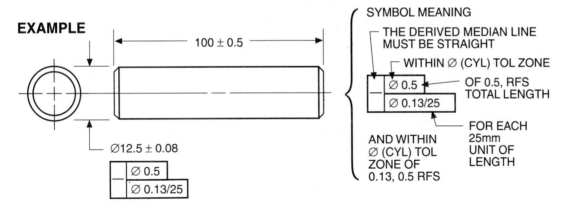

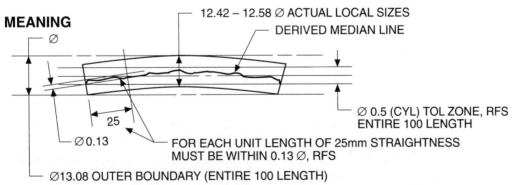

EACH ACTUAL LOCAL SIZE OF THE FEATURE MUST BE WITHIN THE SPECIFIED TOLERANCE OF SIZE. THE DERIVED MEDIAN LINE OF THE FEATURE MUST LIE WITHIN A CYLINDRICAL TOL ZONE OF 0.5 FOR THE TOTAL 100 LENGTH, AND WITHIN A 0.13 CYLINDRICAL TOL ZONE FOR EACH 25mm LENGTH, RFS

CIRCULARITY ◯

Definition. Circularity is the condition of a surface where:

1.   for a feature other than a sphere, all points of the surface intersected by any plane perpendicular to an axis are equidistant from that axis;

2.   for a sphere, all points of the surface intersected by any plane passing through a common center are equidistant from that center.

CIRCULARITY TOLERANCE

A circularity tolerance specifies a tolerance zone bounded by two concentric circles within which each circular element of the surface must lie and applies independently at any plane as described above.

CIRCULARITY TOLERANCE APPLICATION

Limits of size exercise control of circularity within the size tolerance. Often this provides adequate control. However, where necessary to further refine form control, circularity tolerancing can be used on any figure of revolution or circular cross section.

The example illustrates a part with a circularity tolerance of 0.05 specified on a cylindrical part.

The interpretation shows how one establishes the 0.05 tolerance zone. Note that the tolerance zone is the width of the annular zone between the two concentric circles.

A circularity tolerance zone is established relative to the actual local size of the part when measured at the surface periphery at any cross section perpendicular to the part axis. It should be noted that the circularity tolerance applies only at the cross-sectional point of measurement, and is relative to the *size* at that point. Therefore, a cylindrical part with circularity tolerance control could taper or otherwise vary in its surface contour within its size tolerance range, yet still meet circularity requirements if it is within the circularity tolerance at that point.

The part size in this example has been assumed to measure 12.55 at its largest point at the cross section selected for measurement. The 0.05 circularity tolerance zone is then established by two theoretically perfect concentric circles, one at the 12.55 diameter and the other 0.1 *smaller* at the 12.45 diameter. This establishes the tolerance zone of 0.05 *width* between the concentric circles. To be acceptable, the part surface at that cross section must fall within the 0.5 wide tolerance zone.

As is seen, the tolerance zone is established relative to the part actual local size wherever it may fall in its size tolerance range. That is, the part *size* is first determined and its circularity is then defined as a refinement of the part *form* relative to that *size*. Unless otherwise specified, any established size at any point along the surface can be used to determine the circularity tolerance zone. It is therefore seen that the circularity tolerance may be based on *different* sizes on the same part. The circularity tolerance zone, however, remains constant.

Author Advisory—as a "rule-of-thumb," in determining a circularity tolerance, the calculated value should be equal to, or less than, one-half of the feature's total size tolerance; i.e., in the illustrated figure, the total size tolerance is 0.16, the 0.05 determined is less than 0.08. This rule-of-thumb would obviously be applicable only to rigid parts not subject to free state variation.

Note again that the circularity tolerance must always be contained *within* the part *size* tolerance range. The circularity tolerance cannot exceed the size tolerance limits. Furthermore, a circularity tolerance cannot be modified to an MMC application since it controls surface elements only.

A circularity tolerance is a form control of a single part element as it compares to a perfect counterpart of itself. Therefore, no datum references are required nor proper. A circularity tolerance does have a reference center, but this is considered only as a part of the perfect frame of reference (concentric circles and their common center) for measurement, just as a straightness tolerance refers to a perfect line of reference for its measurement. You may compare a circularity tolerance to a straightness tolerance zone curled around a circle.

CIRCULARITY OF A CYLINDER

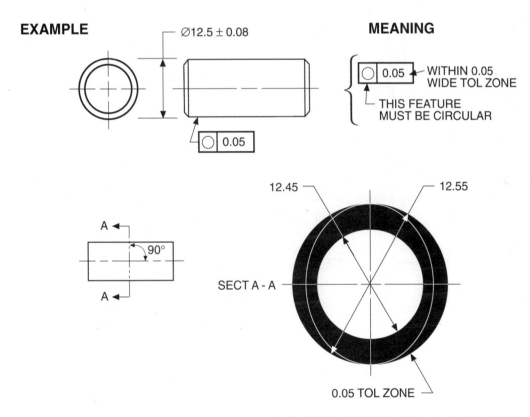

SURFACE PERIPHERY AT ANY CROSS SECTION PERPENDICULAR TO THE FEATURE AXIS MUST BE WITHIN THE SPECIFIED TOLERANCE OF SIZE AND MUST LIE BETWEEN TWO CONCENTRIC CIRCLES, ONE HAVING A RADIUS 0.05 LARGER THAN THE OTHER.

(ABOVE SIZES ARBITRARILY SELECTED FOR ILLUSTRATION)

CIRCULARITY OF A CONE

The example below illustrates a cone-shaped part for which a circularity tolerance of 0.03 is specified. As previously discussed, the periphery at any cross section perpendicular to the axis must be within the specified tolerance of size and must lie between the two concentric circles (one having a radius 0.03 larger than the other).

EXAMPLE

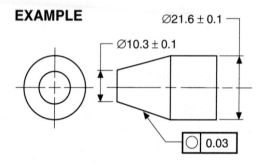

MEANING

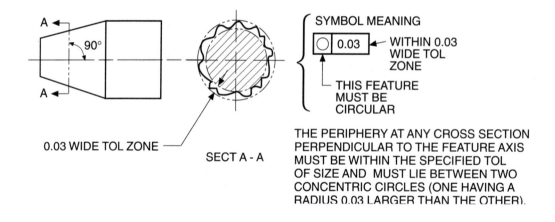

0.03 WIDE TOL ZONE

SECT A - A

SYMBOL MEANING

⊙ | 0.03 ← WITHIN 0.03 WIDE TOL ZONE

└ THIS FEATURE MUST BE CIRCULAR

THE PERIPHERY AT ANY CROSS SECTION PERPENDICULAR TO THE FEATURE AXIS MUST BE WITHIN THE SPECIFIED TOL OF SIZE AND MUST LIE BETWEEN TWO CONCENTRIC CIRCLES (ONE HAVING A RADIUS 0.03 LARGER THAN THE OTHER).

CIRCULARITY OF A SPHERE

Circularity of a spherical part is based upon the same principles as those for a cylinder or cone preceding, except that the tolerance control reference is to *any cross section passing through a common center* rather than to *any cross section perpendicular to the axis,* as in the conventional application of circularity tolerancing.

EXAMPLE 2

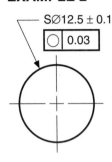

SØ12.5 ± 0.1

| ◯ | 0.03 |

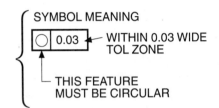

SYMBOL MEANING

| ◯ | 0.03 | ← WITHIN 0.03 WIDE TOL ZONE

└ THIS FEATURE MUST BE CIRCULAR

THE PERIPHERY AT ANY CROSS SECTION PASSING THROUGH A COMMON CENTER MUST BE WITHIN THE SPECIFIED TOLERANCE OF SIZE AND MUST BE BETWEEN TWO CONCENTRIC CIRCLES (ONE HAVING A RADIUS 0.03 LARGER THAN THE OTHER).

MEANING

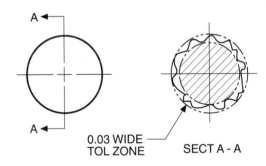

0.03 WIDE
TOL ZONE

SECT A - A

CYLINDRICITY ⌭

Definition. Cylindricity is a condition of a surface of revolution in which all points of the surface are equidistant from a common axis.

CYLINDRICITY TOLERANCE

A cylindricity tolerance specifies a tolerance zone bounded by two concentric cylinders within which the surface must lie.

CYLINDRICITY TOLERANCE APPLICATION

Limits of size exercise control of cylindricity within the size tolerance. This control is often adequate. However, where more refined form control is required, cylindricity tolerancing can be used. Note that in cylindricity, unlike circularity, the tolerance applies simultaneously to both circular and longitudinal elements of the entire surface.

The example illustrates a part with a cylindricity tolerance of 0.05 A cylindricity tolerance is interpreted as "on the radius" or, as in this case, within the 0.05 wide tolerance zone defined by two concentric cylinders 0.05 apart. A cylindricity tolerance can be considered as circularity tolerancing extended to control the *entire* surface of a cylinder.

EXAMPLE

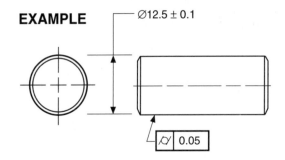

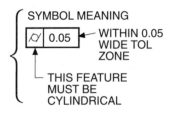

SYMBOL MEANING

⌭ | 0.05 ← WITHIN 0.05 WIDE TOL ZONE

└ THIS FEATURE MUST BE CYLINDRICAL

THE FEATURE MUST BE WITHIN THE SPECIFIED TOLERANCE OF SIZE AND MUST LIE BETWEEN TWO CONCENTRIC CYLINDERS (ONE HAVING A RADIUS 0.05 LARGER THAN THE OTHER).

MEANING

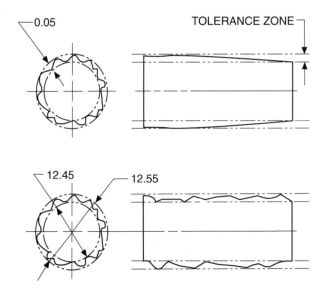

Cylindricity tolerancing can be applied *only* to cylindrical forms. The leader from the feature control symbol may be directed to either view.

It should be noted that a cylindricity tolerance simultaneously controls circularity, straightness, and parallelism of the elements of the cylindrical surface.

A cylindricity tolerance applies to the *entire* cylindrical surface as opposed to the cross-sectional or diametral measurement considered in circularity. Also, in measuring cylindricity, the concentric cylinders defining the tolerance zone are always based on the actual mating size of the part produced. The part size in the example has been assumed to measure 12.55 at its largest diameter. This 12.55 is the actual mating size diameter of the largest of the concentric cylinders defining the tolerance zone. The smaller of the concentric cylinders is 12.55 minus the amount of the cylindricity tolerance (0.5 on R = 0.1 on dia), i.e., 12.45. The cylindricity tolerance zone is therefore 0.05 *wide* between the concentric cylinders. The entire part surface must then fall within this tolerance zone to be acceptable.

It should be noted that the cylindricity tolerance must always be contained *within* the part *size* tolerance range. The cylindricity tolerance cannot exceed the size tolerance limits.

As is seen, the cylindricity tolerance zone is established relative to the part actual mating size wherever it may fall in its *size* tolerance range. That is, the part size is first determined and its *cylindricity* is then defined as a refinement of the part *form* relative to that *size*.

If the largest measurement of the produced part had been near the *low* limit, for example at 12.45, the cylindricity tolerance could not have been more than 0.025 on R (= 0.05 on dia). The cylindricity tolerance cannot exceed *size* tolerance limits. Therefore, the form (cylindricity) variation could not be less than the low limit size of 12.4.

A cylindricity tolerance (similar to flatness) is a form control of a surface element as it compares to a perfect counterpart of itself. Therefore, no datum references are required nor proper. One can compare a cylindricity tolerance zone to a flatness tolerance zone by visualizing that flatness zone curled around a cylinder. A cylindricity tolerance does have a reference axis, but this is considered only a part of the perfect frame of reference (concentric cylinders and their common axis) for measurement, just as a flatness tolerance refers to a perfect plane of reference for its measurement.

Since cylindricity is a form tolerance controlling surface elements only, it cannot be modified to an MMC application.

Author advisory—as a rule-of-thumb in determining a cylindricity tolerance, the calculated value should be equal to or less than, one-half of the feature's size tolerance; i.e., in the illustrated figure the total size tolerance is 0.2, the 0.05 determined is less than 0.1. This rule-of-thumb would obviously be applicable only to rigid parts not subject to free state variation.

EVALUATION OF CIRCULARITY AND CYLINDRICITY

Cylindricity is checked or measured by the same basic techniques that are used to check circularity except that cylindricity involves a cylindrical tolerance zone of uniform thickness over the entire surface and is based on a single size reference. This must be considered in any measuring procedure concerned with cylindricity. For example, in the following discussions of vee block, between centers, and polargraph methods, the measurements are made at cross sections only; whereas in cylindricity, one must consider the entire surface as controlled by the one tolerance zone.

Vee block or between centers methods are often used in open set-up inspection of circularity and cylindricity.

The vee block method provides an approximation or a rough check which may be adequate for some applications. Variables, such as the varying number and arrangement of lobes on the part surface and the angle of the vee block, may affect the resultant measurement sufficiently for the method to become rather inaccurate. In addition, the procedure itself contributes to inaccuracies, because one portion of the part surface (in the vee) is used as the basis for checking another portion of the same surface, thus compounding the chances for error. With certain lobe and vee block combinations, a visibly out-of-circularity part can be rotated with no evidence of error on the dial indicator (see next page).

When the vee block method is used, the total error read is roughly the error on the *diameter*; it must be halved, as an approximation, to compare the result with the specified circularity tolerance which is implied as the radial separation or total width between annular circles.

The between centers method more nearly establishes the reference axis for the geometric relationship. Any error in the centers, however, will affect the accuracy of the resultant measurement. Since the part is rotated about its nominal center or axis, the reading will be on the radius and representative of the total width zone between concentric circles. This method, however, is more correctly termed a runout relationship of a surface and centers and is technically not a circularity analysis. This may be used as a method to support the part recognizing that other considerations are necessary to isolate the circularity error.

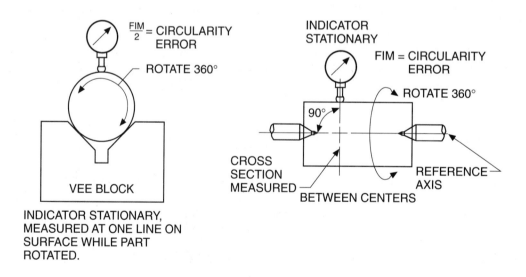

Figures 1 and 2 on the following page demonstrate that certain vee block and lobed-part conditions do not register the correct part circularity or cylindricity error.

Obviously these parts have noticeable form errors. However, when the parts are rotated in the 60° vee block, the true error will not register. In fact, in these hypothetical examples of a five-lobed part and an oval shape, *no* error registers in any position. When the number of lobes of the part is known, certain vee block angles can be used to obtain accurate diametral readings. However, the above indicated variables and the difficulty of accurately predicting conditions on any given part make this method impractical.

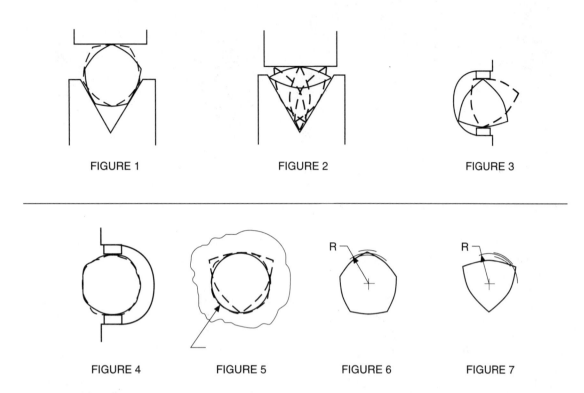

FIGURE 1   FIGURE 2   FIGURE 3

FIGURE 4   FIGURE 5   FIGURE 6   FIGURE 7

Figures 3 and 4 illustrate how "miking" a lobed part (particularly a part with an uneven number of lobes) will not pick up error. The parts shown, although obviously not true in form, register the same measurements at any diametrical location. Suppose the parts in Fig. 1 (five lobes) and Fig. 3 (three lobes) were "miked" at 49.98 since they are intended to go through the 50 ⌀ hole shown in Fig. 5. The parts will not pass through the hole although they were "miked" at the lower size measurement of 49.98.

Figures 6 and 7 illustrate the parts shown in Figs. 1 and 3 evaluated on the basis of radial values. Using appropriate measuring techniques, the error can be determined directly.

PRECISION METHOD OF EVALUATING CIRCULARITY AND CYLINDRICITY

To evaluate circularity and cylindricity precisely, one must relate the part surface periphery to the geometry of a perfectly round or cylindrical form as constructed from a reference axis. Several kinds of special gaging equipment utilizing optical, mechanical, electronic, and pneumatic principles are available. One method uses an electronic probe which travels around the periphery of the part while the part is chuck-mounted on an extremely accurate spindle and transcribes an enlarged profile of the part periphery on a polargraph. This profile is then compared with a transparent overlay gage which contains circles at various increments. Note that the final basis for comparison is the part profile only and the reference axis is merely a means of constructing the geometry for measurement. Although more costly and time consuming than other methods, this method utilizes geometric relationships which more directly evaluate circularity and cylindricity where necessary.

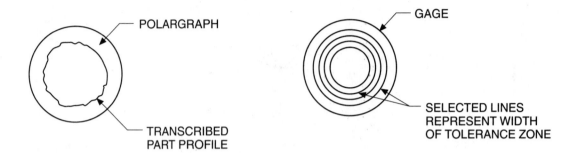

OTHER METHODS OF EVALUATING CIRCULARITY AND CYLINDRICITY

Due to the increasing sophistication of design requirements, manufacturing processes, and measuring processes, such control as circularity and cylindricity may be accomplished by yet further methods.

Such methods as "Least Mean Square," which determines the center of a circular form by mathematical formula, or a computer program based on ordinate and radius measurements may be used. Other methods, such as "minimum circumscribed circle," "maximum inscribed circle," and "minimum radial separation," which utilize precision spindle techniques can be used in appropriate circumstances.

More specific callout of the drawing requirement indicating one of the above methods (or stylus tip radius, cycles per revolution, etc.) may occasionally be necessary.

DATUMS
BASIS FOR RELATING FEATURES TO ONE ANOTHER

MEANING OF DATUM

Flatness, straightness, circularity, cylindricity (and occasionally profile) tolerancing are characteristics or controls which are applied to single features. It might be said that in order to define these controls, the feature is compared to a perfect counterpart of itself with a stated tolerance to indicate the variation permissible from perfect for that feature. It might also be said that the feature serves as its own "datum"; a datum being a feature from which relationships originate. The following definition will further clarify the meaning and intention for a datum.

Definition. A datum is a theoretically exact point, axis, or plane derived from the true geometric counterpart of a specified datum feature. A datum is the origin from which the location or geometric characteristics of features of a part are established.

DATUM VS. DATUM SURFACE AND DATUM FEATURE

A distinction must be made between a "datum" (theoretically exact—as on a drawing), which represents design requirement and function, and the actual "datum surface" or "datum feature" on the produced part.

Definition. A datum feature is an actual feature of a part that is used to establish a datum.

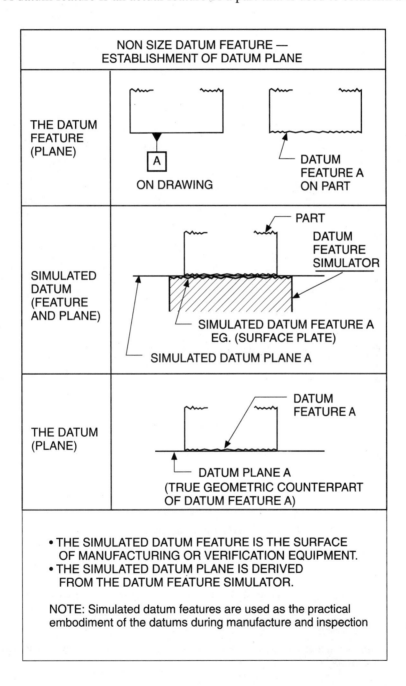

NOTE: See also Author Advisory on page 59.

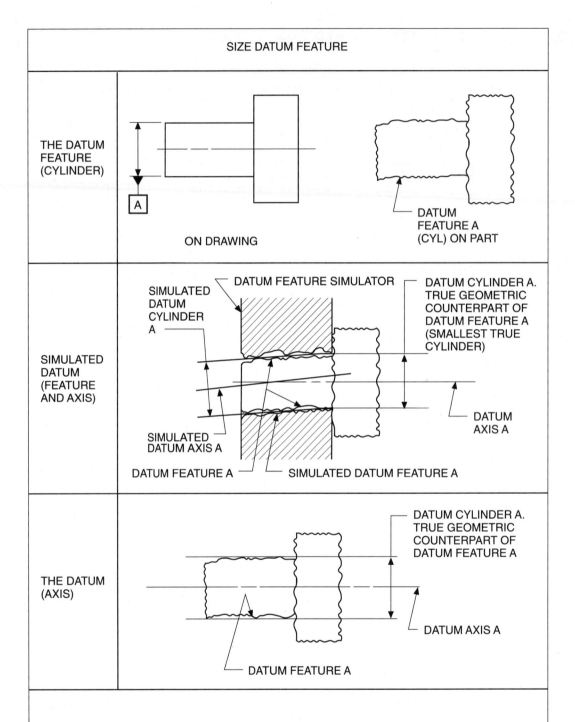

SIZE DATUM FEATURE

THE DATUM FEATURE (CYLINDER)

ON DRAWING

DATUM FEATURE A (CYL) ON PART

SIMULATED DATUM (FEATURE AND AXIS)

SIMULATED DATUM CYLINDER A

DATUM FEATURE SIMULATOR

DATUM CYLINDER A. TRUE GEOMETRIC COUNTERPART OF DATUM FEATURE A (SMALLEST TRUE CYLINDER)

DATUM AXIS A

SIMULATED DATUM AXIS A

DATUM FEATURE A

SIMULATED DATUM FEATURE A

THE DATUM (AXIS)

DATUM CYLINDER A. TRUE GEOMETRIC COUNTERPART OF DATUM FEATURE A

DATUM AXIS A

DATUM FEATURE A

- THE SIMULATED DATUM FEATURE IS THE SURFACE OF MANUFACTURING OR VERIFICATION EQUIPMENT.
- THE SIMULATED DATUM AXIS IS DERIVED FROM THE DATUM FEATURE SIMULATOR.

NOTE: Simulated datum features are used as the practical embodiment of the datums during manufacture and inspection.

NOTE: See also Author Advisory on page 59.

Author advisory: Any physical differences between the "true geometric counterpart of the datum feature," the "simulated datum feature" and the "simulated datum plane" represent a necessary compromise between unpredictable part, tool and gage surface precision variables versus practical application within human capabilities to approach such finite exactness. To attempt to bridge this gap between the theory and the realities involved, the "simulated datum feature," the "simulated datum plane" and the "true geometric counterpart of the datum feature" are, for all practical purposes, normally considered consolidated into one composite result when applied under typical every day conditions. This means that when the "simulated datum feature (via the datum feature simulator)" and the part "datum feature" are brought into contact, "the datum (point, axis or plane)" is considered to be, thus, established about as well as it can be done under normal operating conditions and constraints. Where sophisticated computerized electronic verification or evaluation methods are undertaken, it may be possible and/or necessary to detect and respect such minute differences. Under these conditions, results are based upon recognition, detection and distinction of these values. Because of the above referenced electronic options and more sophisticated requirements arising in industry, the standards must recognize this more detailed nomenclature to adequately describe the datum concepts where necessary.

DATUMS REQUIRED TO RELATE FEATURES

When relating form, orientation, profile, runout, or location tolerances to features in relation to one another, a datum (or datums), must be used. Datum precedence also must be considered when necessary. (See below and Datum section.)

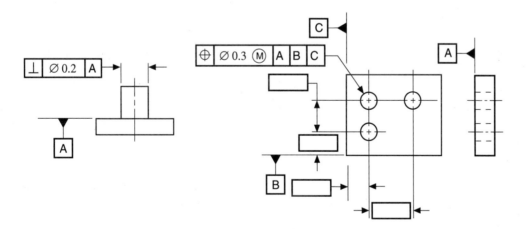

Further examples of datum application using form, orientation, profile, runout, and location characteristics are given in the DATUM section and in following examples throughout this text.

ORIENTATION TOLERANCES—
RELATED FEATURES USING DATUMS

Orientation tolerances used on related features and which require a datum involve the characteristics below:

(PERPENDICULARITY)   (ANGULARITY)   (PARALLELISM)

These characteristics are used to describe orientation tolerances of *single surface, element,* or *size* features and are always related to a *datum.*

Orientation tolerance when applied to a plane surface, is a refinement added to the design which would not be controlled by the size dimensions. Orientation tolerance when applied to a plane surface controls the "flatness" as well as its orientation.

Orientation tolerance when applied to a size feature (hole or pin) is a refinement of orientation (perpendicularity, etc.) within the stated locational tolerance; therefore, it must be *less than* the governing locational tolerance of the features involved. Orientation tolerance and the values determined are based upon design requirements for part relationships, fit, function, strength, appearance, etc., as deemed necessary by the designer.

See following pages for details of application.

PERPENDICULARITY ⊥
(SQUARENESS, NORMALITY)

Definition. Perpendicularity is the condition of a surface, center plane, or axis at a right angle (90°) to a datum plane or axis.

PERPENDICULARITY TOLERANCE

A perpendicularity tolerance specifies one of the following:

1.   a tolerance zone defined by two parallel planes perpendicular to a datum plane or axis within which

    a.   the surface of a feature must lie (see Fig. 1);

    b.   the center plane of a feature must lie (see Fig. 2);

2.   a tolerance zone defined by two parallel planes perpendicular to a datum axis within which the axis of the considered feature must lie (see Fig. 3);

3.   a cylindrical tolerance zone perpendicular to a datum plane within which the axis of the considered feature must lie (see Fig. 4);

4.   a tolerance zone defined by two parallel lines perpendicular to a datum plane or axis within which a line element of the surface must lie (see Fig. 5—radial perpendicularity).

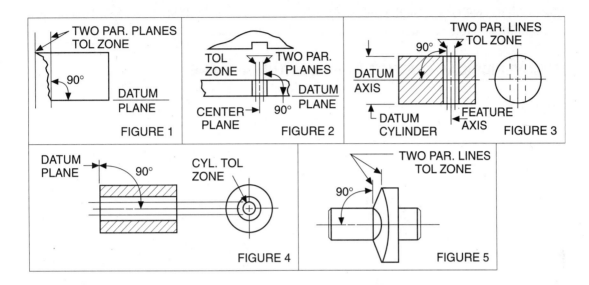

PERPENDICULARITY APPLICATION

The example below illustrates perpendicularity tolerance as applied to a surface.

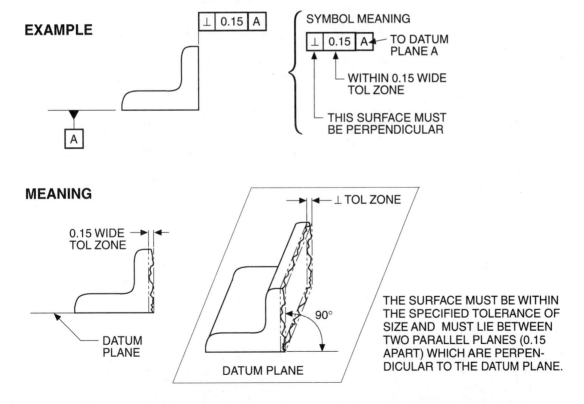

Note that the perpendicularity tolerance applied to a plane surface controls flatness if a flatness tolerance is not specified (that is, the flatness will be at least as good as the perpendicularity).

The examples below and on the following pages show perpendicularity tolerancing under various conditions. Note also how MMC applications permit greater tolerance.

NONCYLINDRICAL FEATURE AT RFS, DATUM A PLANE

EXAMPLE

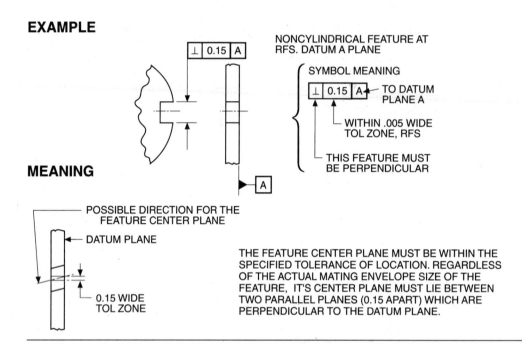

NONCYLINDRICAL FEATURE AT RFS. DATUM A PLANE

SYMBOL MEANING

⊥ 0.15 A — TO DATUM PLANE A

WITHIN .005 WIDE TOL ZONE, RFS

THIS FEATURE MUST BE PERPENDICULAR

MEANING

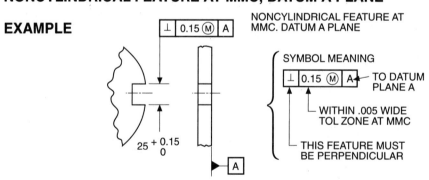

POSSIBLE DIRECTION FOR THE FEATURE CENTER PLANE

DATUM PLANE

0.15 WIDE TOL ZONE

THE FEATURE CENTER PLANE MUST BE WITHIN THE SPECIFIED TOLERANCE OF LOCATION. REGARDLESS OF THE ACTUAL MATING ENVELOPE SIZE OF THE FEATURE, IT'S CENTER PLANE MUST LIE BETWEEN TWO PARALLEL PLANES (0.15 APART) WHICH ARE PERPENDICULAR TO THE DATUM PLANE.

NONCYLINDRICAL FEATURE AT MMC, DATUM A PLANE

EXAMPLE

⊥ 0.15 Ⓜ A

NONCYLINDRICAL FEATURE AT MMC. DATUM A PLANE

SYMBOL MEANING

⊥ 0.15 Ⓜ A — TO DATUM PLANE A

WITHIN .005 WIDE TOL ZONE AT MMC

THIS FEATURE MUST BE PERPENDICULAR

$25 \; {}^{+\,0.15}_{0}$

MEANING

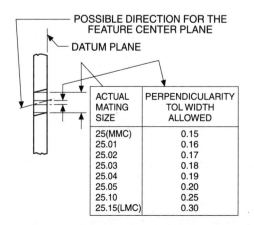

POSSIBLE DIRECTION FOR THE FEATURE CENTER PLANE

DATUM PLANE

ACTUAL MATING SIZE	PERPENDICULARITY TOL WIDTH ALLOWED
25(MMC)	0.15
25.01	0.16
25.02	0.17
25.03	0.18
25.04	0.19
25.05	0.20
25.10	0.25
25.15(LMC)	0.30

THE FEATURE CENTER PLANE MUST BE WITHIN THE SPECIFIED TOLERANCE OF LOCATION. WHEN THE FEATURE IS AT MAXIMUM MATERIAL CONDITION (25), THE MAXIMUM PERPENDICULARITY TOLERANCE IS 0.15 WIDE. WHERE THE ACTUAL MATING ENVELOPE SIZE OF THE FEATURE IS LARGER THAN ITS MAXIMUM MATERIAL CONDITION SIZE, AN INCREASE IN THE PERPENDICULARITY TOLERANCE IS ALLOWED EQUAL TO THAT AMOUNT. VIRTUAL CONDITION IS 24.85.

CYLINDRICAL FEATURE AT RFS, DATUM A CYLINDER RFS

EXAMPLE

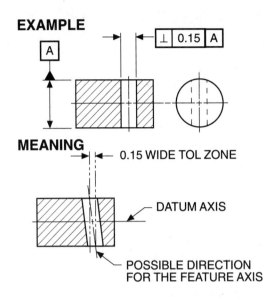

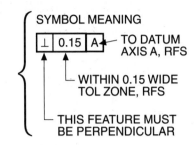

SYMBOL MEANING

⊥ 0.15 A → TO DATUM AXIS A, RFS

└ WITHIN 0.15 WIDE TOL ZONE, RFS

└ THIS FEATURE MUST BE PERPENDICULAR

MEANING

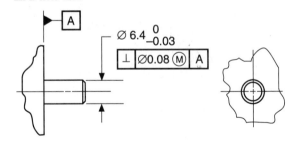

THE FEATURE AXIS MUST BE WITHIN THE SPECIFIED TOLERANCE OF LOCATION. REGARDLESS OF THE ACTUAL MATING ENVELOPE SIZE, ITS AXIS MUST LIE BETWEEN TWO PLANES (0.15 APART) WHICH ARE PERPENDICULAR TO THE DATUM AXIS.

NOTE: This tolerance applies only to the view on which it is specified.

CYLINDRICAL FEATURE AT MMC, DATUM A PLANE

Note that the ⌀ symbol is required to indicate a diameter (cylindrical) tolerance zone.

EXAMPLE

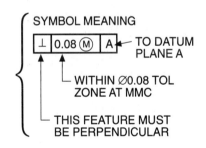

SYMBOL MEANING

⊥ ⌀0.08 Ⓜ A → TO DATUM PLANE A

└ WITHIN ⌀0.08 TOL ZONE AT MMC

└ THIS FEATURE MUST BE PERPENDICULAR

MEANING

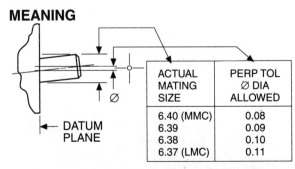

ACTUAL MATING SIZE	PERP TOL ⌀ DIA ALLOWED
6.40 (MMC)	0.08
6.39	0.09
6.38	0.10
6.37 (LMC)	0.11

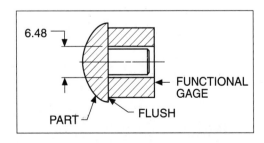

THE FEATURE MUST BE WITHIN THE SPECIFIED TOLERANCE OF LOCATION. WHEN THE FEATURE IS AT MMC 6.4, THE PERPENDICULARITY TOLERANCE IS 0.08 DIAMETER. AS THE FEATURE ACTUAL MATING ENVELOPE SIZE DEPARTS FROM MMC (GETS SMALLER), AN INCREASE IN TOLERANCE IS PERMITTED EQUAL TO THAT AMOUNT OF THAT DEPARTURE. VIRTUAL CONDITION IS ⌀6.48.

CYLINDRICAL FEATURE AT MMC, DATUM A PLANE

Note that the ⌀ symbol is required to indicate a diameter (cylindrical) tolerance zone.

EXAMPLE

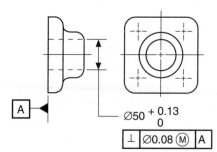

⌀50 $^{+0.13}_{0}$

⊥ ⌀0.08 Ⓜ A

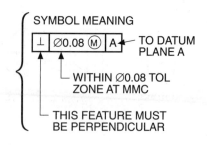

SYMBOL MEANING

⊥ ⌀0.08 Ⓜ A → TO DATUM PLANE A

— WITHIN ⌀0.08 TOL ZONE AT MMC

— THIS FEATURE MUST BE PERPENDICULAR

MEANING

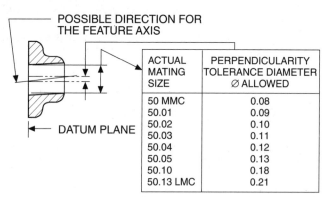

POSSIBLE DIRECTION FOR THE FEATURE AXIS

DATUM PLANE

ACTUAL MATING SIZE	PERPENDICULARITY TOLERANCE DIAMETER ⌀ ALLOWED
50 MMC	0.08
50.01	0.09
50.02	0.10
50.03	0.11
50.04	0.12
50.05	0.13
50.10	0.18
50.13 LMC	0.21

WHEN THE FEATURE IS AT MMC (50), THE MAXIMUM PERPENDICULARITY TOLERANCE FOR ITS AXIS IS 0.08 DIAMETER. AS THE FEATURE ACTUAL MATING ENVELOPE SIZE DEPARTS FROM MMC (GETS LARGER), AN INCREASE IN TOLERANCE IS PERMITTED EQUAL TO THE AMOUNT OF THAT DEPARTURE. VIRTUAL CONDITION IS ⌀49.92

RADIAL PERPENDICULARITY

EXAMPLE

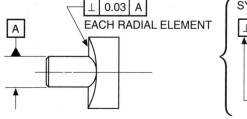

⊥ 0.03 A

EACH RADIAL ELEMENT

A

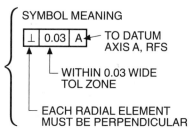

SYMBOL MEANING

⊥ 0.03 A → TO DATUM AXIS A, RFS

— WITHIN 0.03 WIDE TOL ZONE

— EACH RADIAL ELEMENT MUST BE PERPENDICULAR

MEANING

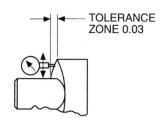

TOLERANCE ZONE 0.03

TRAVEL OF THE INDICATOR IS IN A RADIAL DIRECTION WITH PART HELD STATIONARY. EACH RADIAL ELEMENT OF THE SURFACE MUST BE WITHIN THE SPECIFIED TOLERANCE OF SIZE AND MUST LIE BETWEEN TWO PARALLEL LINES (0.03 APART) WHICH ARE PERPENDICULAR TO THE AXIS OF DIAMETER

Where "element" control is required and the perfect form at MMC requirement of Rule 1 is to be removed, the method below may be used.

EXAMPLE

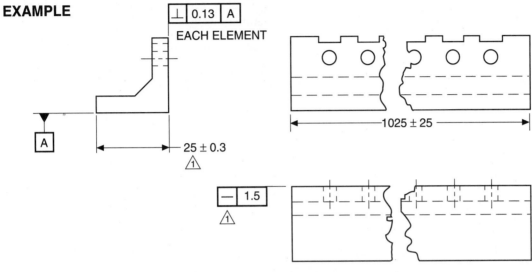

⚠️ PERFECT FORM AT MMC NOT REQUIRED

MEANING

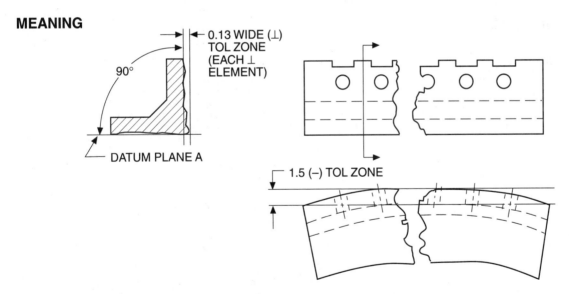

Each perpendicular element of the surface at any location along the entire length of the surface must lie between two parallel lines (0.13 apart) which are perpendicular to datum plane A. Each longitudinal element of the surface must lie between two parallel straight lines (1.5 apart). The part perfect form of the MMC envelope may be exceeded but the stated tolerances are the maximum form error permissible on the surface.

When perpendicularity tolerancing is critical, it may be necessary to limit the *tolerance* deviation to an amount equal to the feature *size* deviation from MMC. This assumes that the part form must be perfect at MMC size and that the virtual condition (size) can be no greater than that at MMC. The only permissible form tolerance must be acquired from the variation in part size (see Example below) in the increase of the feature hole size.

Example 2 limits the tolerance acquired from the feature size increase to a MAX amount.

ZERO TOLERANCE AT MAXIMUM MATERIAL CONDITION

Note that the $\oslash$ symbol is required to indicate a diameter (cylindrical) tolerance zone.

EXAMPLE 1

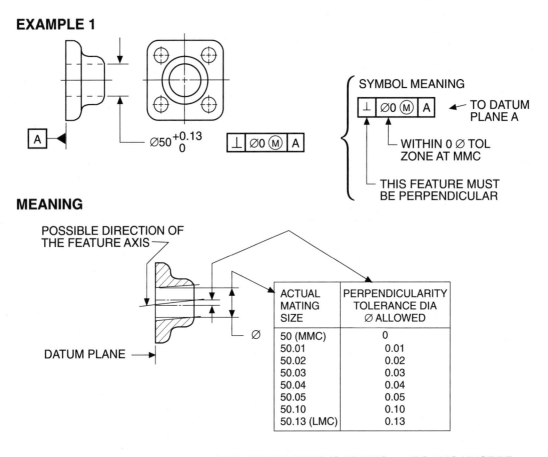

MEANING

ACTUAL MATING SIZE	PERPENDICULARITY TOLERANCE DIA $\oslash$ ALLOWED
50 (MMC)	0
50.01	0.01
50.02	0.02
50.03	0.03
50.04	0.04
50.05	0.05
50.10	0.10
50.13 (LMC)	0.13

WHEN THE FEATURE IS AT MMC 50, ITS AXIS MUST BE PERPENDICULAR TO THE DATUM PLANE. AS THE FEATURE ACTUAL MATING ENVELOPE SIZE DEPARTS FROM MMC (GETS LARGER), AN INCREASE IN TOL IS PERMITTED EQUAL TO THE AMOUNT OF THAT DEPARTURE. VIRTUAL CONDITION AND MMC ARE BOTH 50 IN THIS CASE.

ZERO TOLERANCE AT MMC, MAX DEVIATION

EXAMPLE 2

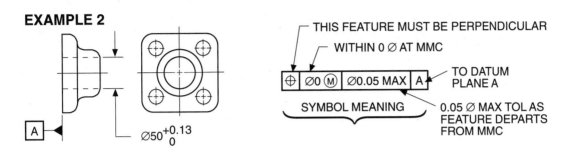

THIS FEATURE MUST BE PERPENDICULAR
WITHIN 0 ∅ AT MMC

⊥ | ∅0 Ⓜ | ∅0.05 MAX | A

SYMBOL MEANING

TO DATUM
PLANE A

0.05 ∅ MAX TOL AS
FEATURE DEPARTS
FROM MMC

A

∅50 +0.13 / 0

MEANING

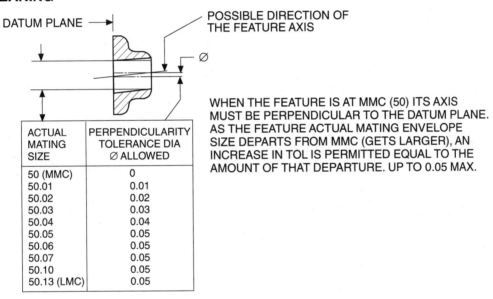

DATUM PLANE

POSSIBLE DIRECTION OF
THE FEATURE AXIS

∅

WHEN THE FEATURE IS AT MMC (50) ITS AXIS
MUST BE PERPENDICULAR TO THE DATUM PLANE.
AS THE FEATURE ACTUAL MATING ENVELOPE
SIZE DEPARTS FROM MMC (GETS LARGER), AN
INCREASE IN TOL IS PERMITTED EQUAL TO THE
AMOUNT OF THAT DEPARTURE. UP TO 0.05 MAX.

ACTUAL MATING SIZE	PERPENDICULARITY TOLERANCE DIA ∅ ALLOWED
50 (MMC)	0
50.01	0.01
50.02	0.02
50.03	0.03
50.04	0.04
50.05	0.05
50.06	0.05
50.07	0.05
50.10	0.05
50.13 (LMC)	0.05

ANGULARITY ∠

Definition. Angularity is the condition of a surface, axis, or median plane which is at a specified angle (other than 90°) from a datum plane or axis.

ANGULARITY TOLERANCE

Angularity is the condition of a surface, center plane, or axis at a specified angle (other than 90°) from a datum plane or axis. An angularity tolerance specifies one of the following:

1. a tolerance zone defined by two parallel planes at the specified basic angle from one or more datum planes, or a datum axis, within which the surface or center plane of the considered feature must lie.

2. a tolerance zone defined by two parallel planes at the specified basic angle from one or more datum planes, or a datum axis, within which the axis of the considered feature must lie.

3. a cylindrical tolerance zone whose axis is at the specified basic angle from one or more datum planes, or a datum axis, within which the axis of the considered feature must lie.

4. a tolerance zone defined by two parallel lines at the specified basic angle from a datum plane, or axis, within which a line element of the surface must lie.

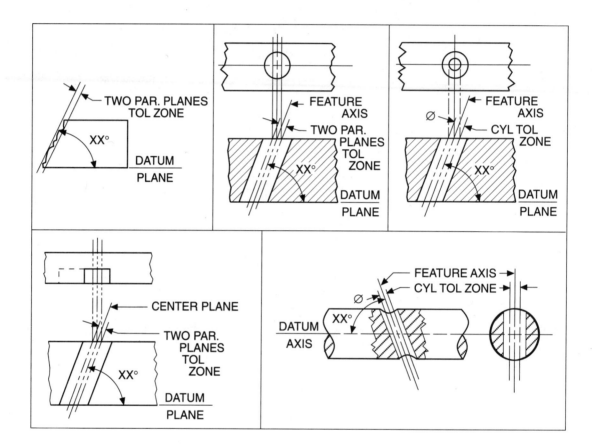

ANGULARITY APPLICATION

The example shows a part with a surface angular requirement. Note that the symbol is interpreted as "This surface must be at 45° angle in relation to datum plane A within 0.13 wide tolerance zone in relation to datum plane A."

The interpretation shows how the tolerance zone is established. Note that the angular tolerance zone is at 45° BASIC (exact) from the datum plane A. To be acceptable, the entire angular surface must fall within this tolerance zone. The angular surface must be contained within the limits of part size.

Note in the lower portion of the interpretation that the tolerance zone is also affected and established by the surface itself. That is, the surface extremities actually determine one plane of the tolerance zone as it bottoms out while the plane is inclined at an exact 45° angle with reference to the datum plane.

PLANE SURFACES

EXAMPLE

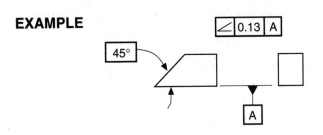

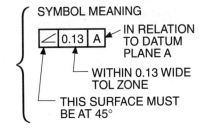

SYMBOL MEANING

∠ 0.13 A — IN RELATION TO DATUM PLANE A

— WITHIN 0.13 WIDE TOL ZONE

— THIS SURFACE MUST BE AT 45°

MEANING

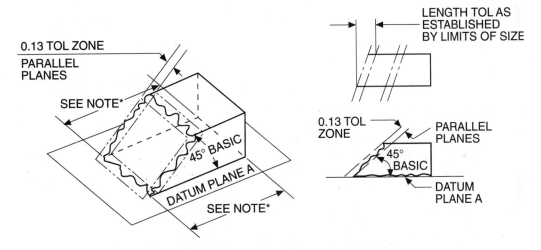

0.13 TOL ZONE PARALLEL PLANES

SEE NOTE*

45° BASIC

DATUM PLANE A

SEE NOTE*

LENGTH TOL AS ESTABLISHED BY LIMITS OF SIZE

0.13 TOL ZONE — PARALLEL PLANES

45° BASIC

DATUM PLANE A

NOTE* Part must be within *size* tolerance limits. The angular tolerance zone, composed of two parallel planes 0.13 apart, is 45° BASIC to the datum plane A. This tolerance zone is established by contact of the outermost of the two planes with the extremities of the angular surface and with the other plane parallel and inward at 0.13 distance. The entire surface must fall within this tolerance zone to be acceptable. Actually, this also controls the *flatness* of the surface to 0.13. Note that the angular surface extremities must be within both size and angular tolerance. See the examples below.

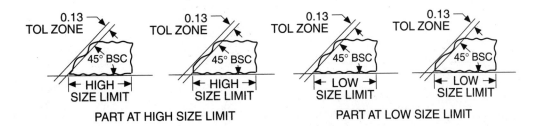

0.13 TOL ZONE — 45° BSC — HIGH SIZE LIMIT

0.13 TOL ZONE — 45° BSC — HIGH SIZE LIMIT

PART AT HIGH SIZE LIMIT

0.13 TOL ZONE — 45° BSC — LOW SIZE LIMIT

0.13 TOL ZONE — 45° BSC — LOW SIZE LIMIT

PART AT LOW SIZE LIMIT

CYLINDRICAL FEATURE AT MMC, DATUM A PLANE

Note that the ⌀ symbol is required to indicate a diameter (cylindrical) tolerance zone. In this example an angularity tolerance has been used to control the axis of a hole at the specified angle. A cylindrical tolerance zone is used. Such an application would normally be used as a refinement of a locational tolerance to ensure that the angular relationship of the hole is maintained to the datum plane surface (in the view shown). The use of the cylindrical tolerance zone requires that the hole axis also remains within the angularity tolerance in a 360° relationship as well as within positional tolerance specified. MMC has been used as based upon design requirements and could gain additional tolerance as the feature is produced (see tabulation).

If it was desired to control the hole in its angularity relationship to the datum feature but only within a total wide zone, the diameter symbol (⌀) would be omitted in the feature control frame. The positional tolerance on such a part would then provide the feature control in the other directions.

EXAMPLE

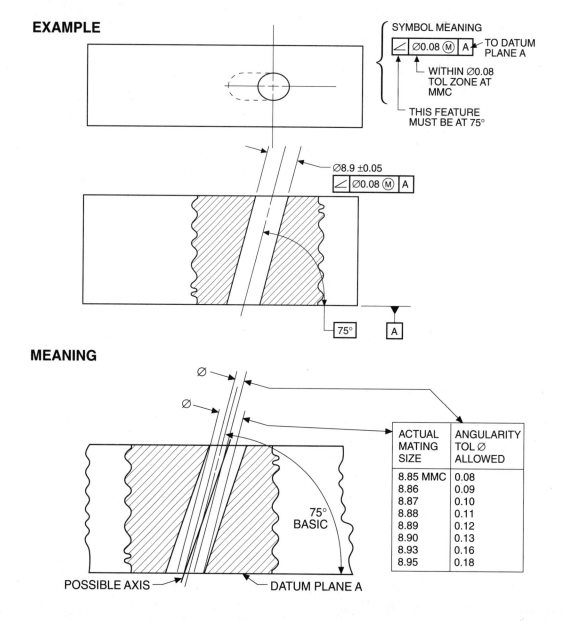

MEANING

ACTUAL MATING SIZE	ANGULARITY TOL ⌀ ALLOWED
8.85 MMC	0.08
8.86	0.09
8.87	0.10
8.88	0.11
8.89	0.12
8.90	0.13
8.93	0.16
8.95	0.18

The feature must be within the specified tolerance of location. When the feature is at MMC 8.85, the angularity tolerance is ⌀ 0.08 diameter. As the feature actual mating envelope size departs from MMC (gets larger), an increase in tolerance is permitted equal to the amount of that departure. Virtual condition is ⌀ 8.77. Note: This tolerance applies only to the view on which it is specified.

PARALLELISM //

Definition. Parallelism is the condition of a surface or center plane, equidistant at all points from a datum plane; or an axis, equidistant along its length from one or more datum planes or a datum axis.

PARALLELISM TOLERANCE

A parallelism tolerance specifies:

1.   a tolerance zone defined by two parallel planes parallel to a datum plane or axis within which the surface or center plane of the considered feature must lie (see Fig. 1);

2.   a tolerance zone defined by two parallel planes parallel to a datum plane or axis, within which the axis of the considered feature must lie (see Fig. 2);

3.   a cylindrical tolerance zone parallel to one or more datum planes or a datum axis, within which the axis of the feature must lie (see Fig. 3);

4.   a tolerance zone defined by two parallel lines parallel to a datum plane or axis, within which the line element of the surface must lie.

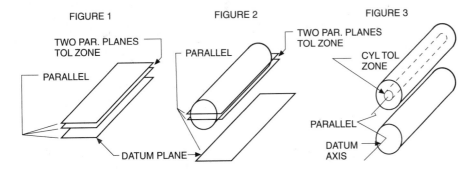

PARALLELISM APPLICATION

Note in the following example that the bottom surface has been selected as the datum and the top surface is to be parallel to datum plane A within 0.05.

The Meaning beneath the example clarifies the symbol: It reads, "This feature must be parallel within 0.05 to datum plane A."

The lower example illustrates the tolerance zone and the manner in which the surface must fall within the tolerance zone to be acceptable. Note that the tolerance zone is established parallel to the datum plane A. Note also that the parallelism tolerance, when applied to a plane surface, controls flatness if a flatness tolerance is not specified (that is, the implied flatness will be *at least* as good as the parallelism).

SURFACE TO DATUM PLANE

EXAMPLE

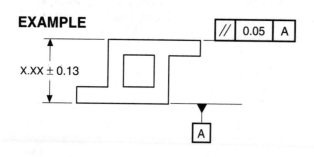

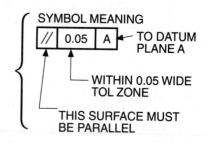

SYMBOL MEANING

→ TO DATUM PLANE A

WITHIN 0.05 WIDE TOL ZONE

THIS SURFACE MUST BE PARALLEL

MEANING

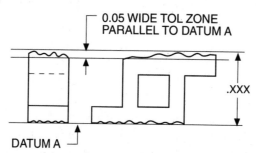

0.05 WIDE TOL ZONE PARALLEL TO DATUM A

.XXX

DATUM A

THIS SURFACE MUST BE WITHIN THE SPECIFIED TOLERANCE OF SIZE AND MUST LIE BETWEEN TWO PLANES 0.05 APART WHICH ARE PARALLEL TO THE DATUM PLANE.

EXAMPLE

MEANING

0.05 TOL ZONE

8.8

PART AT HIGH SIZE LIMIT

0.05 TOL ZONE

8.5

PART AT NOMINAL SIZE

0.05 TOL ZONE

8.2

PART AT LOW SIZE LIMIT

8.5 ± 0.3

(POSSIBLE VARIATION OF PARALLELISM TOLERANCE ZONE WITHIN SIZE TOLERANCE RANGE)

PARALLELISM—FEATURES OF SIZE

This page and the one following cover the application of parallelism tolerancing to features of size. Note how the MMC applications give the possibility of greater tolerance.

FEATURE AT RFS, DATUM A PLANE

EXAMPLE

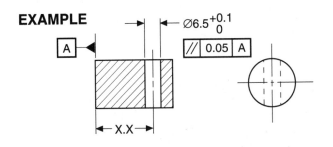

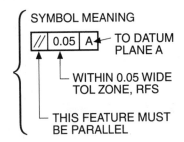

SYMBOL MEANING

// | 0.05 | A ← TO DATUM PLANE A

└ WITHIN 0.05 WIDE TOL ZONE, RFS

└ THIS FEATURE MUST BE PARALLEL

MEANING

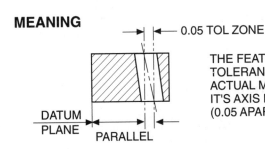

— 0.05 TOL ZONE

THE FEATURE AXIS MUST BE WITHIN THE SPECIFIED TOLERANCE OF LOCATION. REGARDLESS OF THE ACTUAL MATING ENVELOPE SIZE OF THE FEATURE, IT'S AXIS MUST LIE BETWEEN TWO PARALLEL PLANES (0.05 APART) WHICH ARE PARALLEL TO THE DATUM PLANE A.

FEATURE AT MMC, DATUM A PLANE

EXAMPLE

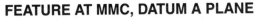

SYMBOL MEANING

// | 0.05 Ⓜ | A ← TO DATUM PLANE A

└ WITHIN 0.05 WIDE TOL ZONE AT MMC

└ THIS FEATURE MUST BE PARALLEL

MEANING

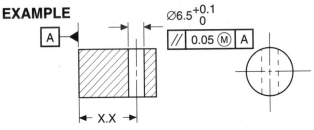

WIDE TOL ZONE

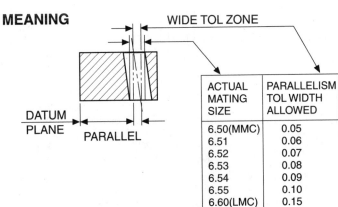

ACTUAL MATING SIZE	PARALLELISM TOL WIDTH ALLOWED
6.50(MMC)	0.05
6.51	0.06
6.52	0.07
6.53	0.08
6.54	0.09
6.55	0.10
6.60(LMC)	0.15

FEATURE AXIS MUST BE WITHIN SPECIFIED TOL OF LOCATION. HOLE (AT MMC) AXIS MUST LIE BETWEEN TWO PARALLEL PLANES (0.05 APART) WHICH ARE PARALLEL TO THE DATUM PLANE A.

WHEN FEATURE ACTUAL MATING ENVELOPE SIZE DEPARTS FROM MMC (GETS LARGER), AN INCREASE IN THE PARALLELISM TOL IS ALLOWED EQUAL TO THAT DEPARTURE.

FEATURE AT RFS, DATUM FEATURE AT RFS

EXAMPLE

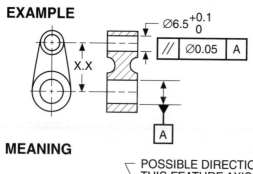

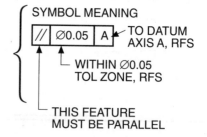

SYMBOL MEANING

TO DATUM AXIS A, RFS

WITHIN Ø0.05 TOL ZONE, RFS

THIS FEATURE MUST BE PARALLEL

MEANING

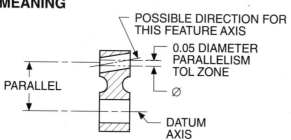

POSSIBLE DIRECTION FOR THIS FEATURE AXIS

0.05 DIAMETER PARALLELISM TOL ZONE

DATUM AXIS

THE FEATURE AXIS MUST BE WITHIN THE SPECIFIED TOL OF LOCATION. REGARDLESS OF THE ACTUAL MATING ENVELOPE SIZE OF THE FEATURE, ITS AXIS MUST LIE WITHIN A 0.05 DIAMETER (CYLINDRICAL) TOLERANCE ZONE WHICH IS PARALLEL TO THE DATUM AXIS.

FEATURE AT MMC, DATUM FEATURE AT RFS

(Note that the $\oslash$ symbol is required to indicate a diameter (cylindrical) tolerance zone.)

EXAMPLE

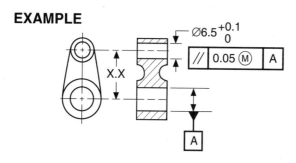

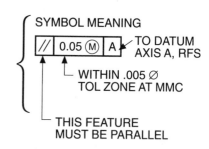

SYMBOL MEANING

TO DATUM AXIS A, RFS

WITHIN .005 $\oslash$ TOL ZONE AT MMC

THIS FEATURE MUST BE PARALLEL

MEANING

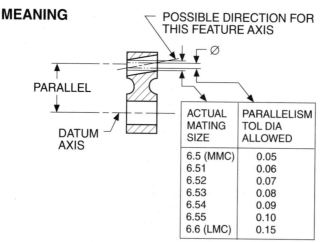

POSSIBLE DIRECTION FOR THIS FEATURE AXIS

ACTUAL MATING SIZE	PARALLELISM TOL DIA ALLOWED
6.5 (MMC)	0.05
6.51	0.06
6.52	0.07
6.53	0.08
6.54	0.09
6.55	0.10
6.6 (LMC)	0.15

THE FEATURE AXIS MUST BE WITHIN THE SPECIFIED TOL OF LOCATION. WHEN THE FEATURE IS AT MAXIMUM MATERIAL CONDITION, 6.5, THE MAXIMUM PARALLELISM TOLERANCE IS 0.05 DIAMETER. WHEN THE FEATURE ACTUAL MATING ENVELOPE SIZE DEPARTS FROM MMC (GETS LARGER), AN INCREASE IN THE PARALLELISM TOL IS ALLOWED EQUAL TO THE AMOUNT OF THAT DEPARTURE.

TOLERANCES OF FORM—PROFILE TOLERANCING

APPLIED TO INDIVIDUAL FEATURES (NO DATUM) OR RELATED FEATURES (USING DATUMS)

Profile tolerancing is of two varieties and involves the characteristics below. According to the design requirement, these characteristics may be applied to an individual feature, such as a *single surface* or *element,* or to related features, such as a *single surface* or *element* relative to a datum or datums.

(PROFILE OF A LINE)   (PROFILE OF A SURFACE)

See following pages for details of application.

PROFILE
METHOD OF SPECIFYING

Definition. Profile tolerancing specifies a uniform boundary along the true profile within which the elements of the surface must lie.

PROFILE TOLERANCE

A profile tolerance (either bilateral or unilateral) specifies a tolerance zone, always* intended and measured normal to the basic profile (applicable to the view in which drawn) at all points of the profile, within which the true part surface profile or line profile must lie.

PROFILE TOLERANCE APPLICATION

Profile tolerancing is an effective method for controlling lines, arcs, irregular surfaces, or other unusual part profiles. Profile tolerances are usually applied to surface features but may also be applied to a line (element on a feature surface). Profile tolerances can also be used as a locational control to size features to specify their shape, form, orientation and relationship to datum features. In any case, these requirements must be specified in association with the desired profile in a plane of projection (view) on the drawing as follows:

a.   An APPROPRIATE VIEW or section is drawn which shows the desired basic profile in true shape.

b.   The profile is defined by basic dimensions. This dimensioning may be in the form of located radii and angles, coordinate dimensioning, formulas, or undimensioned drawing.

c.   Depending on design requirements, the tolerance may be divided bilaterally to both sides of the true profile or applied unilaterally to either side of the true profile.

d.   To control location of a size feature relative to datum planes.

---

*Unless otherwise specified.

Where an equally disposed bilateral tolerance is intended, it is only necessary to show the feature control symbol with a leader directed to the surface. For an unequally disposed or unilateral tolerance, phantom lines are drawn parallel to the true profile to clearly indicate the tolerance zone inside or outside the true profile. One end of a dimension line is extended to the feature control symbol.

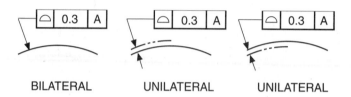

BILATERAL   UNILATERAL   UNILATERAL

e.   Other appropriate dimensions, as well as the applicable feature control symbol, are added. The SYMBOL should be applied in a view in which the surface or lines to be controlled is shown as a profile and which pictorially represents the desired feature orientation and relationship.

f.   If some limits on a drawing are expressed by a profile tolerance and others by regular tolerance dimensions, the extent or limitation of the profile tolerance must be clearly indicated by reference letters applied at the extremities of the profile controlled portion and a symbolic notation added beneath the profile tolerance feature control frame. Where a profile tolerance applies *all around* the profile of the part, the symbolic representation ⊕ is specified or the notation ALL AROUND is placed beneath the feature control frame (see pages 78 and 79).

Common surface profile tolerancing includes a combination of FORM, ORIENTATION, and SIZE controls, with the profile basic dimensions amounting to overall dimensions as well. In this situation, the standard limits of size interpretation, which require that "form" control must be contained within "size" control (i.e., Rule 1 on page 26), do not apply.

Surface profile tolerancing may also be associated with conventional size dimensions and tolerances. The profile itself, however, must be controlled by basic dimensions and the profile tolerance zone. Under this condition the standard limits of size interpretation (Rule 1) do apply and the profile tolerance zone must be contained *within* the *size* tolerance zone. If, for example, the bilateral profile method is used, some portions of the profile tolerance zone may be sacrificed if the controlled feature is at its extreme size limit at that point of the profile.

Since profile tolerancing is a control of *surface form* or *orientation,* no modifier (MMC) can be used on the feature controlled. Where a datum reference is used, it is usually intended to apply RFS. However, there may be unique applications permitting the use of an MMC datum to expedite gaging provided there is no detrimental effect on the design requirement (e.g., a functional shaft size as related to a cam profile).

GAGING

The example meaning on page 78 shows the basic gaging principle of measurement *normal to the basic profile.* Gage traverse or part rotation techniques could be used. Where overall size permits, optical comparator (or similar) techniques are usually the most economical and effective method of inspecting this type of part. These techniques make it possible to compare a blown-up (10X, 20X, etc.) projected shadow or profile of the part with an optical gage chart containing accurately scribed profile tolerance zone limits.

Note, however, that shadow image of the part profile which is seen on the comparator screen will be the extreme profile and may not represent the entire surface. Focus adjustment and surface illumination methods (available on some comparators) may be able to account for the entire surface. If necessary, further inspection may be required to account for surface irregularities which are below the extreme profile or not made visible by the method used.

A larger part exceeding the normal limitations of the optical chart size might be checked in segments or with the use of tracer and reticle.

Computerization techniques, either independently applied or in conjunction with optical methods, also provide possible verification options.

TWO TYPES OF PROFILE TOLERANCE

In practice, a profile tolerance may be applied to either an entire surface or to individual line element profiles taken at various cross sections through the part. The two types or methods of controlling profile are:

a. Profile of a surface ⌓
    The tolerance zone established by profile of a surface tolerance is a three-dimensional zone or total control across the entire length and width or circumference of the feature; it may be applied to parts having a constant cross section or to parts having a surface of revolution. Usually profile of a surface requires datum references.

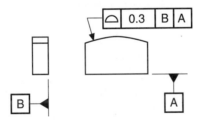

b. Profile of a line ⌒
    The tolerance zone established by profile of a line tolerance is a two-dimensional zone extending along the length of the considered feature; it may apply to the profiles of parts having a varying cross section such as a propeller, aircraft wing, nose cone, or to random cross sections of parts where it is not desired to control the entire surface as a single entity. Profile of a line may, or may not, require datum references.

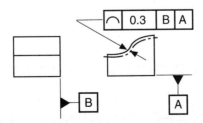

⌒ PROFILE OF A SURFACE

EXAMPLE

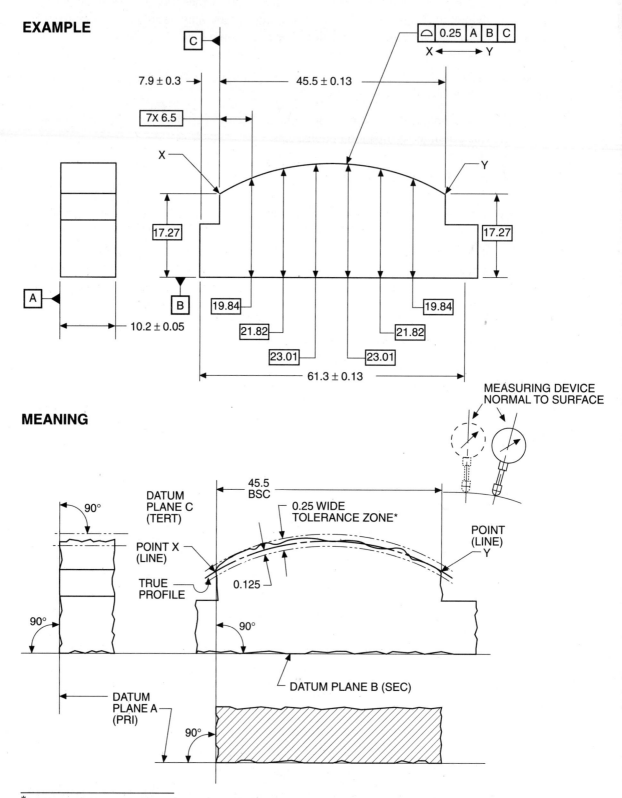

MEANING

*0.25 total wide zone (0.125 each side of true profile). The surface is between the two profile boundaries 0.25 apart, equally disposed about the true profile, which are perpendicular to datum plane A and positioned with respect to datum planes B and C.

PROFILE
PROFILE OF A SURFACE ⌒

Where the profile tolerance is to be used in controlling the entire surface profile, the symbolic representation may be specified or the words "ALL AROUND" should be added below the profile tolerance.

ZONE TOLERANCE AROUND ENTIRE PROFILE

EXAMPLE

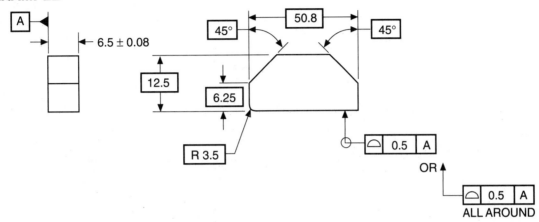

MEANING

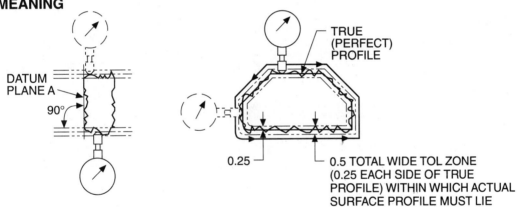

GAGING

The example illustrates the basic gaging principle of measurement *normal to the true profile* as oriented from datum A. Where overall size permits, optical "comparator" (or similar) techniques are usually the most economical and effective method of inspecting this type of part (see also GAGING of preceding example).

Where the profile tolerance is to be used in controlling all, or large portions, of the surface profile, specific notations (such as angular degrees on a cam) should be added with the profile tolerance symbol or note.

This type of tolerance control is particularly useful on cams or similar parts.

DIFFERENT ZONE TOLERANCE AROUND ENTIRE PROFILE

EXAMPLE

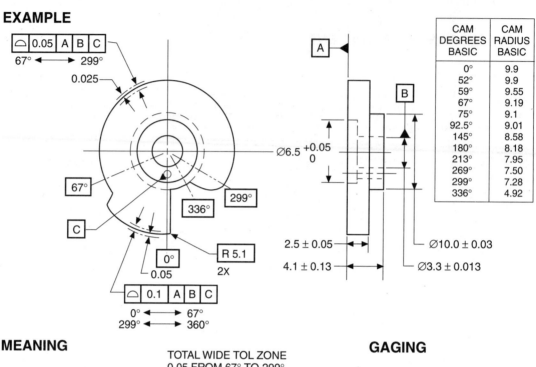

	CAM DEGREES BASIC	CAM RADIUS BASIC
	0°	9.9
	52°	9.9
	59°	9.55
	67°	9.19
	75°	9.1
	92.5°	9.01
	145°	8.58
	180°	8.18
	213°	7.95
	269°	7.50
	299°	7.28
	336°	4.92

MEANING

TOTAL WIDE TOL ZONE
0.05 FROM 67° TO 299°
0.1 FROM 0° TO 67°
0.1 FROM 299° TO 360°
(0.025 & 0.05 RESPECTIVELY
EACH SIDE OF TRUE PROFILE)
WITHIN WHICH ACTUAL
SURFACE PROFILE MUST LIE

GAGING

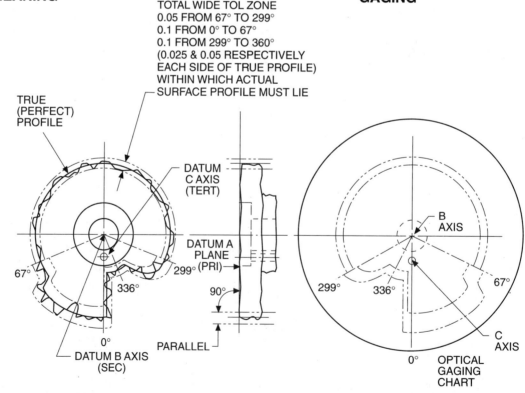

Optical comparator (or similar) techniques are ideal for this part. A transparent chart made to 20X, 50X, etc., size with accurately scribed lines defining the tolerance zones permits a direct surface profile check of the magnified part with the chart. If the surface profile of the part lies within the scribed tolerance zones when oriented to the datum B and C axes and while bottoming on datum A, it is acceptable.

PROFILE
PROFILE OF A LINE

The profile of a surface controls the entire surface, or all the elements of the surface, within a uniform tolerance zone as established from the true profile. However, where line elements of a surface are to be specifically controlled or controlled as a refinement of size or surface profile control, the profile of a line characteristic may be used. Line profile control is applied in a manner similar to the application of surface profile.

The example below illustrates line profile control as a refinement of size. As with surface profile, line profile must be shown in the drawing view in which the control applies. The tolerance zone is established in the same way as surface profile. However, its tolerance zone is disposed about each element of the surface. Therefore the tolerance zone applies for the full length of each element of the surface (in the view in which it is shown), but only for the width or height at a cutting plane bisecting the element. This cutting plane must be considered to be perpendicular or parallel to the part orientation within its size tolerances. Where datum references are specified, the orientation is with respect to the datums to give a more specific relationship. Each actual line element of the controlled profile may vary within its prescribed tolerance zone, but the element-to-element control may vary within the entire size tolerance.

Line profile control is illustrated in the example below and on page 82. Accuracy and consistency of the individual elements at the part surface are critical, but variation from element to element is less critical. More lenient control within the size tolerances is adequate in the direction perpendicular to the line profile. The requirement to hold an accurate line profile on an irregularly shaped part within a more lenient size control is similar to holding straightness within a more lenient size control on a square, rectangular, or cylindrical part.

Surface profile and line profile may be applied to the same feature, where applicable—for example, in situations in which the line elements are to be controlled more closely than the surface as a whole. Combined surface and line profile control may be used and specified as shown in the figure below.

EXAMPLE

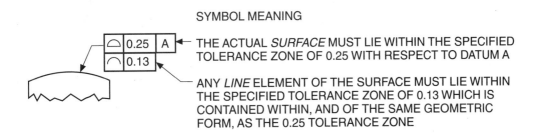

SYMBOL MEANING

THE ACTUAL *SURFACE* MUST LIE WITHIN THE SPECIFIED TOLERANCE ZONE OF 0.25 WITH RESPECT TO DATUM A

ANY *LINE* ELEMENT OF THE SURFACE MUST LIE WITHIN THE SPECIFIED TOLERANCE ZONE OF 0.13 WHICH IS CONTAINED WITHIN, AND OF THE SAME GEOMETRIC FORM, AS THE 0.25 TOLERANCE ZONE

PROFILE OF A LINE

EXAMPLE

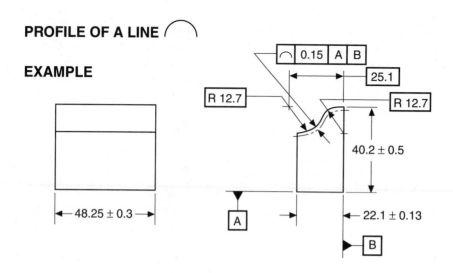

MEANING

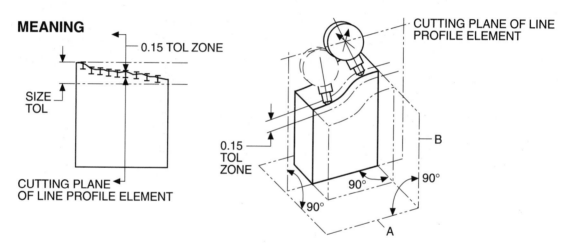

Any profile line element of the surface (elements in the view in which the symbol is shown and elements represented by the true profile) must be within the specified 0.15 tolerance zone which is unilaterally disposed from the true profile. The *profile* tolerance zone must be contained within the *size* tolerance zone.

GAGING

The illustration above shows the basic gaging principle of measurement *normal to the basic profile* and in an imaginary cutting plane which bisects the line element while oriented basically perpendicular and parallel within the size tolerances.

Dial indicator, master template, and optical comparator (or similar) techniques can be used (see also GAGING in preceding examples).

COPLANAR SURFACES

Profile tolerancing may be used to specify coplanarity of two or more surfaces where it is desired to treat these surfaces as a single interrupted or noncontinuous surface. In such an application, a control is provided similar to that achieved by a flatness tolerance applied to a single plane surface. Coplanarity is the condition of two or more surfaces having all elements in one plane (see page 84).

As shown in Example 1, the profile of a surface tolerance establishes a tolerance zone defined by two parallel planes within which the considered surfaces must lie. No datum reference is stated (as in the case of flatness) since the orientation of the tolerance zone is established from the contact of the part against a reference standard; the datum is established by the considered surfaces themselves.

Where more than two surfaces are involved, it may be desirable to identify which specific surface (or surfaces) is to be used as a datum and to establish the tolerance zone. Where necessary, datum target methods could also be used. Where a datum (or datums) is used, it is understood that the tolerance zone established applies to all coplanar surfaces, including the datum surfaces, unless otherwise specified as shown in Example 2.

OFFSET SURFACES

As a variation of coplanar surfaces, the principles described above may be extended to offset surfaces as shown in the example. In this instance, the variation from coplanar is a desired amount (2.5) and is stated by a basic dimension.

COPLANAR SURFACES

EXAMPLE 1

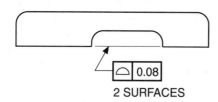

2 SURFACES

MEANING

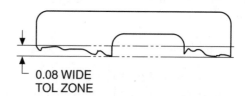

0.08 WIDE
TOL ZONE

THE SURFACES MUST BE WITHIN
THE SPECIFIED TOL OF SIZE AND
MUST BOTH LIE BETWEEN TWO
PARALLEL PLANES 0.08 APART.

EXAMPLE 2

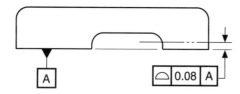

MEANING

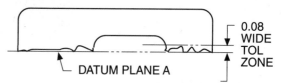

0.08
WIDE
TOL
ZONE

DATUM PLANE A

THE SURFACE MUST BE WITHIN
THE SPECIFIED TOL OF SIZE AND
MUST LIE BETWEEN TWO
PARALLEL PLANES 0.08 APART
AS ESTABLISHED RELATIVE TO
DATUM PLANE A

OFFSET SURFACES

EXAMPLE

MEANING

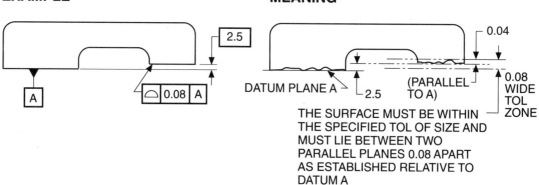

0.04

DATUM PLANE A

2.5

(PARALLEL
TO A)

0.08
WIDE
TOL
ZONE

THE SURFACE MUST BE WITHIN
THE SPECIFIED TOL OF SIZE AND
MUST LIE BETWEEN TWO
PARALLEL PLANES 0.08 APART
AS ESTABLISHED RELATIVE TO
DATUM A

CONICITY

Profile of a surface tolerance can be applied to conical parts to control conicity. It may be applied as an independent control of form (Example 1, next page) or as a combined control of form and orientation with respect to a datum feature and datum axis (Example 2, next page). It could also be used, when necessary, as profile ALL AROUND control, which can control size and form simultaneously using basic dimensions to define the conical shape.

In Example 1, profile of a surface is used as a *form* control (of the 20° cone) as a refinement within the size control. This is comparable to a cylindricity tolerance controlling *form* of a cylinder as a refinement within its size control. No datum is required nor proper in this case.

In Example 2, profile of a surface is used as a *form* and *orientation* control (of the 15° cone) as a refinement within the size control and relative to datums A and B. The added notation ALL AROUND (or symbolic version) could be used if desired for clarification but is not necessary here. In this case, profile tolerance is applied to a surface of revolution as a design preference (hypothetically) over other methods (i.e., runout).

See DATUM section for more information on datums and datum precedence (primary, secondary, tertiary datums).

CONICAL SURFACES

EXAMPLE 1

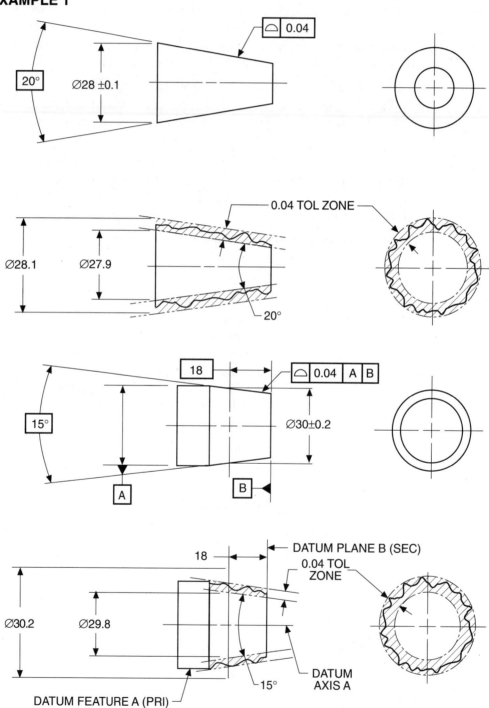

RUNOUT TOLERANCES—
RELATED FEATURES USING DATUMS

A runout tolerance is a relationship between surfaces or features, therefore, a datum (or datums) is required.

Runout tolerance may be applied in two different ways using the characteristics shown below.

(CIRCULAR RUNOUT)   (TOTAL RUNOUT)

See following pages for details of application.

RUNOUT ⟋ ⟋⟋ (CIRCULAR AND TOTAL)

Definition. Runout is the composite deviation from the desired form and orientation of a part surface of revolution during full rotation (360°) of the part on a datum axis.

RUNOUT TOLERANCE

Runout tolerance states how far an actual surface or feature is permitted to deviate from the desired form and orientation implied by the drawing during full rotation (360°) of the part on a datum axis.

RUNOUT APPLICATION

Runout tolerancing is a method used to control the composite surface effect of one or more features of a part relative to a datum axis. Runout tolerance is applicable to rotating parts in which this composite surface control is based on the part function and design requirement. A runout tolerance always applies on an RFS basis; namely, size variation has no effect upon the runout tolerance compliance.

Each considered feature must be within its individual runout tolerance when rotated 360° about the datum axis. The tolerance specified for a controlled surface is the total tolerance or full indicator movement (FIM) in terms of common inspection criteria. Former terms, full indicator reading (FIR) and total indicator reading (TIR), have the same meaning as FIM.

As is seen on the following page, the basis of runout tolerance control is the datum axis of the part. Surfaces controlled may be those constructed *around* a datum axis, or those constructed at *right angles* to a datum axis. As is also seen, the datum axis is established from a datum feature.

NOTE On runout tolerance, where it is desired to consider a worst case "outer boundary" or "inner boundary," it is determined by the maximum material condition of the feature plus (if outside diameter) or minus (if inside diameter) the runout tolerance.

BASIS OF CONTROL OF RUNOUT TOLERANCE

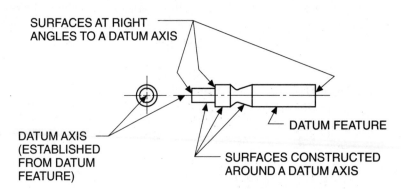

SURFACES AT RIGHT
ANGLES TO A DATUM AXIS

DATUM FEATURE

DATUM AXIS
(ESTABLISHED
FROM DATUM
FEATURE)

SURFACES CONSTRUCTED
AROUND A DATUM AXIS

TYPES OF FEATURES APPLICABLE TO A RUNOUT TOLERANCE

↗ CIRCULAR RUNOUT

EXAMPLE

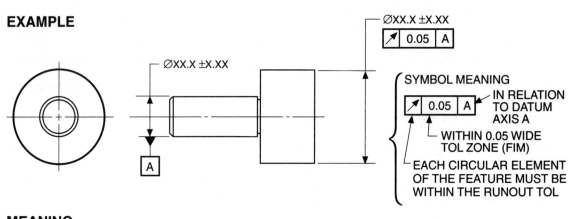

ØXX.X ±X.XX

| ↗ | 0.05 | A |

ØXX.X ±X.XX

A

SYMBOL MEANING

| ↗ | 0.05 | A |

IN RELATION
TO DATUM
AXIS A

WITHIN 0.05 WIDE
TOL ZONE (FIM)

EACH CIRCULAR ELEMENT
OF THE FEATURE MUST BE
WITHIN THE RUNOUT TOL

MEANING

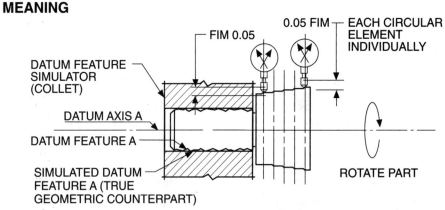

FIM 0.05

0.05 FIM

EACH CIRCULAR
ELEMENT
INDIVIDUALLY

DATUM FEATURE
SIMULATOR
(COLLET)

DATUM AXIS A

DATUM FEATURE A

SIMULATED DATUM
FEATURE A (TRUE
GEOMETRIC COUNTERPART)

ROTATE PART

↗↗ TOTAL RUNOUT

EXAMPLE

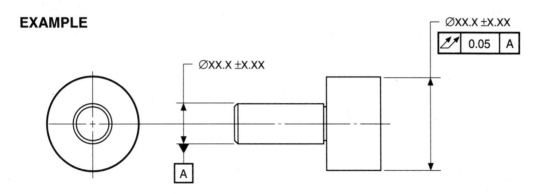

MEANING

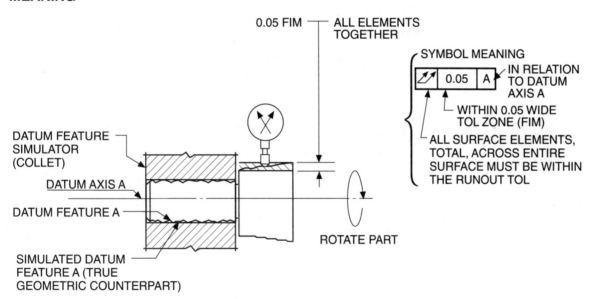

The datum axis may be established by a single diameter (cylinder) of sufficient length, two diameters with sufficient axial separation, or a diameter and a face surface which is at right angles to it. Features selected as datums should, as much as possible, be functional to the part requirement (e.g., bearing mounting diameters, etc.).

TYPES OF RUNOUT CONTROL

The two types of runout control are *circular* runout and *total* runout. Selection of the proper type is based upon the design requirement and manufacturing considerations. The fundamental difference between the two types are illustrated on page 88 and above with detailed examples following in this section. Note that with either method, the collective or composite control of various form and orientation variations of the part provides a more direct representation of part functions, integrates manufacturing operations, and minimizes inspection setup requirements. If necessary to predict a worst case possibility on such parts, an "outer boundary" or "inner boundary" is determined; see definitions.

COAXIAL FEATURES—
SELECTION OF PROPER CONTROL

There are four characteristics for controlling interrelated coaxial features:

1.   RUNOUT TOLERANCE (circular or total)   (RFS)

2.   POSITION TOLERANCE   (MMC or RFS)

3.   CONCENTRICITY TOLERANCE   (RFS)

4.   PROFILE OF A SURFACE   (RFS DATUM)

Any of the above methods provides effective control. However, it is important to select the *most appropriate* one to both meet the design requirements and provide the most economical manufacturing conditions. (See also details of preceding and following sections.)

Below are recommendations to assist in selecting the proper control:

If the need is to control only CIRCULAR cross-sectional elements in a composite relationship to the datum axis, RFS, e.g., multidiameters on a shaft, use:

CIRCULAR RUNOUT ↗   **EXAMPLE** | ↗ | 0.1 | A – B |

(This method controls any composite error effect of circularity, concentricity, and circular cross-sectional profile variations.)

If the need is to control the TOTAL cylindrical or profile surface in composite relative to the datum axis RFS, e.g., multi-diameters on a shaft, bearing mounting diameters, etc., use:

TOTAL RUNOUT ↗↗   **EXAMPLE** | ↗↗ | 0.1 | A – B |

(This method controls any composite error effect of circularity, cylindricity, straightness, coaxiality, angularity, and parallelism.)

NOTE   Runout is always implied as an RFS application. It cannot be applied on an MMC basis, since an MMC situation involves functional interchangeability or assemblability (probably of mating parts), in which case POSITION tolerance would be used. See below.

If the need is to control the total cylindrical or profile surface and its actual mating envelope axis relative to the datum axis on an MMC or RFS basis, e.g., on mating parts to assure interchangeability or assemblability, use:

POSITION ⊕   **(IF MMC)**   **EXAMPLE** | ⊕ | ⌀ 0.1 Ⓜ | A Ⓜ |

                        (IF RFS)   **EXAMPLE** | ⊕ | ⌀ 0.1 | A |

                        OR RFS DATUM | ⊕ | ⌀ 0.1 Ⓜ | A |

If the need is to control the *axis* of one or more features in composite relative to a *datum axis,* RFS, e.g., to control such as balance of a rotating part, use:

CONCENTRICITY ◎   **EXAMPLE** | ◎ | ⌀0.1 | A – B |

NOTE   Concentricity is always implied as an RFS application. Variations in size (departure from MMC size, out-of-circularity, out-of-cylindricity, etc.) do not in themselves conclude *axis* error.

If the need is to control the total cylindrical or profile surface simultaneously with the size dimension(s) (using basic dimensions for both), relative to a datum axis, e.g., precise fit, multi-diameters, etc., use:

PROFILE OF A SURFACE ◠   **EXAMPLE** | ◠ | 0.1 | A |

PART MOUNTED ON FUNCTIONAL DIAMETER (DATUM)

Circular runout provides a composite control of circular elements of a surface. Circular runout is normally a less complex requirement than total runout. The tolerance is applied independently at any circular cross section or measuring position on the part as it rotates through 360°. Circular runout should be considered when the part function and manufacturing requirements are satisfied by this type of control. Where more complete control of all elements in composite is necessary, total runout should be considered.

Circular runout controls composite variations of circularity and cross-sectional size and form variations of the surface at each circular element where applied to surfaces constructed around a datum axis. It controls circular elements of the surface (wobble) where applied to surfaces constructed at right angles (perpendicular) to a datum axis.

The example below utilizes circular runout control. It applies circular runout to the angular surface controlling the individual circular elements of the part as it rotates.

When circular runout is to be applied at specified locations, it must be so stated on the drawing. The example on the next page illustrates this application.

Please see further pages and examples in this section for further applications of and considerations on circular runout and for circular runout and total runout use on the same part.

EXAMPLE

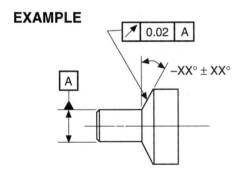

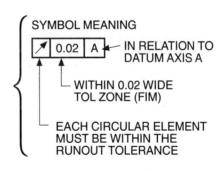

SYMBOL MEANING

IN RELATION TO DATUM AXIS A

WITHIN 0.02 WIDE TOL ZONE (FIM)

EACH CIRCULAR ELEMENT MUST BE WITHIN THE RUNOUT TOLERANCE

MEANING

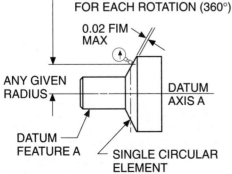

FIXED POSITION OF INDICATOR FOR EACH ROTATION (360°)

0.02 FIM MAX

ANY GIVEN RADIUS

DATUM AXIS A

DATUM FEATURE A

SINGLE CIRCULAR ELEMENT

THE FEATURE MUST BE WITHIN THE SPECIFIED TOLERANCE OF SIZE AT ANY MEASURING POSITION. EACH CIRCULAR ELEMENT OF THE SURFACE MUST BE WITHIN 0.02 FULL INDICATOR MOVEMENT WHEN THE PART IS ROTATED ONE FULL ROTATION ABOUT THE SPECIFIED DATUM AXIS WITH THE INDICATOR FIXED IN A POSITION NORMAL TO THE SURFACE. (THIS DOES NOT CONTROL FORM OF THE TOTAL SPECIFIED SURFACE AREA, BUT ONLY CONTROLS THE RUNOUT OF EACH CIRCULAR ELEMENT.)

EXAMPLE

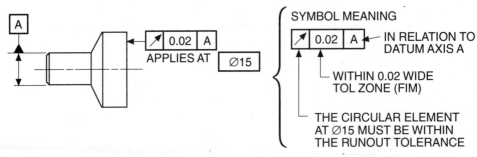

SYMBOL MEANING

⬈ 0.02 A ← IN RELATION TO DATUM AXIS A

└─ WITHIN 0.02 WIDE TOL ZONE (FIM)

└─ THE CIRCULAR ELEMENT AT Ø15 MUST BE WITHIN THE RUNOUT TOLERANCE

MEANING

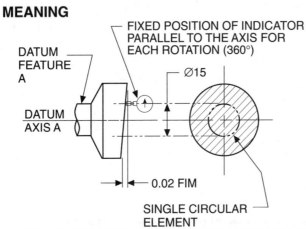

FIXED POSITION OF INDICATOR PARALLEL TO THE AXIS FOR EACH ROTATION (360°)

DATUM FEATURE A

DATUM AXIS A

0.02 FIM

SINGLE CIRCULAR ELEMENT

Ø15

THE CIRCULAR ELEMENT OF THE SURFACE AT Ø15 MUST BE WITHIN 0.02 FULL INDICATOR MOVEMENT WHEN THE PART IS ROTATED ONE FULL ROTATION ABOUT THE SPECIFIED DATUM AXIS WITH THE INDICATOR FIXED IN A POSITION PARALLEL TO THE AXIS. (THIS DOES NOT CONTROL PERPENDICULARITY, BUT CONTROLS ONLY THE LATERAL RUNOUT (WOBBLE) OF EACH CIRCULAR ELEMENT AT THE SPECIFIED SURFACE LOCATION.)

PART MOUNTED ON TWO FUNCTIONAL DIAMETERS (DATUMS)

Where multiple diameters of a rotating part are to be controlled relative to a datum axis, the datum axis can be established by the features which will provide the functional mounting of the part at assembly. These are called multiple datums.

Two diameters (cylinders) are used on the part at right to establish the datum axis of relationship. These diameters could represent the bearing mounting features and thus they establish the datum axis of rotation. Note that two datum features are selected (C and D). By stating the two datums simultaneously with a dash line between them, the relationship of all of the other features so designated relate to their common datum axis C–D. The precedence of datums C and D are equal in this case.

Any two surfaces given runout tolerances about the datum axis are to be individually within their stated runout tolerance; collectively they are related to each other within the sum of their indicator readings.

A runout tolerance specified for the datum feature has no effect on the considered features related to it.

PART MOUNTED ON TWO FUNCTIONAL DIAMETERS (DATUMS)

EXAMPLE

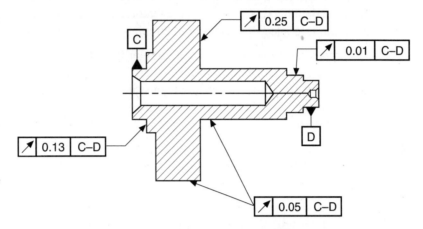

(SIZE DIMENSIONS & TOLERANCES NOT SHOWN FOR SIMPLICITY)

MEANING

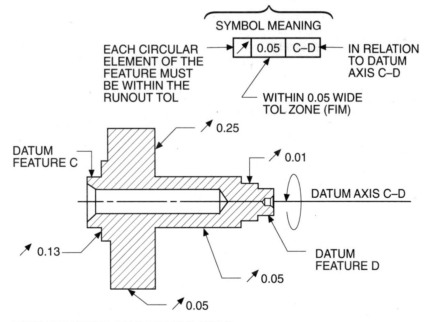

WHEN MOUNTED ON DATUMS C AND D.
DESIGNATED SURFACES MUST BE WITHIN CIRCULAR
RUNOUT (⟋) TOLERANCE SPECIFIED, RFS

PART MOUNTED ON TWO FUNCTIONAL DIAMETERS (DATUMS) INCLUDING RUNOUT TOLERANCE ON DATUMS

The part shown at right extends the principles of the previous example. It is a shaft of multi-diameters about a common datum axis C–D with each feature, *including the datum features*, stating an individual runout tolerance. It utilizes both circular runout and total runout control.

Total runout controls composite surface variations of circularity, cylindricity, parallelism, straightness, angularity, taper, and profile of a surface where applied to features constructed around a datum axis. Total runout applied to surfaces constructed perpendicular to a datum axis controls composite variations of perpendicularity and flatness. The tolerance is applied *simultaneously* at all circular and profile measuring positions as the part is rotated through 360°.

Due to the design requirements involved in this example, certain diameters (the datums) must be given total runout control, whereas the remaining diameters and face surfaces may be controlled with circular runout.

Since in the end assembly of this part, (mount to bearings), the relationship to both datums simultaneously is the result desired, all the features including each datum (C and D) must meet their individual runout tolerances. Note that the runout tolerance of each datum (C and D) individually is relative to their common axis C–D.

Multiple leaders may be used in controlling two or more features with a common runout tolerance as is shown. Runout tolerance may be specified individually or in groups, as is convenient, without affecting the runout tolerance.

PART MOUNTED ON TWO FUNCTIONAL DIAMETERS

EXAMPLE

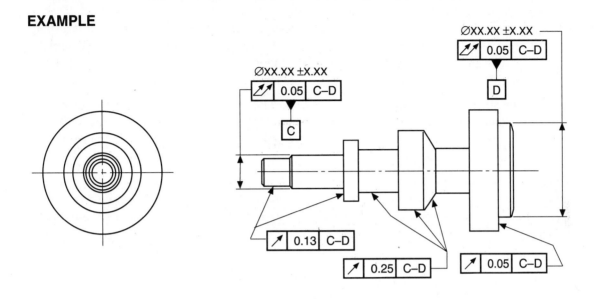

MEANING

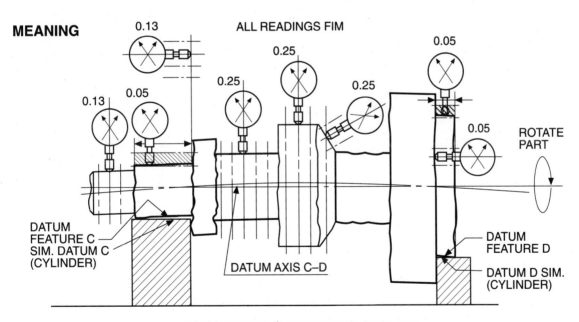

—EACH CIRCULAR ELEMENT INDIVIDUALLY—
—DATUM DIAMETERS C AND D ALL ELEMENTS
    TOGETHER (TOTAL)—

PART MOUNTED ON TWO FUNCTIONAL DIAMETERS (DATUMS) INCLUDING RUNOUT AND CYLINDRICITY TOLERANCE ON DATUM

The part shown at the right further extends the principles of the previous examples. It is a shaft of multi-diameters about a common datum axis C–D, with each feature including the datum features stating an individual runout tolerance. In addition to having both circular and total runout control, each of the datum features has a cylindricity tolerance.

It should be noted that the datum features of this part have controls of three orders of magnitude: size, total runout, and cylindricity (form). In this example, an accurate fit to bearings as controlled by size and form was required in addition to the runout relationship of the datum features to the datum axis (C–D) of rotation.

Features with a specific relationship to another feature rather than to the common datum axis may be so indicated by addition of appropriate datum references (such as datum E on page at right). Note that the inside diameter on the right end of the part relates the runout requirement to datum E.

PART MOUNTED ON TWO FUNCTIONAL DIAMETERS (DATUMS)

EXAMPLE

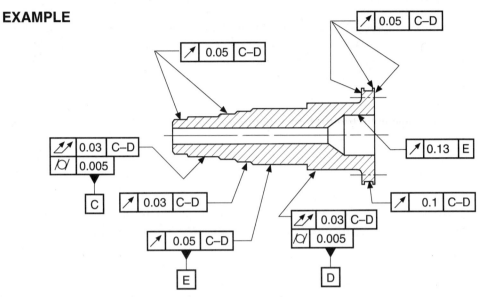

(SIZE DIMENSIONS & TOLERANCES NOT SHOWN FOR SIMPLICITY)

MEANING

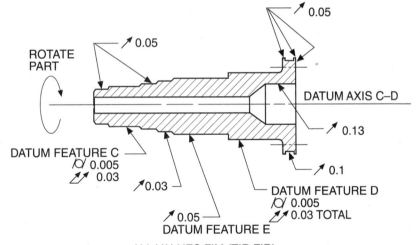

WHEN MOUNTED ON DATUMS C AND D, DESIGNATED SURFACES
MUST BE WITHIN CIRCULAR RUNOUT (⟋) TOLERANCE SPECIFIED.
DATUMS C AND D MUST ALSO BE WITHIN TOTAL RUNOUT (⟍⟋)
TOLERANCE SPECIFIED AND CYLINDRICAL WITHIN 0.005.

WHEN MOUNTED ON DATUM E, DESIGNATED SURFACE MUST BE
WITHIN CIRCULAR RUNOUT (⟋) TOLERANCE SPECIFIED.

PART MOUNTED ON FUNCTIONAL FACE SURFACE (DATUM) AND DIAMETER (DATUM)

Runout tolerancing may be applied to features of rotation where the feature datum references are an axis and a face surface perpendicular to the axis. In such an application, datum precedence is usually considered necessary. The influence of the appropriate feature as a primary datum is determined relative to the design requirement.

In the example at the right, the face surface perpendicular to the axis of rotation is considered the functionally important surface (bottoms on the inner race of bearing) and is thus selected as the primary datum A. Secondary datum diameter (cylinder) B is in contact with the extremities of datum feature B. With the part in this orientation and located functionally on its datums, all related features must meet their individual runout tolerances when the part is rotated about datum axis B.

Note that the primary datum A and secondary datum B features are indicated with that precedence by separate enclosures in the feature control frame (reading left to right).

PART MOUNTED ON FUNCTIONAL FACE SURFACE (DATUM) AND DIAMETER (DATUM)

EXAMPLE

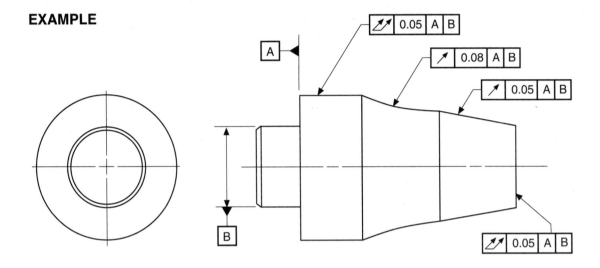

MEANING

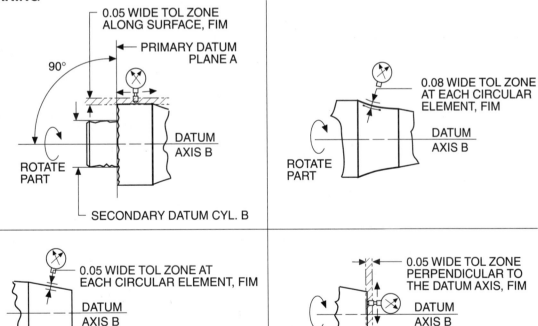

PART MOUNTED ON FUNCTIONAL FACE SURFACE (DATUM) AND DIAMETER (DATUM)

The part shown at right extends the principles of runout tolerancing. It illustrates a situation in which the large flat mounting surface requires flatness control to assure desired accuracy of the primary datum surface as it mounts and orients the remainder of the part to proper functional relationships. In this instance it can be assumed that this method was most representative of part end function.

The remainder of the part relates to the datum reference frame established by datum features C and D, with C being the primary datum for attitude of the part, and D the secondary datum for establishing axis of rotation. Note again that the other features *including datum D* relate to the datum system C and D.

PART MOUNTED ON LARGE FLAT SURFACE
(DATUM) AND DIAMETER (DATUM)

EXAMPLE

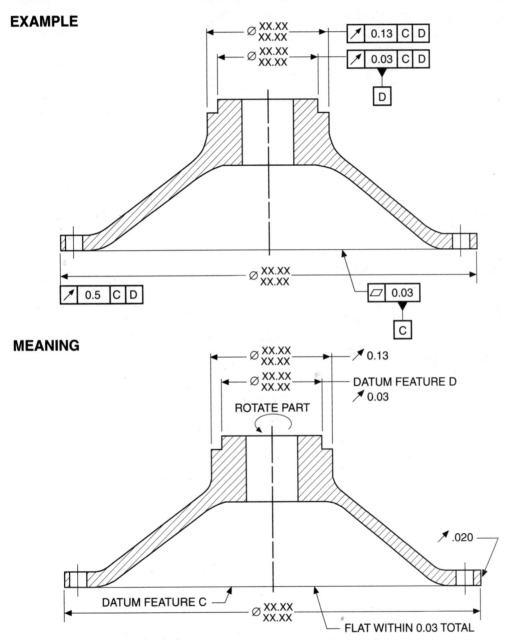

MEANING

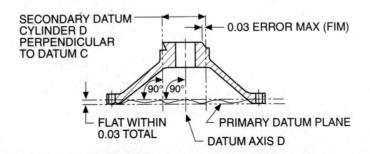

WHEN MOUNTED ON DATUMS C AND D, DESIGNATED SURFACES MUST BE
WITHIN CIRCULAR RUNOUT (⚯) TOLERANCE SPECIFIED.

The following example illustrates a functional design requirement where the part mounts upon live centers in assembly. Suppose components such as a gear, pulley, clutch, and brake mount upon the four diameters shown. The datum axis A–B is established for the runout relationships. The size and runout tolerance are determined by the calculated values permissible as based upon the design requirement. Another possibility might be where the datums A and B are used as "temporary datums" so that the ⌀ 12.49 and ⌀ 8.89 features could be established as shown with these features then specified as datums (e.g. C and D) for the ⌀ 23.88 and ⌀ 18.92 relationship to the new datums axis "C–D."

Typical total runout applications, inspection analysis, and meaning are discussed below and in the following text. An understanding of the fundamentals involved in these examples will provide the reader with the basis for total runout characteristic and use. The methods and principles shown may be readily adapted to circular runout as well, except for the substitution, where appropriate in the text and illustrations, of the circular elements (only) principles. In checking total runout, the datum feature, or features, is normally centered or positioned in an appropriate inspection device (e.g., between centers, collet, chuck, mandrel, vee block, or other centering device) to establish the datum axis from which the total runout relationships are to occur.

PART MOUNTED ON MACHINING CENTERS

EXAMPLE

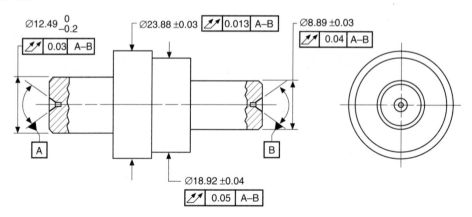

MEANING

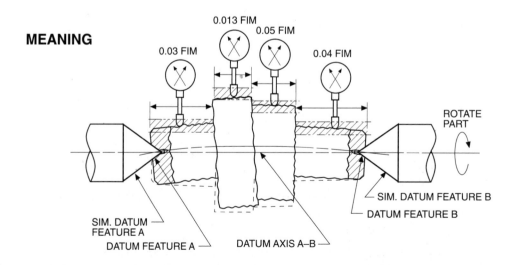

PART MOUNTED ON TWO FUNCTIONAL DIAMETERS

EXAMPLE

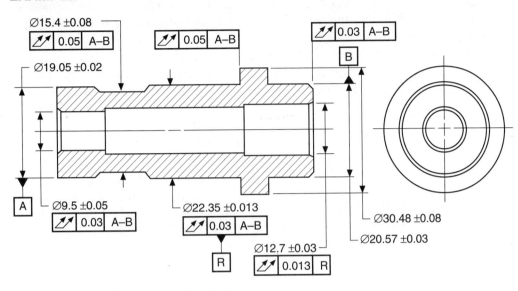

MEANING 1

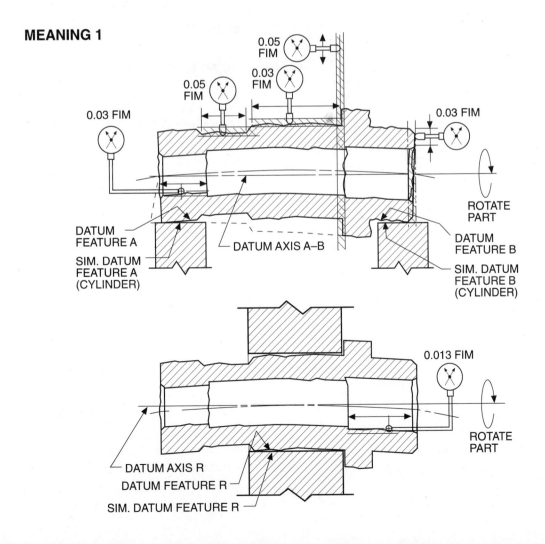

TOTAL SURFACE RUNOUT (WOBBLE)

Note that when flat or face surface runout (wobble) is part of the specification as in the case of the example on the preceding page, the runout analysis must consider the need for restricting end movement of the part while checking it. If collets or other holding devices are applied to establish the datum axis, end movement of the part is controlled. Also, specified secondary or tertiary datums (specified faces, shoulders, etc.) may indicate the end location or "stop" surface. However, in the absence of the above conditions and when the flat or face surface runout is to be checked, one should, if possible, "stop" on the surface being checked (see MEANING 1 on the preceding page and continued below), in order to avoid adding the error of *another* surface to the reading. When this is done, however, dependent on the location or placement of the stop, the dial indicator runout reading may need to be *halved* (divided by 2) before comparison with the stated part drawing runout tolerance to determine whether the requirement has been met. Referring to the illustration below, note that to use the fixed factor 2 the stop should be as close as possible to the O.D. of the surface being checked. As the placement of the stop approaches the axis of the mounting datum diameters (see also MEANING 2 and 3), the factor approaches a 1:1 ratio.

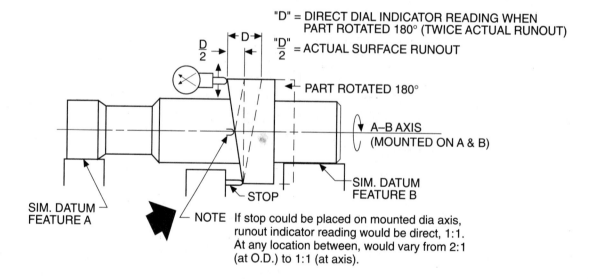

"D" = DIRECT DIAL INDICATOR READING WHEN PART ROTATED 180° (TWICE ACTUAL RUNOUT)

$\dfrac{"D"}{2}$ = ACTUAL SURFACE RUNOUT

PART ROTATED 180°

A–B AXIS
(MOUNTED ON A & B)

SIM. DATUM FEATURE B

STOP

SIM. DATUM FEATURE A

NOTE If stop could be placed on mounted dia axis, runout indicator reading would be direct, 1:1. At any location between, would vary from 2:1 (at O.D.) to 1:1 (at axis).

MEANING 2

As an alternative method a full end stop may be used. The end stop must, however, be extremely accurate at 90° (exact within gage tolerance) to the axis of the mounting diameters. See below.

In this case the dial indicator reading taken is direct and can be compared immediately with the stated part drawing runout (wobble) tolerance.

Note also that no error of the part surface or face which is used for "stopping" is introduced into the indicator reading, since the high point or extremity of this surface remains in consistent contact with the full stop gage as the part is rotated.

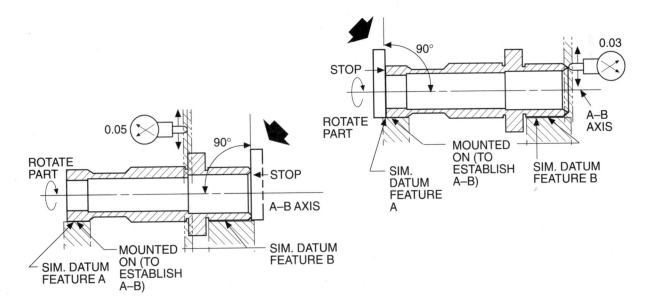

MEANING 3

If the part has a *closed end* and a stop can be placed at the *axis* of the mounting datum diameters (see illustration below), the runout requirement may be checked without regard for extreme accuracy of the stop. The dial indicator reading taken is direct and can be compared immediately with the stated part drawing runout (wobble) tolerance.

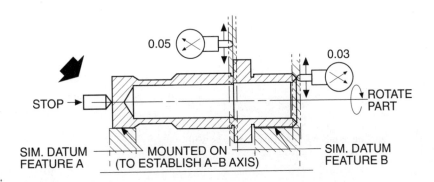

RUNOUT ⬈
PARTIAL SURFACE*

Where it is desirable to indicate that a runout tolerance is applicable only at a specific place or portion of a surface feature, it may be shown as indicated in the example below. Basic (exact) dimensions are shown indicating the portion of the surface to which the runout tolerance applies.

A thick chain line is shown slightly off the part with the leader from the feature control frame extended to the concerned surface. The thick chain line means that the stated tolerance applies only at the designated area. The basic dimensions invoke standard inspection accuracy in verifying the requirements. Otherwise, standard interpretation is given to the total runout requirement shown (see previous pages).

EXAMPLE

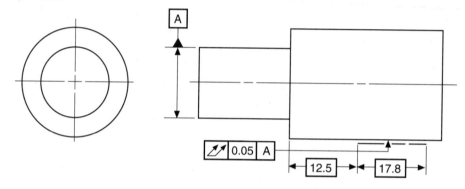

MEANING

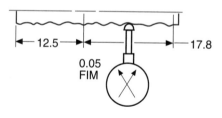

---

*This principle may be applied to other geometric controls as well.

TOLERANCES OF LOCATION

Tolerances of location state the permissible variation in the specified location of a feature in relation to some other feature or datum. Tolerances of location refer to the geometric characteristics: position, symmetry, and concentricity.

In the course of the discussion on location tolerancing, more detail on maximum material condition, datums, basic dimensioning, and the interrelationship of location and orientation tolerancing will be introduced.

Location tolerances involve features of size and relationships to center planes and axes. At least two features, one of which is a size feature, are required before location tolerancing is valid. Where function or interchangeability of mating part features is involved, the MMC principle may be introduced to great advantage. Perhaps the most widely used and best example of the application of this principle is position tolerancing.

The use of the position concept in conjunction with the maximum material condition concept provides some of the major advantages of the geometric tolerancing system.

POSITION ⌖

Definition. Position is a term used to describe the perfect (exact) location of a point, line, or plane of a feature in relationship with a datum reference or other feature.

POSITION TOLERANCE

A position tolerance is the total permissible variation in the location of a feature about its exact true position. For cylindrical features (holes and bosses) the position tolerance is the *diameter (cylinder)* of the tolerance zone within which the axis of the feature must lie, the center of the tolerance zone being at the exact true position. For other features (slots, tabs, etc.) the position tolerance is the total width of the tolerance zone within which the center plane of the feature must lie, the center plane of the zone being at the exact true position.

POSITION THEORY

The illustration at right examines the position theory as typically applied to a part for purposes of function or interchangeability. As a means of describing this theory it is helpful to first compare the position system with the bilateral or coordinate system.

Imagine a part with four holes in a pattern which must line up with a mating part to accept screws, pins, rivets, etc., to accomplish assembly, or four holes in a pattern to accept the pins, dowels, or studs of a mating part to accomplish assembly.

The top figure at the right shows the part with a hole pattern dimensioned and toleranced using a coordinate system. The bottom figure shows the same part dimensioned using the position system. Comparing the two approaches, the following differences are noted:

1.   The derived tolerance zones for the hole centers are square in the coordinate system and round in the position system.

2.   The hole center location tolerance in the top figure is part of the coordinates (the 50.8 and 44.5 dimensions). In the bottom figure, however, the location tolerance is associated with the hole size dimension and is shown in the feature control frame at the right. The 50.8 and 44.5 coordinates are retained in the position application, but are stated as BASIC or exact values.

For this comparison, the 0.1 square coordinate tolerance zone has been converted to an equivalent 0.14 position tolerance zone. The two tolerance zones are superimposed on each other in the enlarged detail.

The black dots represent possible inspected centers of this hole on eight separate piece parts. Note that if the coordinate zone is applied, only three of the eight parts are acceptable. However, with the position zone applied, six of the eight parts appear immediately acceptable.

COORDINATE SYSTEM

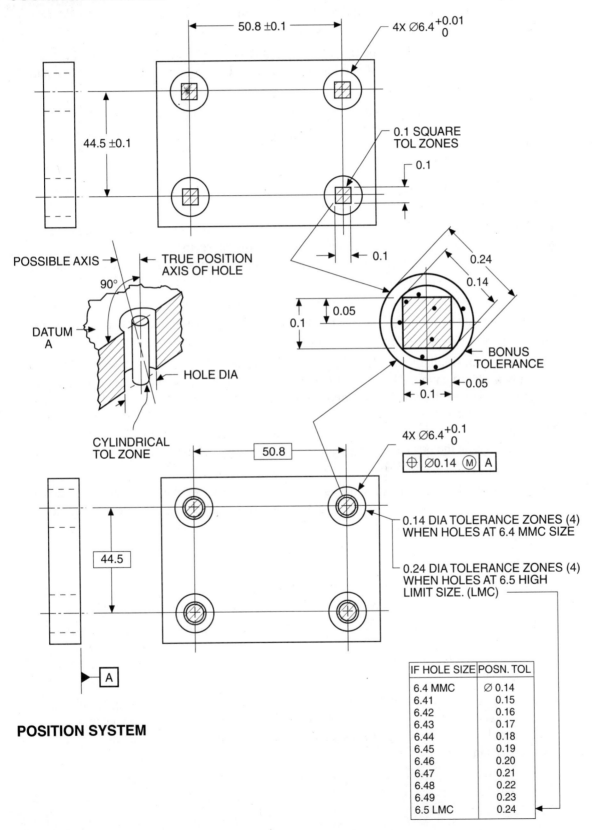

50.8 ±0.1

4X Ø6.4 +0.01 / 0

44.5 ±0.1

0.1 SQUARE
TOL ZONES

0.1

0.1

POSSIBLE AXIS → | ← TRUE POSITION
AXIS OF HOLE

90°

DATUM
A

HOLE DIA

CYLINDRICAL
TOL ZONE

0.24

0.14

0.05

0.1

0.1

BONUS
TOLERANCE

0.05

0.1

4X Ø6.4 +0.1 / 0

⌖ | Ø0.14 Ⓜ | A

50.8

44.5

0.14 DIA TOLERANCE ZONES (4)
WHEN HOLES AT 6.4 MMC SIZE

0.24 DIA TOLERANCE ZONES (4)
WHEN HOLES AT 6.5 HIGH
LIMIT SIZE. (LMC)

A

POSITION SYSTEM

IF HOLE SIZE	POSN. TOL
6.4 MMC	Ø 0.14
6.41	0.15
6.42	0.16
6.43	0.17
6.44	0.18
6.45	0.19
6.46	0.20
6.47	0.21
6.48	0.22
6.49	0.23
6.5 LMC	0.24

The position diameter shaped zone can be justified by recognizing that the 0.14 diagonal is unlimited in orientation. Also, a cylindrical hole should normally have a cylindrical tolerance zone.

A closer analysis of the representative black dots and their position with respect to the desired exact location clearly illustrates the fallacies of the coordinate system when applied to a part such as that illustrated.

The dot in the upper left diagonal corner of the square zone and the dot on the left outside the square zone are in reality at nearly the same distance from the desired exact center. However, in terms of the square coordinate zone, the hole on the left is unacceptable by a wide margin, whereas the upper left hole is acceptable.

Note, then, that a hole produced off center under the coordinate system has *greater* tolerance if the shift is on the diagonal and not in the horizontal or vertical direction.

Realizing that the normal function of a hole relates to its mating feature in *any* direction (i.e., a hole vs. a round pin), we see that the square zone restriction seems unreasonable and incorrect. Thus the position tolerance zone, which recognizes and accounts for unlimited orientation of round or cylindrical features as they relate to one another, is more realistic and practical.

In normal applications of position principles, the tolerance is derived, of course, from the design requirement, *not* from converted coordinates. The maximum material sizes of the features (hole and mating component) are used to determine this tolerance.

Thus the 0.14 position tolerance of the example on the preceding page (see also Fig. 1) would normally be based on the MMC size of the hole (6.4). As the hole size deviates from the MMC size, the position of the hole is permitted to shift off its "true position" beyond the original tolerance zone to the extent of that departure. The "bonus tolerance" of 0.24 illustrates the possible position tolerance should the hole be produced, for example, to its high limit size of 6.5. The tabulation on the preceding page shows the enlargement of the position tolerance zone as the hole size departs from MMC.

Although we have considered only one hole to this point in the explanation, the same reasoning applies to all the holes in the pattern. Note that position tolerancing is also a noncumulative type of control in which each hole relates to its own true or exact position and no error is accumulated from the other holes in the pattern.

Position tolerancing is usually applied on mating parts in cases where fit, function, and interchangeability are the considerations. It provides greater production tolerances, ensures design requirements, and provides the advantages of functional inspection practices as desired.

Functional gaging techniques, familiar to a large segment of industry through many years of application, are fundamentally based on the MMC position concept. It should be clearly understood, however, that functional gages are not mandatory in fulfilling MMC position requirements.

Functional gages are used and discussed in this text for the dual purpose of explaining the principles involved in position tolerancing and of introducing the functional gage technique as a valuable tool. A functional gage can be considered as a simulated master mating part at its worst condition.

Position, although a locational tolerance, also includes form and orientation tolerance elements in composite. For example, as shown in the illustration on page 109, perpendicularity is invoked as part of the control to the extent of the diameter zone, actually as a "cylindrical" zone, for the depth of the hole. Further, the holes in the pattern are parallel to one another within the position tolerance. Various other elements of form are included as a part of the composite functional control provided by position tolerancing. Use of datum references and the relation of the hole pattern to specific surfaces or other features further ensures that the holes (and their tolerance zones) are related to part function and will be uniformly interpreted. For purposes of simplicity in explaining the theory, only *one* datum feature was specified in this example. A complete specification of this requirement would include *two* additional datums (edges of the part). For a more complete discussion on the datum reference frame, see the DATUM section and numerous examples following in this text.

POSITION SYSTEM

The example below further clarifies the position theory; two of the holes on the part shown in the example are enlarged to illustrate the actual effect of feature *size* variation on the previous positional location of the features.

POSITION THEORY

Figure 1 shows the two $6.4^{+0.1}_{0}$ holes at MMC size (or the low limit of their size tolerance) of 6.4 and with their centers perfectly located in the 0.14 diameter position tolerance zone. The drawing illustrates the mating part situation represented by a functional or fixed pin. The gage pins are shown undersize an amount equal to the positional tolerance of 0.14, i.e., at 6.26 diameter. This represents the maximum permissible offset of the holes within their stated positional tolerance when the hole is at MMC size of 6.4.

FIGURE 1

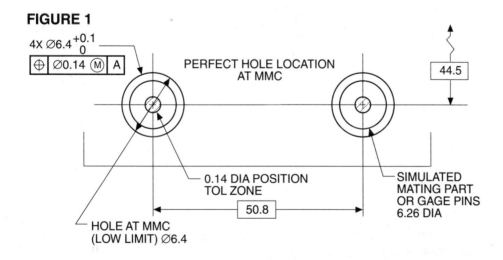

Figure 2 shows the two 6.4 MMC holes offset in opposite directions to the maximum permissible limits of the 0.14 position tolerance zone. Note that we illustrate the worst condition: the edges of the holes are tangent to the diameters of the simulated mating part or gage pins. The holes are within tolerance and, as can be seen, would satisfactorily pass the simulated mating part condition as represented by the gage pins.

In Fig. 3 the 6.4 $^{+0.1}_{0}$ holes have been produced to the opposite, or *high* limit (least material condition) size of 6.5. It can now be seen that when we retain the same offset and tangency of the holes and mating part of the gage pins as shown in Fig. 2, the produced centers of the holes are allowed to shift *beyond* the original 0.14 tolerance zone to a resulting 0.24 diameter tolerance zone still providing an acceptable situation.

FIGURE 2

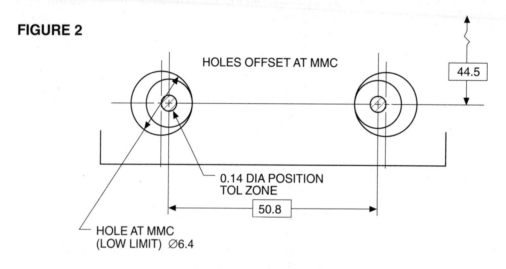

HOLES OFFSET AT MMC

44.5

0.14 DIA POSITION
TOL ZONE

50.8

HOLE AT MMC
(LOW LIMIT) ⌀6.4

FIGURE 3

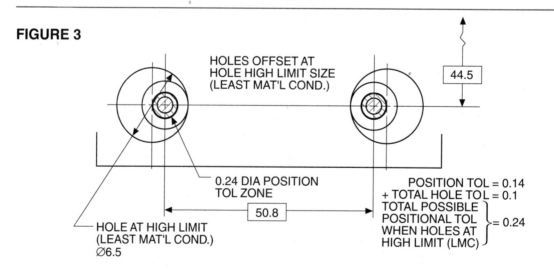

HOLES OFFSET AT
HOLE HIGH LIMIT SIZE
(LEAST MAT'L COND.)

44.5

0.24 DIA POSITION
TOL ZONE

50.8

HOLE AT HIGH LIMIT
(LEAST MAT'L COND.)
⌀6.5

POSITION TOL = 0.14
+ TOTAL HOLE TOL = 0.1
TOTAL POSSIBLE
POSITIONAL TOL   }= 0.24
WHEN HOLES AT
HIGH LIMIT (LMC)

The foregoing illustrates the interrelationship of size and location tolerances which is utilized in position dimensioning and tolerancing.

Although in this example we have used only two of the holes, the same reasoning applies to *all* the holes in the pattern; similarly, each individual hole could be offset within its tolerance zone in any direction around 360° and provide an acceptable situation.

It should be noted that a functional or fixed pin gage such as the one used here to explain the position theory can be used *only* to check the *positional* location of the holes. *Positional* tolerance can be added as the holes increase in size or depart from MMC size within their size tolerance range. Hole size tolerance, however, must be held within the tolerances specified on the drawing and must be checked individually and separately from the positional check.

MATING PARTS—FLOATING FASTENER

Position tolerancing techniques are most effective and appropriate in mating part situations. The illustrations on page 115, in addition to demonstrating the calculations required, also emphasize the importance of decisions at the design stage to recognize and initiate the position principles.

The mating parts shown in the illustration are to be interchangeable. Thus, the calculation of their position tolerances should be based on the two parts and their interface with the fastener in terms of MMC sizes.

The two parts are to be assembled with four M5 screws. The holes in the two parts are to line up sufficiently to pass the four screws at assembly. Since the four screws ("fasteners") are separate components, they are considered to have some "float" with respect to one another. The colloquial term, "floating fastener" application, has been popularly used to describe this situation.

The calculations are shown in the upper right corner of the illustration. Also, note that, in this case, the same basic dimensions and position tolerances are used on both parts. They are, of course, separate parts and are on separate drawings.

The position tolerance calculations are based on the MMC sizes of the holes and the screws. The maximum material basis then sets the stage for maximum producibility, interchangeability, functional gaging (if desired), etc., at production. As seen from the illustration, part acceptance tolerances will increase as the hole actual mating sizes in the parts are actually produced and vary in size as a departure from MMC. From the 0.4 diameter tolerance calculated, the tolerance may increase to as much as 0.55 dependent upon the actually produced hole size. It should be noted that clearance between the mating features (in this case hole and screw) is the criterion for establishing the position tolerances.

Simultaneously with these production advantages, the design is protected since it has been based upon the realities of the hole and screw sizes as they interrelate at assembly and in their function. Thus, as parts are produced, assembly is ensured, and the design function is carried out specifically as planned.

A possible functional gage is also shown in the illustration on page 115. The 5.0 gage pin diameters are determined by the MMC size of the hole, 5.4, minus the stated position tolerance of 0.4. In our example, the same functional gage can be used on both parts. Functional gages are, of course, not required with position application, but they do, however, provide an effective method of evaluation where desired. Functional gaging principles may be carried out by the use of coordinate measuring machines (CMM's), and the variables data derived, through the use of calculators, computer programs or tabulated mathematical equivalents (see pages 322, 323). This technique is often referred to as "soft gaging."

MATING PARTS—FLOATING FASTENER

Referring to the position tolerance calculations, if more than two parts are assembled in a floating fastener application, we must determine the position tolerance to ensure that any two parts and the fastener will mate properly. Calculate each part to mate with the fastener using the illustrated formula and MMC sizes.

Only the primary datums (the interface surfaces) are shown on the illustrations for simplicity of explanation of this method. The hole pattern alignment is emphasized between the two parts. A complete specification of these requirements on both parts would include two additional datums to locate the holes and the pattern from the part outside surfaces (datums).

See page 213 for methods of centralizing the hole pattern on the part. See also pages 121, 123 following, pages 210–222, "Datums" (beginning at page 223) and other sections of this text describing application of datums.

The clearance holes on these parts are all specified as same size. Where they are specified as *different sizes,* each part can be calculated separately for the allowable positional displacement based on the difference between the MMC of the hole and fastener. If one part in our example had $5.1 \, ^{+0.05}_{0}$ –4HOLES specified, the method below would be used:

MMC hole	$\varnothing 5.1$
MMC fastener	$(-) \; \varnothing 5.0$
	$\varnothing 0.1$

$$\boxed{\; \oplus \; | \; \varnothing \, 0.1 \; \text{ⓜ} \; | \; A \;}$$

Other part	
MMC hole	$\varnothing 5.4$
MMC fastener	$(-) \; \varnothing 5.0$
	$\varnothing 0.4$

$$\boxed{\; \oplus \; | \; \varnothing \, 0.4 \; \text{ⓜ} \; | \; A \;}$$

The position tolerance calculation method illustrated assumes the possibility of a zero interference-zero clearance condition of the mating part features at extreme tolerance limits. Additional compensation of the calculated tolerance values should be considered as necessary relative to the particular application.

Formulas used as a basis for the position floating fastener calculations are:

To calculate position tolerance with fastener and hole size known:

$$T = H - F$$

where T = tolerance, H = MMC hole, and F = MMC fastener.

Where the hole size or fastener size is to be derived from an established position tolerance, the formula is altered to:

$$H = T + F$$

$$F = H - T$$

NOTE   Where mating parts are to be assembled with rivets, dowel pins, press fits, etc., other considerations or methods may be necessary; such as, drilling or fabricating of the holes, pins or other features at assembly.

MATING PARTS—FLOATING FASTENER

EXAMPLE

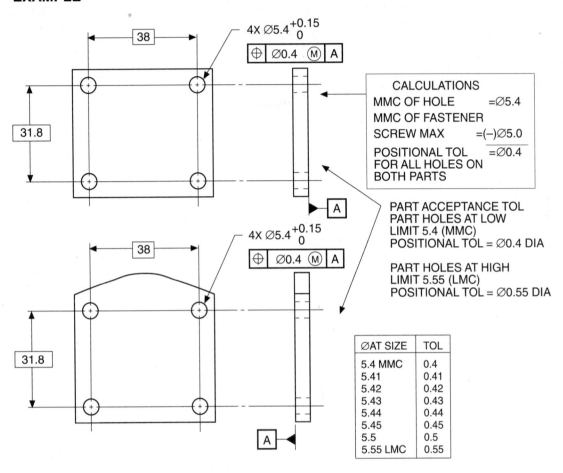

4X ⌀5.4 $^{+0.15}_{0}$

| ⊕ | ⌀0.4 | Ⓜ | A |

CALCULATIONS
MMC OF HOLE   =⌀5.4
MMC OF FASTENER
SCREW MAX   =(–)⌀5.0
POSITIONAL TOL   =⌀0.4
FOR ALL HOLES ON
BOTH PARTS

PART ACCEPTANCE TOL
PART HOLES AT LOW
LIMIT 5.4 (MMC)
POSITIONAL TOL = ⌀0.4 DIA

PART HOLES AT HIGH
LIMIT 5.55 (LMC)
POSITIONAL TOL = ⌀0.55 DIA

⌀AT SIZE	TOL
5.4 MMC	0.4
5.41	0.41
5.42	0.42
5.43	0.43
5.44	0.44
5.45	0.45
5.5	0.5
5.55 LMC	0.55

GAGE FOR ABOVE PARTS

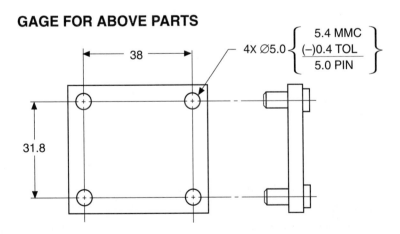

4X ⌀5.0 { 5.4 MMC / (–)0.4 TOL / 5.0 PIN }

MATING PARTS—FIXED FASTENER

When one of two mating parts has "fixed" features, such as the threaded studs in this example, the "fixed fastener" method is used in calculating position tolerances.

The term "fixed fastener" is a colloquialism popularly used to describe this application. Both the term and the technique are applied to numerous other manufacturing situations such as locating dowels and holes, tapped holes, etc.

The advantages of the MMC principle as described in the foregoing "floating fastener" application also apply here. However, with a "fixed fastener" application, the difference between the MMC sizes of mating features must be divided between the two features, since the total position tolerance must be shared by the two mating features. In this example, the two mating features (actually four of each in each pattern) are the studs and the clearance holes. The studs must fit through the holes at assembly.

Again, we see that the clearance of the mating features as they relate to each other at assembly determines the position tolerances. When one feature is to be assembled *within* another on the basis of the MMC sizes and "worst" condition of assembly, the clearance, or total tolerance, must be divided for assignment to *each* of the mating part features. In this case, the derived 0.4 was divided equally, with 0.2 diameter position tolerance assigned to each mating part feature (stud and hole). The total tolerance of 0.4 can be distributed to the two parts as desired, so long as the total is 0.4 (e.g., 0.3 + 0.1, 0.25 + 0.15, etc.). This decision is made at the design stage, however, and must be fixed on the drawing before release to production.

Only the primary datums (the interface surfaces) are shown on the illustrations for simplicity of explanation of the fixed fastener method. See page 213 for methods of centralizing the hole and pin patterns on the parts. See also pages 121,123 following, pages 213–222, "Datums" (beginning at page 223) and other sections of this text describing application of datums.

Application of the MMC principle to situations of this type guarantees functional interchangeability, design integrity, maximum production tolerance, functional gaging (if desired), and uniform understanding of the requirements.

As the part features of both parts are produced, any departure in actual mating size from MMC will increase the calculated position by an amount equal to that departure. Thus, for example, the position tolerance of the upper part could possibly increase up to 0.35 and that of the lower part up to 0.33 dependent upon the amount of departure from their MMC sizes. However, parts must actually be produced and sizes established before the *amount* of increase in tolerance can be determined.

Functional gages (shown below each part in the illustration) can be used for checking, and, although their use is not a must, they provide a very effective method of evaluation if desired. Note that the functional gages resemble the mating parts; as a matter of fact, functional gages simulate mating parts at their worst condition.

The functional gage pins of the upper part are determined by the MMC hole size minus the stated position tolerance. Gage tolerances are not shown, although they may be imagined to be on the order of $5.202 \, ^{+0.005}_{0}$ for pin size, and ± 0.005 on between pin locations. Local gage practices would prevail using a minimum amount of part tolerance (up to 10%) for gage making tolerances as assigned.

The functional gage on the lower part of the illustration contains holes instead of pins. The gage hole sizes are determined by the MMC (O.D.) size of the 5.0 pins plus the stated position tolerance. The tolerances are similar to those of the above pin gage. Tolerances on the order of $5.198-^{0}_{0.005}$ for hole size, and ± 0.005 between holes could be applied, depending on local gage practices using a minimum amount of part tolerance (up to 10%) for gage making tolerances as assigned.

EXAMPLE

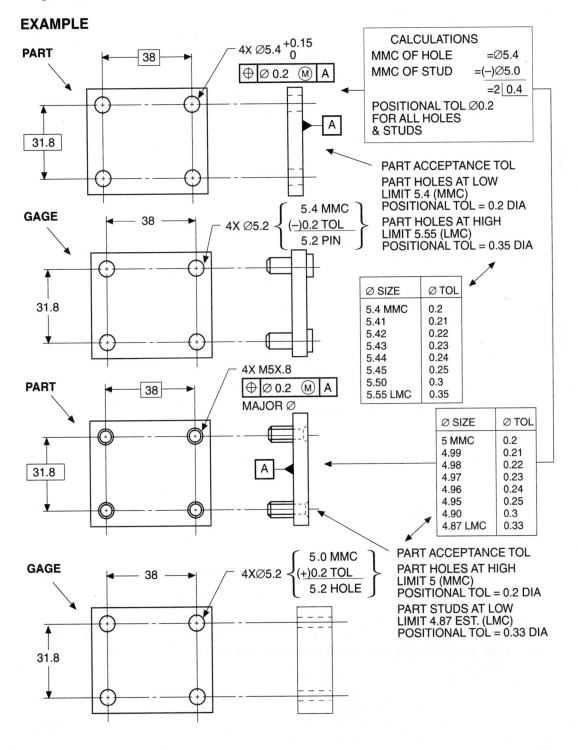

CALCULATIONS
MMC OF HOLE   =Ø5.4
MMC OF STUD   =(−)Ø5.0
                 =2|0.4

POSITIONAL TOL Ø0.2
FOR ALL HOLES
& STUDS

PART ACCEPTANCE TOL

PART HOLES AT LOW
LIMIT 5.4 (MMC)
POSITIONAL TOL = 0.2 DIA

PART HOLES AT HIGH
LIMIT 5.55 (LMC)
POSITIONAL TOL = 0.35 DIA

Ø SIZE	Ø TOL
5.4 MMC	0.2
5.41	0.21
5.42	0.22
5.43	0.23
5.44	0.24
5.45	0.25
5.50	0.3
5.55 LMC	0.35

Ø SIZE	Ø TOL
5 MMC	0.2
4.99	0.21
4.98	0.22
4.97	0.23
4.96	0.24
4.95	0.25
4.90	0.3
4.87 LMC	0.33

PART ACCEPTANCE TOL

PART HOLES AT HIGH
LIMIT 5 (MMC)
POSITIONAL TOL = 0.2 DIA

PART STUDS AT LOW
LIMIT 4.87 EST. (LMC)
POSITIONAL TOL = 0.33 DIA

It should be noted that the term MAJOR $\oslash$ is used beneath the position callout on the lower part on page 117. In the absence of this special notation of exception, the ANSI Y14.5, Pitch Diameter Rule would have invoked the tolerance on the basis of the pitch diameter of the threads. The major diameter (or O.D.) of the thread was the desired criterion in this example. See POSITION—EXTENDED PRINCIPLES section for additional examples of fixed fastener applications.

The calculations on these parts illustrate a balanced tolerance application in which the total permissible position tolerance of the two parts is equally divided, for example, 0.2 on each part. The total position tolerance can, however, be distributed as desired, as discussed earlier.

If more than two parts are assembled in a fixed fastener application, each part containing clearance holes must be calculated to mate with the part with the fixed features.

The position tolerance calculation method illustrated assumes the possibility of a zero interference-zero clearance condition of the mating part features at extreme tolerance limits. Additional compensation of the calculated tolerance values should be considered as necessary relative to the particular application.

Formulas used as a basis for the position fixed fastener (or locator) calculations are:

$$T = \frac{H - F}{2}$$

MMC hole	= H
MMC fastener	= F
(or pin, dowel, etc.)	
Tolerance	= T

Where the hole size or fastener (or pin, dowel, etc.) size is to be derived from an established tolerance, the formula is altered to:

$$H = F + 2T$$
$$F = H - 2T$$

The illustrations on page 119 show position tolerancing applied to two mating parts with a circular hole pattern. The same reasoning applies here as in the preceding examples except that the basic dimensions are angular (45° angles, 8 places) and a diameter (the 38 diameter).

These two parts again are of the fixed fastening type, the studs of the lower part being the fixed elements. To determine the positional tolerances for each part, the MMC of the hole and the MMC of the stud are used to determine the total positional tolerance. This is divided by two to give the positional tolerance value for each part. The total value may be divided as desired, as previously described.

Note again how the positional tolerance *increases* as the actual mating sizes of the holes in the upper part and the studs in the lower part depart from their MMC sizes, that is, when the holes get larger and the pins get smaller during the production process.

Functional gages are shown in the illustration for each of the parts. Note that the pins in the upper gage are calculated to the MMC or low limit of the holes in the part (which is 4.8 in this case) *minus* the positional tolerance (0.15), resulting in the 4.65 gage pin size.

EXAMPLES

8X ⌀4.8 +0.16 / 0

| ⊕ | ⌀ 0.15 Ⓜ | A |

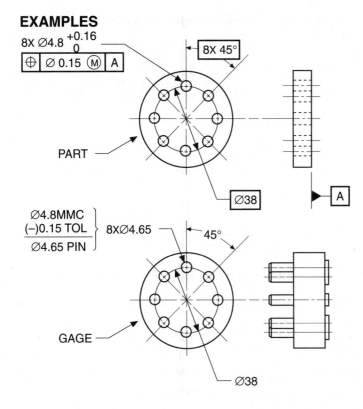

8X 45°

PART

⌀4.8MMC
(–)0.15 TOL
──────────
⌀4.65 PIN } 8X⌀4.65

45°

GAGE

⌀38

CALCULATIONS	
MMC OF HOLE	= ⌀4.8
MMC OF STUD	= (–)⌀4.5
	2|0.3
POSITIONAL	= ⌀0.15
TOL FOR ALL	
HOLES AND STUDS	

PART ACCEPTANCE TOL

PART HOLES AT LOW
 LIMIT 4.8 (MMC)
 POSITIONAL TOL = ⌀0.15

PART HOLES AT HIGH
 LIMIT 4.96
 POSITIONAL TOL = ⌀0.31

⌀ SIZE	⌀ TOL
4.8 MMC	0.15
4.81	0.16
4.82	0.17
4.83	0.18
4.84	0.19
4.85	0.20
4.9	0.25
4.96 LMC	0.31

8X ⌀4.5 0 / –0.1

| ⊕ | ⌀0.15 Ⓜ | A |

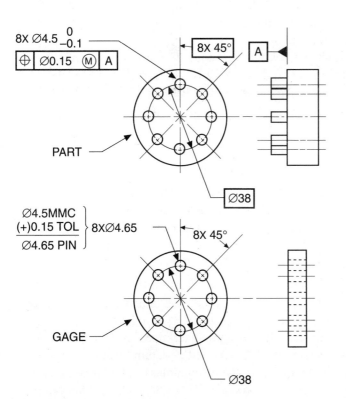

8X 45° | A |

PART

⌀38

⌀4.5MMC
(+)0.15 TOL
──────────
⌀4.65 PIN } 8X⌀4.65

8X 45°

GAGE

⌀38

PART ACCEPTANCE TOL

PART STUDS AT HIGH
 LIMIT 4.5 (MMC)
 POSITIONAL TOL = ⌀0.15

PART STUDS AT LOW
 LIMIT 4.4
 POSITIONAL TOL = ⌀0.25

⌀ SIZE	⌀ TOL
4.5 MMC	0.15
4.49	0.16
4.48	0.17
4.47	0.18
4.46	0.19
4.45	0.20
4.4 LMC	0.25

The lower gage is calculated in reverse, using the MMC on high limit of the stud, 4.5, plus the positional tolerance (0.15), resulting in the 4.65 gage hole size.

These calculations illustrate a balanced tolerance application in which the total permissible position tolerance of the two parts is equally divided, for example, 0.15 on each part. The total position tolerance can, however, be distributed as desired, for example, 0.1 on one part, 0.2 on the other, etc., so long as it totals the tolerance calculated (in this case 0.3).

The position tolerance calculation method illustrated here and in preceding examples assumes the possibility of a zero interference-zero clearance condition of the mating part features at extreme tolerance limits. Additional compensation of the calculated tolerance value should be considered as necessary relative to the particular application.

A primary datum (A) only is shown in the examples for simplicity of explanation. A secondary datum (i.e., the outside diameter) would be required to complete the specifications on both parts.

RELATION TO SPECIFIED DATUM SURFACES

Datum planes or surfaces as the basis for position relationships must be *specified* on the drawing. This illustration introduces the datum reference frame and datum precedence as it relates to hole patterns. The primary datum A placed first in the feature control frame is probably established because it is the most important stabilizing feature for the part interface, the corresponding feature, with a mating part. The secondary datum B is placed next in the feature control frame indicating the next most important orientation of the part to a mating surface and the third, or tertiary, datum C is then added in the third segment of the feature control frame to complete the hole pattern relationship to the datum planes, the three mutually perpendicular datum planes. Very often the selection of which features is to be selected as the secondary datum is arbitrary. That is, the part function and interface with a mating assembly may suggest that either of the second or third necessary datums could be equally selected as secondary datum. However, to give uniform meaning to design intent and manufacturing and verification follow-through, a decision is made to remove the either/or ambiguity.

The extremities or high points of the datum surfaces from which the ⊡19⊡ dimensions are taken establish the datum planes from which the position hole pattern is oriented. These datum planes are functional to the part requirement and may also be used for tooling or fixturing reference and in establishing measuring planes to inspect the part.

The position tolerance zone is 0.25 when the holes are produced at MMC or the low limit of 6.4. The tolerance zone increases up to 0.45, if the hole actual mating envelope size is produced larger to the high (LMC) limit of 6.6. Note that the position tolerance applies while the part is in contact with the datum surface; thus the position tolerance stated also controls the pattern location from the edges and provides the tolerance for the ⊡19⊡ pattern locating dimensions.

Note that in this example, the precision of the datum feature surfaces will be the result of manufacturing processes used to produce the rather lenient part overall size dimensions. The question is then raised, are these surfaces of sufficient precision to ensure the part relationship to the mating part situation is adequate. It is the decision of the design to determine this. Note also that the squareness (perpendicularity) of the two outside surfaces selected as datum features (B and C) have no ensured specific perpendicularity requirement. The result, unless further refined by form and/or orientation controls will be "whatever is produced." This depends on what can be considered the "four factors", i.e., good workmanship, discretion of the production operations, probabilities and any unless otherwise specified controls (such as title block, industry specs, milspecs, etc.). See the following example on page 122 and 123 for further options.

EXAMPLE

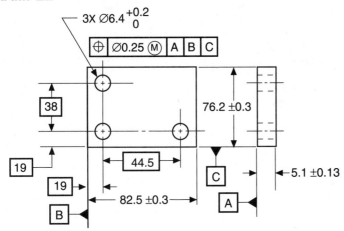

MEANING

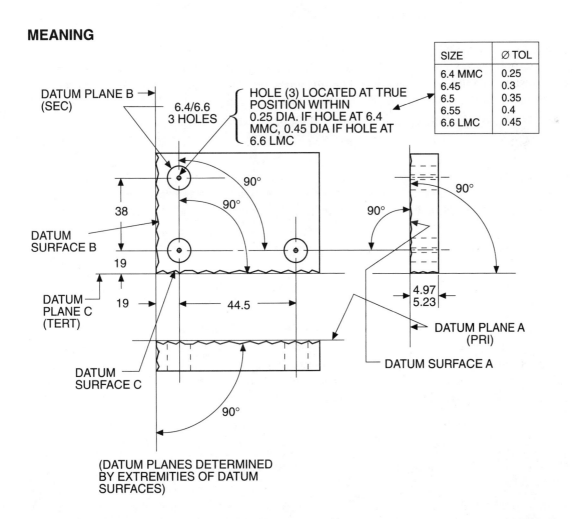

SIZE	Ø TOL
6.4 MMC	0.25
6.45	0.3
6.5	0.35
6.55	0.4
6.6 LMC	0.45

HOLE (3) LOCATED AT TRUE POSITION WITHIN 0.25 DIA. IF HOLE AT 6.4 MMC, 0.45 DIA IF HOLE AT 6.6 LMC

DATUM PLANE B (SEC)

6.4/6.6 3 HOLES

DATUM SURFACE B

DATUM PLANE C (TERT)

DATUM SURFACE C

DATUM PLANE A (PRI)

DATUM SURFACE A

(DATUM PLANES DETERMINED BY EXTREMITIES OF DATUM SURFACES)

RELATION TO SPECIFIED DATUM SURFACES — WITH FORM AND ORIENTATION TOLERANCES

The illustration on the right shows the same part as that shown on the preceding page, but with added form and orientation tolerances specified to control the precision of the datum features.

Where part function, and thus the stated drawing requirements, are more critical, specified datums and greater geometric control are essential to the design.

In this example, it was considered necessary to control the accuracy of the datum surfaces in their specific relationship to each other. To accomplish this, identification of the specific surfaces as datum references was required. Further, since the hole position pattern was critical in its orientation to the surfaces, datum identification was required. With specification of the datums, precedence of the datum surfaces is established, and the part and the hole pattern are stabilized relative to the datum reference frame. Datum precedence is also established.

Datum surface A (top surface of the part) is to be held to a flatness of 0.03 total. Datum surface B is to be perpendicular to datum plane A within 0.03 total. Datum surface C is to be perpendicular to datum plane A within 0.03 total and also perpendicular to datum plane B within 0.05 total.

The extremities or high points of the datum surfaces from which the ⌴19⌴ dimensions are taken establish the secondary and tertiary datum planes (B and C). The primary datum plane (A) is established by the extremities of datum surface A. The part orientation with respect to the position pattern is thus fixed. These datum planes are functional to the part requirement and may also be used for tooling or fixturing reference.

The position tolerance zone is 0.25 when the holes are produced at MMC or the low limit of 6.4. The tolerance zone increases up to 0.45 if the holes are produced larger to the high limit of 6.6. Note that the position tolerance applies while the part is in contact with the datum surfaces according to their stated precedence or sequence. The position tolerance stated also controls the pattern location from the edges and provides the tolerance for the ⌴19⌴ pattern locating dimensions.

In this example all geometric controls are specifically stated, removing all doubt as to design intent and follow-through manufacture and inspection requirements. The tools provided by specified datums and greater geometric control can be very effectively applied and will protect design integrity.

It should be noted that the form and orientation controls shown are not necessarily required when specifying the datum planes (A,B,C). The designer should consider the precision of the surfaces being used as datum features (as controlled by size), and then only as necessary, per the design requirements, add any form or orientation controls. See the preceding illustration and explanation (pages 120 and 121) where no form and orientation tolerances were required.

POSITION TOLERANCE—SPECIFIED DATUMS—FORM AND ORIENTATION TOLERANCES

EXAMPLE

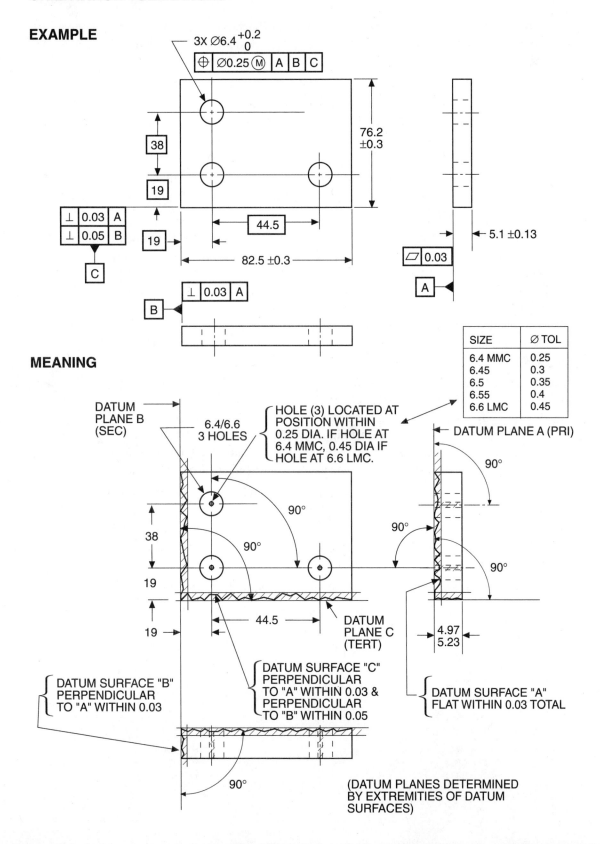

SIZE	⌀ TOL
6.4 MMC	0.25
6.45	0.3
6.5	0.35
6.55	0.4
6.6 LMC	0.45

MEANING

RELATION TO DATUM SURFACES—
COMPOSITE POSITION TOLERANCING

When the location of a pattern of features from datum surfaces is less important than the accuracy required *within* the pattern of features, composite position tolerancing may be used. Composite position tolerancing also extends the use of specified datum relationships and geometric tolerance control.

Composite position tolerancing incorporates a dual feature control frame with two positional controls. One, the upper segment in the symbol, specifies the applicable datums and the pattern locating position tolerance. The lower segment specifies the applicable datum and the feature relating position tolerance. A single position tolerance symbol is used.

The composite position tolerance method utilizes the full advantages of MMC (and RFS or LMC if desired) and extends the principles to control of patterns of features as well as of the individual feature interrelationship.

In the upper right pattern of eight mounting holes in the example, the theoretical center of the pattern is located by the 51.5 and 102 basic dimensions and is related to datums A, B, and C. This center is the axis for the 51 basic diameter. The individual theoretical centers of the eight 4.9 holes are established by the intersection of the 45° basic angles at the 51 basic diameter. The pattern as a unit, yet actually determined by the holes themselves, is to be located within a position tolerance of 0.8 when the holes are at MMC. As the individual holes in the pattern depart from MMC toward LMC, additional tolerance to that hole (and thus to the pattern) is acquired equal to that departure.

Within the hole pattern itself, the feature relating position tolerance is established at 0.13 diameter. Then, as previously described, each hole in the pattern may increase its position tolerance an amount equal to the departure from MMC as it is produced to a maximum of 0.23 at LMC. Note that the hole-to-hole interrelationship in the pattern, as well as the relationship to datum A, is maintained. The attitude (perpendicularity) of each individual hole simultaneously must be within the 0.13 diameter tolerance zone as well as within position of 0.13 at MMC.

As can be seen in the typical hole cross-sectional view, the axes of both the large (pattern) and small (feature) tolerance zones are parallel. The axes of the holes must lie within *both* the larger and smaller tolerance zones. Portions of the smaller zones may fall outside the peripheries of the larger tolerance. However, this portion of the smaller zone is not usable since the axis of each hole must fall within both zones.

The lower left four hole pattern follows the same reasoning as described above and as seen in the illustration.

EXAMPLE

MEANING

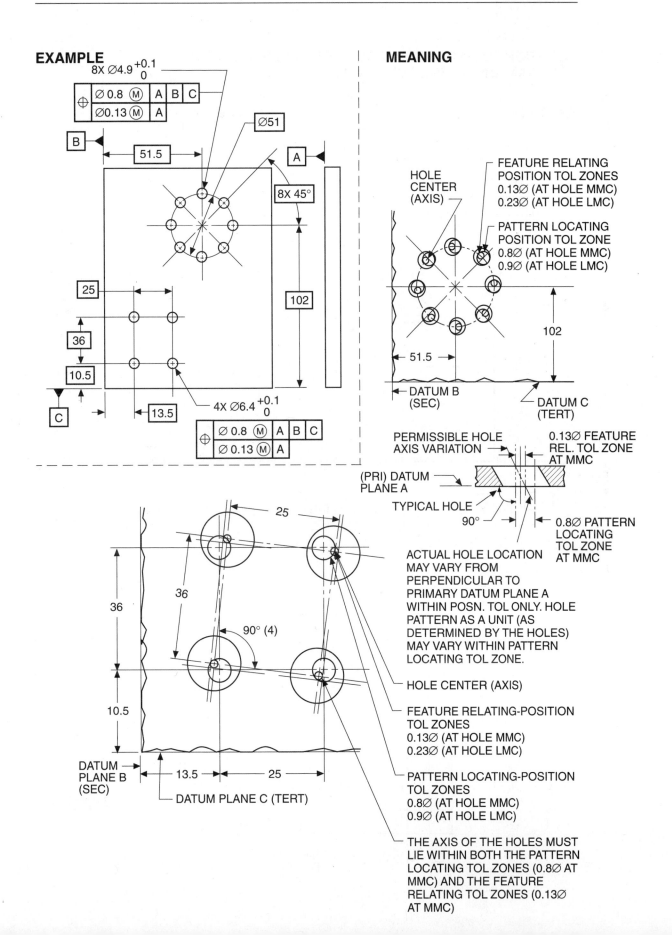

8X Ø4.9 +0.1 / 0

⌖	Ø 0.8 Ⓜ	A	B	C
	Ø0.13 Ⓜ	A		

Ø51

B

51.5

8X 45°

A

25

102

36

10.5

C

13.5

4X Ø6.4 +0.1 / 0

⌖	Ø 0.8 Ⓜ	A	B	C
	Ø 0.13 Ⓜ	A		

HOLE CENTER (AXIS)

FEATURE RELATING POSITION TOL ZONES
0.13Ø (AT HOLE MMC)
0.23Ø (AT HOLE LMC)

PATTERN LOCATING POSITION TOL ZONE
0.8Ø (AT HOLE MMC)
0.9Ø (AT HOLE LMC)

102

51.5

DATUM B (SEC)

DATUM C (TERT)

PERMISSIBLE HOLE AXIS VARIATION

0.13Ø FEATURE REL. TOL ZONE AT MMC

(PRI) DATUM PLANE A

TYPICAL HOLE

90°

0.8Ø PATTERN LOCATING TOL ZONE AT MMC

25

36

90° (4)

36

10.5

DATUM PLANE B (SEC)

13.5

25

DATUM PLANE C (TERT)

ACTUAL HOLE LOCATION MAY VARY FROM PERPENDICULAR TO PRIMARY DATUM PLANE A WITHIN POSN. TOL ONLY. HOLE PATTERN AS A UNIT (AS DETERMINED BY THE HOLES) MAY VARY WITHIN PATTERN LOCATING TOL ZONE.

HOLE CENTER (AXIS)

FEATURE RELATING-POSITION TOL ZONES
0.13Ø (AT HOLE MMC)
0.23Ø (AT HOLE LMC)

PATTERN LOCATING-POSITION TOL ZONES
0.8Ø (AT HOLE MMC)
0.9Ø (AT HOLE LMC)

THE AXIS OF THE HOLES MUST LIE WITHIN BOTH THE PATTERN LOCATING TOL ZONES (0.8Ø AT MMC) AND THE FEATURE RELATING TOL ZONES (0.13Ø AT MMC)

RELATION TO DATUM SURFACES COMPOSITE POSITION TOLERANCING—FUNCTIONAL GAGES

EXAMPLE **GAGES**

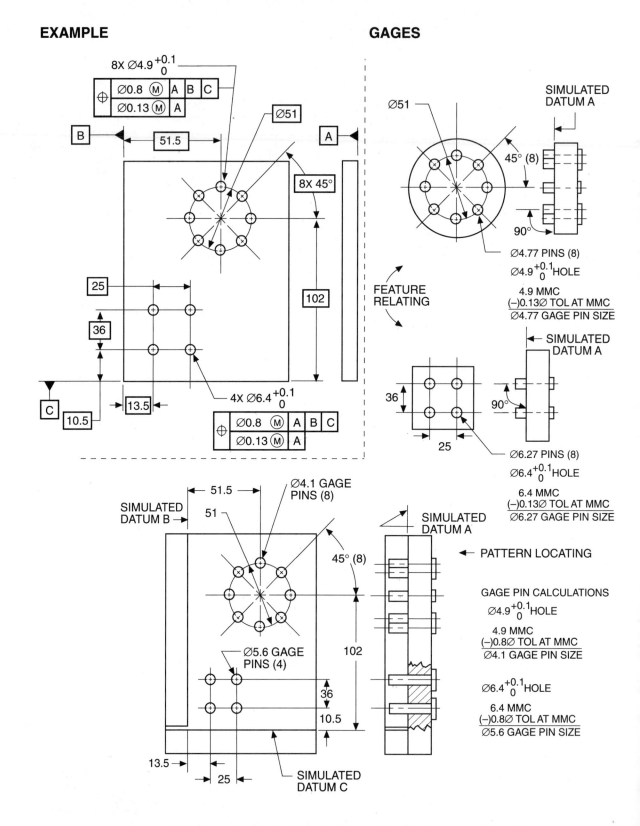

COMPOSITE POSITION TOLERANCING—
FUNCTIONAL AND PAPER GAGING, GRAPHIC ANALYSIS, COMPUTATION

The previous page illustrates functional gage principles for the four hole pattern of the part on page 125. Other methods—paper gaging and mathematical computation—are represented by a sample situation on pages 128, 129, 296, and 297.

Paper gaging methods are shown to demonstrate actual usable techniques which can also be used to quantify position tolerance principles pictorially. The mathematical method may be used in lieu of paper gaging or functional gaging directly as backup verification of borderline cases, or for further analytical work. Paper gaging, also referred to as graphic analysis, provides a useful tutorial tool to explore principles and better understand the mechanics of positional tolerancing. Once understood, mathematical or computer manipulation of the data in such a situation is more readily achieved.

PAPER GAGING, GRAPHIC ANALYSIS

Paper gaging is accomplished through plotting an enlarged scale of coordinately measured feature positions onto a piece of standard graph paper and then plotting the resulting differentials (actual position versus true position) to a selected scale (e.g., one square = 0.05) with a dot on the graph. An overlay chart (gage) of tracing paper or other transparent material containing a series of graph-scale circles of desired increments is placed over the graph to depict the position tolerance zones. Note that the paper gaging method simulates part function and functional gaging. However, the individual tolerance zones are each assumed to be represented by the one exact (true) position on the graph. The exact (basic) dimensions of the pattern are assumed as 0 in the X and Y directions.

Step 1 The four 6.4 holes are located both as a pattern (⌀ 0.8) and hole-to-hole within the pattern (⌀ 0.13). In open set-up inspection methods, two steps are required to determine whether both requirements have been met. Step 1, using coordinate measuring and paper gaging, will determine whether the hole pattern as a unit (as based upon individual hole location) has met the ⌀ 0.8 tolerance requirement. The part is set up according to the datum surfaces and is measured to holes 1, 2, and 3 (see illustration on page 129); only three of the four holes are necessary for this evaluation.

The resulting X and Y measurements are compared to the specified coordinate dimensions on the drawing resulting in a differential (off position) value. This differential is plotted on the paper gage graph according to the above stated scale (i.e., one square = 0.05).

When the actual location of the three holes are plotted, an overlay gage, with the circles, the scale of the graph plot, and a representation of the tolerances, is placed over the graph plot. The center of the overlay must be placed on the center of the plot. If the plotted centers fall within the position zone (in this case the ⌀ 0.8 zone), the pattern, via the holes, meets the requirement. Since hole #1 in our hypothetical example exceeds the ⌀ 0.8 tolerance zone, MMC principles may be invoked. The size of hole #1 is found to be 6.5 which is 0.1 departure from MMC; thus that hole has ⌀ 0.9 tolerance, and as seen in the illustration on the following page, is now acceptable. The hole pattern is also thus found acceptable within the ⌀ 0.8 tolerance.

Step 2 To evaluate the accuracy of the four holes in the pattern (part shown on page 125) relative to the individual 0.13 position tolerances, it is necessary to consider the hole-to-hole relationship in the pattern, exclusively. The four-hole grid pattern from which the individual hole positions can be compared is established by selecting two of the holes as a basis. In our example, hole #1 is selected as the origin for the X- and Y- coordinate measurements.

Since one additional hole must be selected to give orientation or square-up to the pattern, hole #3 is selected for the X orientation.

From the part orientation now established, each hole is measured in X and Y from this set-up. From the illustration on page 129, resultant differentials are derived from our hypothetical measurements. Note, of course, that #1 is zero in X and Y (as the origin for measurement), and hole #3 is zero in X as the square-up for the basic pattern.

The differentials are plotted on the graph paper to the desired scale (in this case every five squares = 0.01) using an origin on the graph as the hole #1 position. Each hole is plotted in the appropriate value and quadrant in X and Y from the hole #1 position.

With the holes plotted on the graph, the overlay gage with the circles representing the tolerances scaled to the graph plot is placed over the graph. The overlay is moved at random to try to encompass all the plotted hole centers simultaneously within the stated tolerance 0.13 circles. After trial and error, let us assume that the illustrated location of the overlay gage can successfully encompass holes #1, #2, and #3 but hole #4 is outside the 0.13 circle. MMC principles may now be invoked. By checking the actual mating size of hole #4, it is determined to be 6.45, which is a departure of 0.05 from MMC; thus hole #4 has 0.18 position tolerance, and as is seen in the illustration, is now acceptable.

All holes in the pattern have met their position tolerance requirements. It is extremely important to note that hole #1, even though used as a zero location origin, is actually assumed as an equal partner in the pattern and must likewise be treated as imperfect in its relationship to its desired position.

It should be noted that other methods of deriving coordinate data could be used. For example, via computer analysis determine the centroid of the hole pattern and then derive X and Y values. Any method chosen is attempting to duplicate a situation which most closely represents the best fit manipulation of the mating part pins (or representative hard gage). The method shown would represent only one position, or rotation, of the pattern of the many possible options.

Paper gaging simulates hard gaging and part function and thus is an effective technique. This method is further advantageous in that it visually detects error trends through periodic inspections, gives a permanent record, and requires no gaging tolerance.

COMPUTATION

Mathematical verification is possible using a calculator or computer program, proving the holes within the $\oslash 0.8$ and $\oslash 0.13$ tolerance zones.

Step 1. Mathematically proven using a calculator or computer program (similar to pages 293–296).

- Hole #1 calculates to $\oslash 0.8602325 - 0.1$ (bonus tol) = $\oslash 0.7602325$. Hole #1 is *good* using the $\oslash 0.1$ bonus tolerance (less than $\varnothing 0.8$).
- Hole #2 calculates to = $\oslash 0.6708204$. Hole #2 is *good* (less than $\oslash 0.8$).
- Hole #3 calculates to = 0.5830952. Hole #3 is *good* (less than $\oslash 0.8$).

Step 2. Mathematically proven using a suitable computer program (similar to pages 296–302).

- Hole—hole location *is* within tolerance as calculated.
- Hole—hole position tolerance calculated for *all four holes,* considering bonus tolerance of hole #4 ($\oslash 0.05$), is $\oslash 0.1222532$. That is, the smallest circle (tolerance zone) which will encompass all the four hole locations (including bonus tol $\oslash 0.05$ of hole #4) is $\oslash 0.1222532$. This is less than $\oslash 0.13$ the stated tolerance. Therefore, all four holes are acceptable.

STEP 1

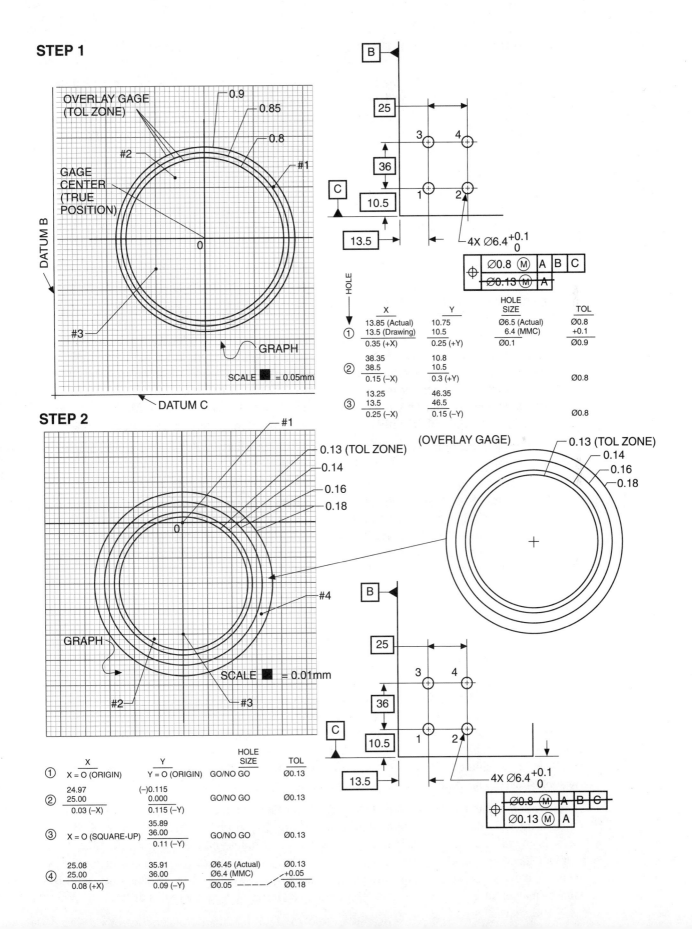

STEP 1 graph labels:

OVERLAY GAGE (TOL ZONE)

0.9
0.85
0.8

#2
#1

GAGE CENTER (TRUE POSITION)

0

#3

DATUM B

GRAPH

SCALE ▪ = 0.05mm

DATUM C

B

25

3   4

36

C

10.5   1   2

13.5

4X Ø6.4 +0.1 / 0

⊕	Ø0.8 Ⓜ	A	B	C
	Ø0.13 Ⓜ	A		

		X	Y	HOLE SIZE	TOL
		13.85 (Actual)	10.75	Ø6.5 (Actual)	Ø0.8
①		13.5 (Drawing)	10.5	6.4 (MMC)	+0.1
		0.35 (+X)	0.25 (+Y)	Ø0.1	Ø0.9
		38.35	10.8		
②		38.5	10.5		
		0.15 (–X)	0.3 (+Y)		Ø0.8
		13.25	46.35		
③		13.5	46.5		
		0.25 (–X)	0.15 (–Y)		Ø0.8

HOLE

STEP 2

#1

0.13 (TOL ZONE)
0.14
0.16
0.18

0

#4

GRAPH

SCALE ▪ = 0.01mm

#2   #3

(OVERLAY GAGE)

0.13 (TOL ZONE)
0.14
0.16
0.18

+

B

25

3   4

36

C

10.5   1   2

13.5

4X Ø6.4 +0.1 / 0

⊕	Ø0.8 Ⓜ	A	B	C
	Ø0.13 Ⓜ	A		

	X	Y	HOLE SIZE	TOL
①	X = O (ORIGIN)	Y = O (ORIGIN)	GO/NO GO	Ø0.13
②	24.97	(–)0.115		
	25.00	0.000	GO/NO GO	Ø0.13
	0.03 (–X)	0.115 (–Y)		
③		35.89		
	X = O (SQUARE-UP)	36.00	GO/NO GO	Ø0.13
		0.11 (–Y)		
④	25.08	35.91	Ø6.45 (Actual)	Ø0.13
	25.00	36.00	Ø6.4 (MMC)	+0.05
	0.08 (+X)	0.09 (–Y)	Ø0.05 ----	Ø0.18

COMPOSITE POSITION TOLERANCING

The composite position tolerancing principle can be applied in numerous ways to meet similar, yet different, design requirements from that shown on previous pages. For example, where the two patterns of features are in fact *one* pattern, or are to be treated as *one* pattern for both the pattern locating and feature relating (upper *and* lower segments), the method shown in the upper example could be used. One callout addressed to both patterns and the number designation 12X (twelve holes) will convey the one pattern meaning.

Functional gaging, coordinate measuring, and evaluation of the upper segment, pattern locating tolerance, would be exactly as shown in previous pages. The lower segment, feature relating tolerance, would also be handled as in previous pages except that it would be put into a *single gage* (accommodating all twelve holes at once) or into a *single measuring process* where all twelve holes are measured and evaluated as *one* pattern.

The lower example illustrates the possibility where one pattern (the four $\oslash$ 6.4 holes) is to serve as the datum reference. Although feasible, this method includes some complications in verification. Where the design requirement is that the diametrical pattern (the eight $\oslash$ 4.9 holes) be located with respect to the four holes (designated datum D) location, this method can be used. Production is so guided. However, verification from the four holes required either some form or functional gage which can pick up the four holes simultaneously (and thus their functional centroid), or open set-up measuring by coordinate methods selecting two of the four holes or mathematically deriving the centroid of the four holes via computer. Methods shown previously can then be applied with some trial and error attempts possibly necessary. Where a pattern of features is used as a single datum reference, the centroid of the pattern establishes the datum axis from which the feature relationship originates.

If a functional gage method is to be used in this application, due consideration must be given to the Datum Virtual Rule as it would apply to the datum D pattern gage pins (i.e., $\oslash$ 6.27). Should the design require more stringent control or relationship of the eight hole pattern to the datum D pattern, zero (0) tolerancing of the feature relating hole pattern could be considered (see below). In such a case, the Datum Virtual Condition Rule would not apply since the gage "pins" would be at MMC size.

i.e.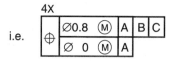

Consideration could be given to whether selection of two of the four holes might better serve as secondary and tertiary datum features (arbitrarily selected by design). This method would remove most of the complexities of using a group of features as a single datum. If RFS reference to the datum as a group is considered, the foregoing method (select two holes) should seriously be considered. Otherwise, the complexities become deeper and can only be resolved by sophisticated analytical or computer methods.

As discussed in later sections of this text, it can be argued whether using a pattern of features (such as holes as shown) as a single datum, and where the datum is referenced on an MMC basis, is any different than if the datum feature pattern of holes and the features related to it were all stated as a single requirement as one pattern. When the datum virtual condition rule is invoked, it can be proven that they are exactly the same. However, where extenuating circumstances may enter in, or surely where RFS datum reference is used, there is a difference. Where

COMPOSITE POSITION TOLERANCING—ONE PATTERN

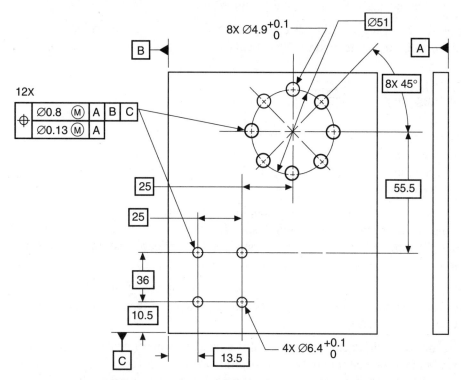

COMPOSITE POSITION TOLERANCING—HOLE PATTERN AS DATUM

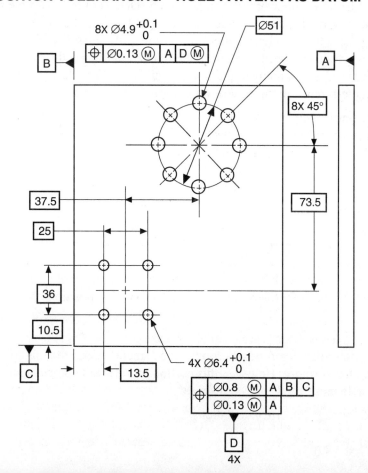

CMM processing is used, the RFS method is automatic in measurement. Whether the MMC latitude is included in the analysis can bring differing results.

See later sections in this text for further discussion on two hole datum applications, coordinate measuring, relationship to gaging, and mathematical (computer) simulation of hole patterns.

RELATION TO DATUM SURFACES
COMPOSITE POSITION TOLERANCING

When the location of a pattern of features from datum surfaces is less important that the accuracy required *within* the pattern of feature composite position tolerancing may be used. Composite position tolerancing also extends the use of specified datum relationships and geometric tolerance control.

Composite position tolerancing incorporates a dual feature control frame with two positional controls. One, the upper segment in the symbol, specifies the applicable datums and the pattern locating position tolerance. The lower segment specifies the applicable datum and the feature relating position tolerance. A single position tolerance symbol is used.

The composite position tolerance method utilizes the full advantages of MMC (and RFS or LMC if desired) and extends the principles to control of patterns of features as well as of the individual feature interrelationship.

Extending the composite tolerancing principles, this example incorporates a *secondary* datum in the features relating tolerance frame. Such a specification permits the patterns to float within the restrictions of the pattern locating tolerance but maintain a parallel orientation of the feature relating tolerance framework to datum plane B. Details of the four hole pattern meaning are shown; the bolt circle pattern would float independently in the same manner. Note that since the purpose is to *orient* each pattern framework only (*location* of the pattern from the datum reference frame is specified in the pattern location frame), the basic dimensions from the datums B and C do not apply to the lower segment feature relating tolerance. The basic dimensions, hole-to-hole, in the pattern yet control the hole locations to one another.

Functional gaging principles would be as described before in previous examples relative to the pattern locating tolerance. Possible functional gages for the feature-relating requirements are shown at right. Note how the gages stabilize against datums A and B with movement permitted of the gage pin pattern parallel to datum plane B. Graphic analysis and CMM processing with computer soft-gaging techniques are again possible.

Within the hole pattern itself, the feature relating position tolerance is established at 0.13 diameter. Then, as previously described, each hole in the patterns may increase its position tolerance an amount equal to the departure from MMC as it is produced to a maximum of 0.23 at LMC. Note that the hole-to-hole interrelationship in the pattern, as well as the relationship to datum A, is maintained. The attitude (perpendicularity) of each individual hole simultaneously must be within the 0.13 diameter tolerance zone as well as within position of 0.13 at MMC.

As can be seen in the typical hole cross-sectional view, the axes of both the large (pattern) and small (feature) tolerance zones are parallel. The axes of the holes must lie within *both* the larger and smaller tolerance zones. Portions of the smaller zones may fall outside the peripheries of the larger tolerance. However, this portion of the smaller zone is not usable since the axis of each hole must fall within both zones.

See page 126 for gage pin size calculations.

COMPOSITE POSITION TOLERANCING

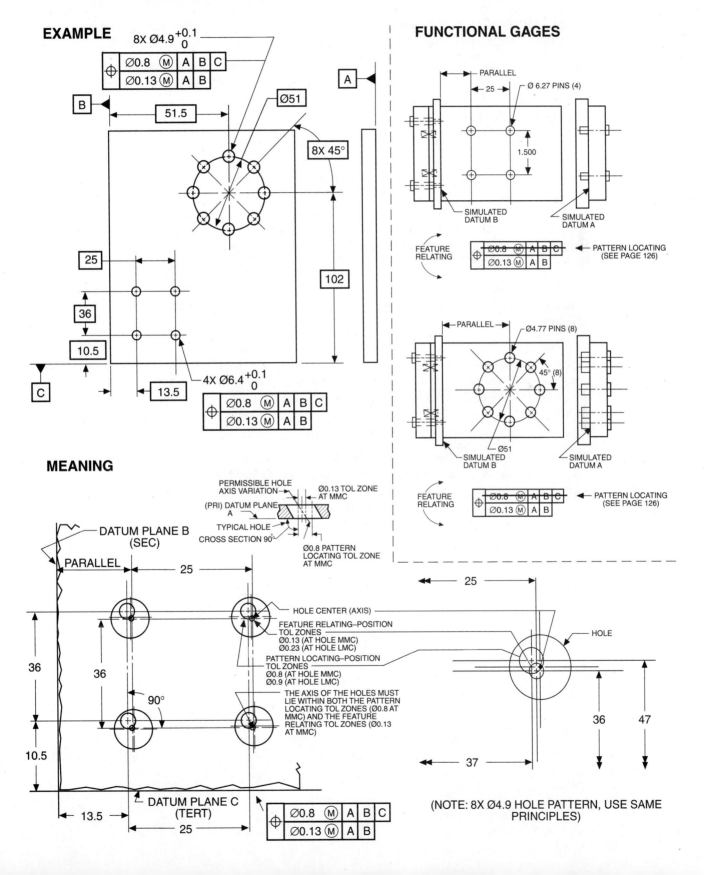

EXAMPLE

FUNCTIONAL GAGES

MEANING

(NOTE: 8X Ø4.9 HOLE PATTERN, USE SAME PRINCIPLES)

MMC RELATED TO MMC DATUM FEATURE

When a pattern of holes is dimensioned relative to the location of another hole, this hole is identified as a datum and the hole pattern is located dimensionally with respect to it.

In the example, the four holes are related to the center datum hole. As the position of the center datum hole shifts, the position of the 4 hole pattern itself must follow as dictated by the function of the part. Imagine that this part has a mating part with a shaft and four pins which must assemble with the five holes in the illustrated part.

Note that the 12.5 datum center hole is also located from datums (surfaces) A, B, and C. It is given a position tolerance of $\oslash$ 0.3 at MMC and a refined orientation, or perpendicularity, tolerance of $\oslash$ 0.08 at MMC. This properly controls the center hole relationship to the edge surfaces with a rather lenient position tolerance while the orientation to the primary datum A is maintained to closer tolerance. The center hole is identified as datum D so that the four 9.5 holes can be located with respect to it. Under MMC principles, the position and perpendicularity tolerances increase an amount equal to the hole size departure from MMC as shown in the illustration.

Wherever the datum hole D position varies in the design considerations, or in actual production, the four 9.5 hole pattern must follow. Note that the positional pattern dimensions originate at datum D to carry out this intent.

The four 9.5 holes are located by a position tolerance of $\oslash$ 0.13 at MMC with respect to datum A (for orientation), datum D at MMC (virtual condition) (for location), and datum B (for rotation). Reference to the lower portion of the illustration under Meaning will assist in the understanding of the effect of the three datums and the importance of datum precedence. Although not illustrated, imagine the four 9.5 hole position tolerances have been calculated relative to a mating part using the "fixed fastener" method previously explained.

The four holes individually with respect to their own true positions can vary in location up to $\oslash$ 0.13 at MMC. While at any actual size in a departure from MMC a hole can vary an additional amount (enlarges positional tolerance) equal to that departure. For example (as shown), if the hole size is produced at 9.53, the position tolerance becomes $\oslash$ 0.16, at 9.55 it becomes $\oslash$ 0.18, etc., up to $\oslash$ 0.21 if the hole is produced at 9.58 LMC.

Since the datum feature is a size feature, its variation in size (12.5 to 12.55) has an effect on the four hole pattern relationship which locates from it. That is, as the actual size of the datum feature increases, its relationship to the mating part corresponding feature (e.g., a shaft) changes; if the imagined mating part shaft and corresponding pins (at MMC) will insert into holes larger than those we used to calculate the position tolerances, greater latitude (off position) of the pattern as a unit can be realized. This latitude is, however, to the hole pattern as a unit and not relative to the four holes individually or hole-to-hole in the four hole pattern.

Please see the following gaging illustrations and text describing evaluation of this part with representative techniques.

EXAMPLES

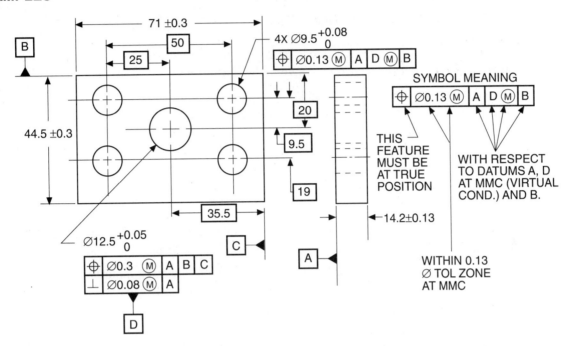

MEANING

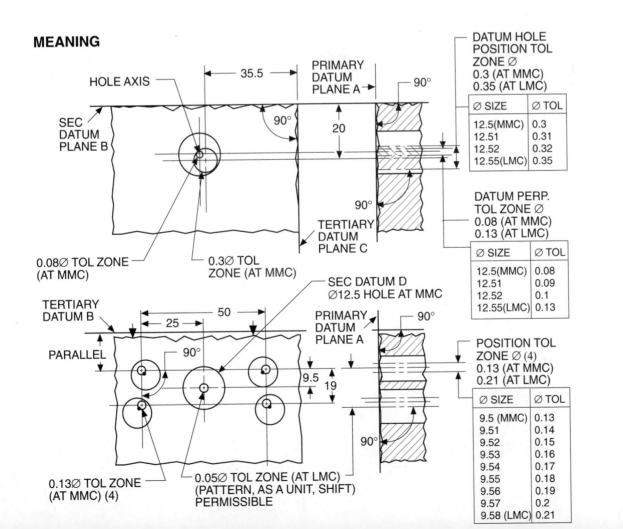

MMC RELATED TO MMC DATUM FEATURE
FUNCTIONAL GAGING PRINCIPLES

Functional gaging principles which can be applied to the preceding example (duplicated at the upper left of the opposite page) are shown in the illustration.

It must first be clarified that functional gaging is not required when position tolerancing is used. Other techniques (i.e., CMM, open set-up coordinate methods, optical methods, etc.) can also be used. For example, the paper gaging and computation methods, illustrated earlier in this section could effectively be applied using the datum D feature as zero and orientation and rotation from datums A and B.

The purpose of this illustration is to depict representative functional gages and show actual methods as well as to demonstrate principle.

The illustration is intended to be self-descriptive and to explain the necessary details. Note that three gages (or operations) are required to fulfill the requirements.

Note specifically the clarity afforded by the clearly specified design requirements via the datums. Also, note the manner in which the gages pattern after the part function and simulate mating part relationships.

Also worthy of note is the manner in which the datum D pick-up pin size virtual condition is established based upon the Datum Virtual Condition Rule. The virtual condition applicable (12.42) is derived from the smallest orientation or position tolerance controlling the hole (i.e., 0.08) as subtracted from MMC size 12.5.

See pages 298-302 for further detail describing graphic analysis and mathematical/computer analysis of this part, i.e., "soft gaging."

EXAMPLE

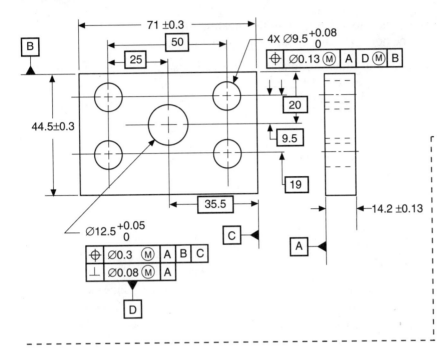

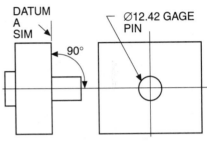

PERPENDICULARITY GAGE PIN CALCULATIONS (⊥)

DATUM A SIM → 90° — Ø12.42 GAGE PIN

FUNCTIONAL GAGES

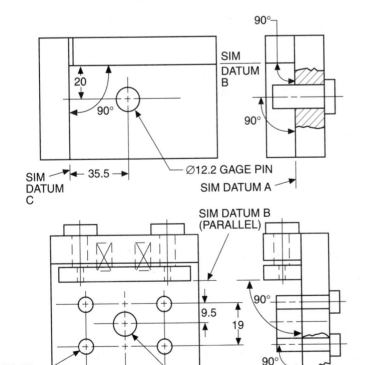

90°

SIM DATUM B

90°

20
90°

SIM DATUM C

35.5

Ø12.2 GAGE PIN

SIM DATUM A

SIM DATUM B (PARALLEL)

9.5

19

90°

90°

Ø9.37 GAGE PIN (4)

SIM DATUM A

DATUM D PIN Ø12.42 (VIRTUAL COND.)

25

50

GAGE PIN CALCULATION

$Ø12.5^{+0.05}_{0}$ HOLE

Ø12.5 MMC
(–)0.08Ø⊥ TOL AT MMC
Ø12.42 GAGE PIN SIZE

Ø12.5 HOLE LOCATING (⊕) GAGE PIN CALCULATIONS

$Ø12.5^{+0.05}_{0}$ HOLE

Ø12.5 MMC
(–)0.3Ø⊕TOL AT MMC
Ø12.2 GAGE PIN SIZE

PATTERN LOCATING (⊕) GAGE PIN CALCULATIONS

$Ø12.5^{+0.05}_{0}$ HOLE (DATUM D)

Ø12.5 MMC
(–)0.08Ø⊥ TOL AT MMC
Ø12.42 DATUM PIN SIZE
(VIRTUAL CONDITION)

$Ø9.5^{+0.08}_{0}$ HOLE

Ø9.5 MMC
(–)0.13Ø⊕TOL AT MMC
Ø9.37 GAGE PIN SIZE

MMC RELATED TO RFS DATUM FEATURE

When a pattern of holes is dimensioned relative to the location of another hole, this hole is identified as a datum and the hole pattern is located dimensionally with respect to it.

In the illustrated example, the four holes are related to the center datum hole. As the position of the center datum hole shifts, the position of the four hole pattern itself must follow as dictated by the function of the part. Imagine that this part has a mating part with a shaft and four pins which must assemble with the five holes in the illustrated part.

Differing from the preceding illustration, assume there is a closer precision fit required between the shaft of the imagined part as it fits into the 12.5 hole. The four pins of the imagined part, however, are to relate to the four 9.5 holes on an MMC basis as in the preceding illustration. Therefore, since in this case the relationship between the four holes and their datum is critical or more precise, the datum D is referenced regardless of feature size (RFS).

Note that the 12.5 datum center hole is located from surface datums A, B and C. It is given a position tolerance relative to these edges of ⌀ 0.3 at MMC since the relationship to the edges on this basis can have a rather lenient position tolerance. The orientation of the datum hole relative to the primary datum A is, however, to be maintained to a closer tolerance. Since datum D position and orientation is controlled on an MMC basis, the tolerances (position and perpendicularity) increase an amount equal to the produced size departure from MMC shown in the illustration. It should be noted here, however, that the reference to datum D in the relationship of the four 9.5 holes is on an RFS basis. This means the four hole pattern takes reference from the exact center (axis) of the datum hole at whatever size it is produced (RFS) within 12.5 to 12.55.

Wherever the datum hole D position varies in the design considerations, or in actual production, the four 9.5 hole pattern must follow. Note that the positional pattern dimensions originate at datum D to carry out this intent.

The four 9.5 holes are located by a position tolerance of ⌀ 0.13 at MMC with respect to datum A (for orientation), datum D at RFS (for location), and datum B (for rotation). Reference to the lower portion of the illustration under Meaning will assist in understanding the effect of the three datums and the importance of datum precedence. Although not illustrated, imagine that the four 9.5 hole position tolerances have been calculated relative to a mating part using the "fixed fastener" method previously explained.

The four holes individually, with respect to their own true or exact positions, can vary in location up to ⌀ 0.13 at MMC. While at any actual size in a departure from MMC, a hole can vary an additional amount (enlarge positional tolerance) equal to that departure. For example (as shown), if the hole size is produced at 9.53 the position tolerance becomes ⌀ 0.16, at 9.55 it becomes ⌀ 0.18, etc., up to ⌀ 0.21, if the hole is produced at 9.58 (LMC).

Although the datum feature is a size feature as in the preceding illustration, its variation in size (12.5 to 12.55) will have *no* effect on the four hole pattern relationship. This is because the pattern relationship is to the center (axis) of the datum D, 12.5 hole, no matter to which size it is produced in its size tolerance range (i.e., RFS).

In this instance, it is seen that a more critical or precise relationship is maintained between the four holes and their datum. The design requirements and mating part functional interface determined this approach in our example.

Please see the following gaging illustrations and text describing evaluation of this part with representative techniques.

EXAMPLE

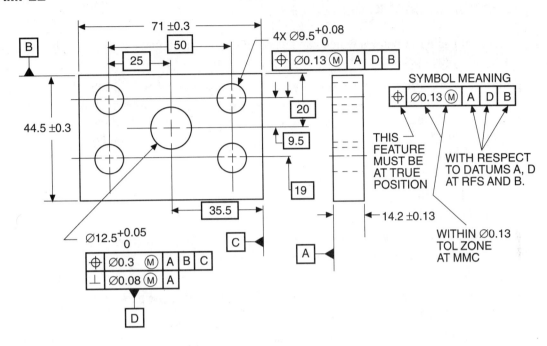

MEANING

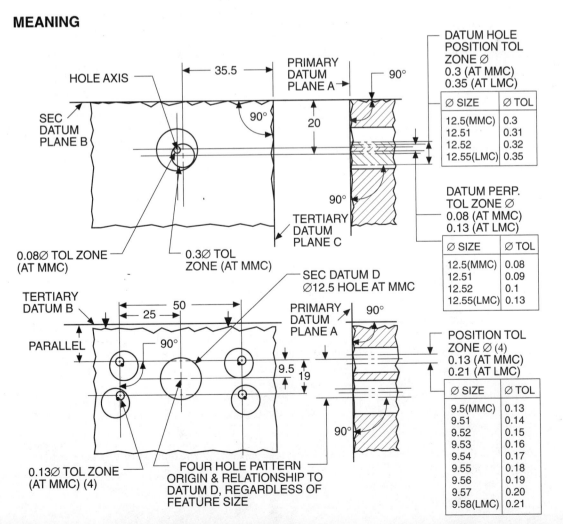

MMC RELATED TO RFS DATUM FEATURE
FUNCTIONAL GAGING PRINCIPLES

Functional gaging principles which can be applied to the preceding example (duplicated at the upper left of the opposite page) are shown in the illustration.

It must first be clarified that functional gaging is not required when position tolerancing is used. Other techniques (i.e., CMM, open set-up coordinate methods, optical methods, etc.) can also be used. For example, the paper gaging and computation methods illustrated earlier in this section could effectively be applied using the datum D feature as zero and orientation and rotation from datums A and B.

The purpose in this illustration is to depict representative functional gages to show actual methods as well as to illustrate principle. Note that functional principles, identical to the preceding part, can be used except for the RFS pickup of the datum feature.

The illustration is intended to be self-descriptive and to explain the necessary details. Note that three gages (or operations) are required to fulfill the requirements.

Note specifically the clarity afforded by the clearly specified design requirements via the datums. Also, note the manner in which the gages pattern after the part function and simulate mating part relationships.

In this example, the use of the RFS datum has tightened requirements. Costs in manufacturing will probably be higher than with total use of MMC as in the preceding example; yet, functional principles can be applied as shown.

EXAMPLE

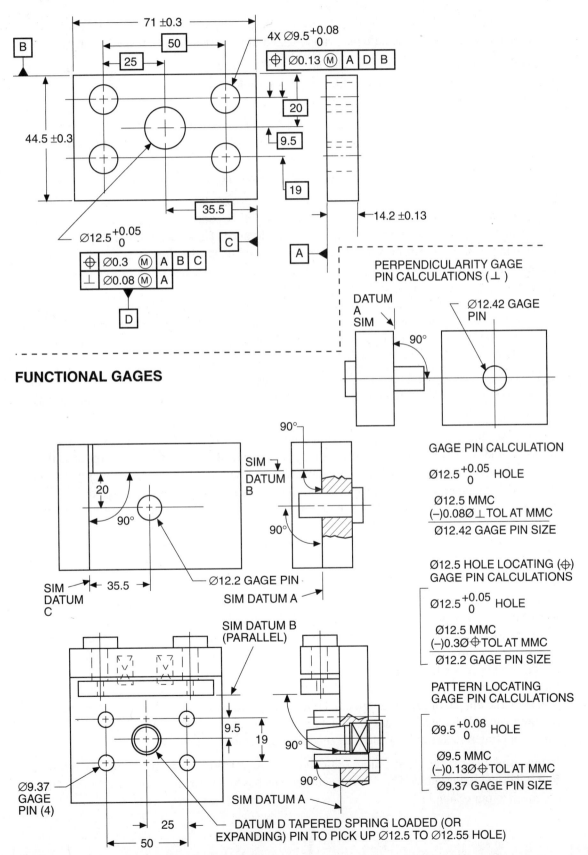

FUNCTIONAL GAGES

RFS RELATED TO RFS DATUM FEATURE

When a pattern of holes is dimensioned relative to the location of another hole, this hole is identified as a datum and the hole pattern is located dimensionally with respect to it.

In the illustrated example, the four holes are related to the center datum hole. As the position of the center datum hole shifts, the position of the four hole pattern itself must follow as dictated by the function of the part.

In this instance, note that the four holes of the pattern are assigned an RFS position tolerance and that they are related to the datum hole D also on an RFS basis. Imagine that this part has precision requirements between the holes either to provide accurate relationship with a mating part or to maintain accuracy for a mating situation, such as a semicritical gear plate mounting.

Note that the 12.5 datum center hole is located from surface datums A, B, and C. It is given a position tolerance relative to these edges of $\oslash$ 0.3 at MMC since the relationship to the edges can be on this basis with a rather lenient position tolerance. The orientation of the datum hole relative to the primary datum A is, however, to be maintained to a closer tolerance. Since datum D position and orientation are controlled on an MMC basis, the tolerances (position and perpendicularity) increase an amount equal to the produced size departure from MMC as shown in the illustration. Note, however, that the reference to datum D in the relationship of the four 9.5 holes is on an RFS basis. The four hole pattern, therefore, takes its positional reference from the exact center (axis) of the datum hole at whatever size it is produced (RFS) within 12.5 to 12.55.

Wherever the datum hole D position varies in the design considerations or in actual production, the four 9.5 pattern must follow. Note that the positional pattern dimensions originate at datum D to carry out this intent.

The four 9.5 holes are located by a position tolerance of $\oslash$ 0.13 at RFS with respect to datum A (for attitude), datum D at RFS (for location), and datum B (for rotation). Reference to the explanation portion of the illustration (under Meaning) will assist in understanding the effect of the three datums and the importance of datum precedence. Although not illustrated, imagine that, as previously stated, the four 9.5 hole position tolerances have been determined to relate to a mating part or to maintain accuracy in a mating situation where other features or components must relate with precision regardless of the produced sizes (RFS) of the 9.5 holes (9.5 to 9.58).

The four holes individually, with respect to their own true or exact positions, can vary in location up to $\oslash$ 0.13. Under the RFS method, however, this tolerance applies to each hole regardless of the size to which it is produced. That is to say (as shown in the lower right corner of the illustration), the $\oslash$ 0.13 position tolerance is the maximum allowable to each hole no matter to which size it is produced (9.5 to 9.58). If, as was shown in previous illustrations, the MMC method had been applied, the position tolerance would increase to the extent of departure of the 9.5 holes from MMC. Not so, however, in this example since the RFS method has been invoked. The choice of proper approach, be it MMC or RFS, is of course decided by the design requirements. MMC methods are obviously recommended wherever possible due to the added functional (interchangeability) and tolerance advantages.

In this instance, it is seen that a more critical or precise relationship is maintained between the four holes and their datum. The design requirements based on part function determined this approach in our example.

EXAMPLE

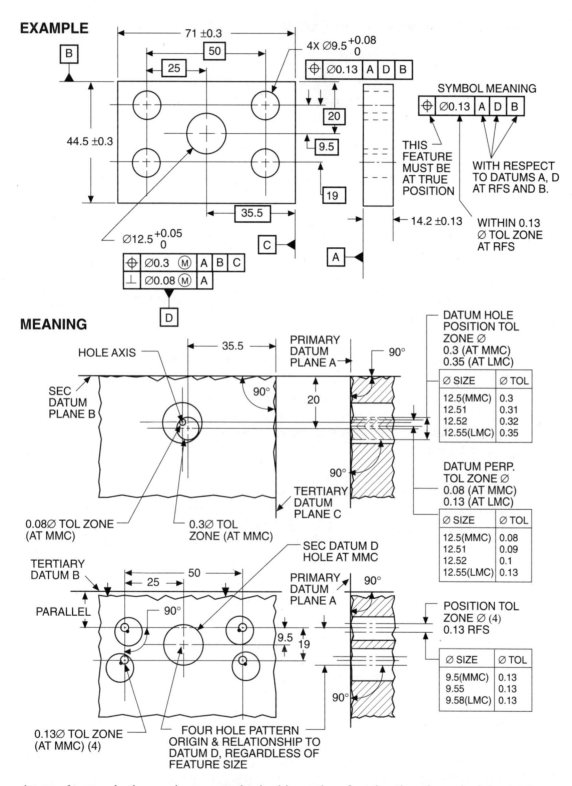

MEANING

As may be seen in the previous examples in this section, functional gaging principles can be used to evaluate the datum hole location and relationships. However, the four hole (9.5) pattern relative to datum D in this example cannot utilize such methods because it is an RFS application; open set-up or CMM methods would be necessary. The techniques of "paper gaging and computation," described earlier in the text could, of course, be used as desired. See also pages 298–301.

MMC RELATED TO MMC DATUM FEATURE
PROJECTED TOLERANCE ZONE[*]

This illustration shows a part with four tapped mounting holes. A cover plate mechanism is assumed as a mating part (not shown). Position tolerancing is used to assure assembly of the mating parts. The projected tolerance zone requirement is also added.

We wish to establish a relationship between the mounting surface, the mounting hole pattern, and the 38.25 counterbored seat diameter (identified as datum B). The mating part has a seat mechanism which must fit within this counterbore and attach and locate with screws to the four mounting holes on the flange surface.

The flange surface itself is established as the primary datum and is identified as datum A to ensure clarity of the hole pattern positional relationship with the top surface.

We calculate the position tolerances using the "fixed fastener" method as based on the M6 screw and the clearance hole in the mating part cover plate. Assuming that the cover plate has 6.16 clearance holes and 6.4 maximum thickness, we calculate the position tolerance and assign 0.08 to each part (0.16/2 = 0.08).

The four M6 holes are thus designated as shown. The feature control symbol specifies "at true position within $\oslash$ 0.08 at MMC size of the holes with respect to datums A and B (at MMC) and to datum C."

Conventional position tolerancing discussed previously reveals that a position (MMC) application is affected by the actual mating size departure of the involved features from MMC. However, in this instance, the peculiarity of thread assembly requires further consideration and may cause an exception to the usually inferred interpretation.

The centering effect of a screw as it tightens in the tapped hole tends to negate or diminish any *added* position tolerance based on greater clearance between the pitch diameters of the screw and tapped hole. The screw seeks center as, in tightening, the flank of the mating thread forms (screw and hole) come into contact or bottom out. Thus additional position tolerance relative to the tapped hole increase in pitch diameter size (departure from MMC) may not be fully realized. There may be some additional tolerance derived in actual assembly but it cannot be predicted.

The same reasoning as in the foregoing paragraph applies to attempts to assign position tolerance to countersinks. Position tolerance would be appropriate to the through hole, but, as a rule, is not practical for the countersink itself.

Added tolerance may be acquired, however, for the tapped hole pattern as a unit relative to datum B (counterbore) as it departs from MMC size (gets larger). Datum B is a straight sided feature, and size deviation from MMC will have the effect previously discussed in other examples.

(Continued on page 146).

---

[*]See also page 198 for further explanation of principle and alternatives to projected tolerance zone methodology.

EXAMPLE

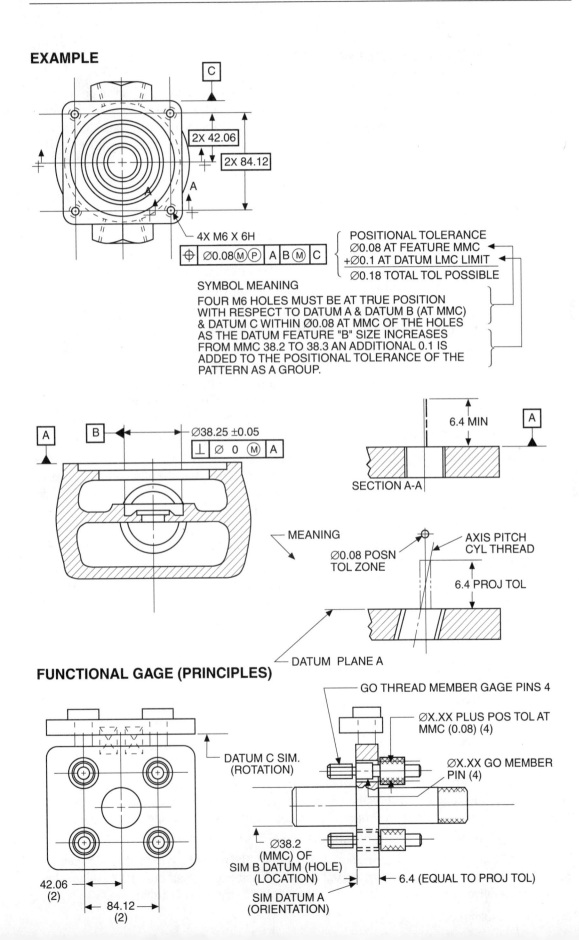

2X 42.06

2X 84.12

4X M6 X 6H

⊕ | ⌀0.08 Ⓜ Ⓟ | A | B Ⓜ | C

POSITIONAL TOLERANCE
⌀0.08 AT FEATURE MMC
+⌀0.1 AT DATUM LMC LIMIT
⌀0.18 TOTAL TOL POSSIBLE

SYMBOL MEANING
FOUR M6 HOLES MUST BE AT TRUE POSITION
WITH RESPECT TO DATUM A & DATUM B (AT MMC)
& DATUM C WITHIN ⌀0.08 AT MMC OF THE HOLES
AS THE DATUM FEATURE "B" SIZE INCREASES
FROM MMC 38.2 TO 38.3 AN ADDITIONAL 0.1 IS
ADDED TO THE POSITIONAL TOLERANCE OF THE
PATTERN AS A GROUP.

A   B   ⌀38.25 ±0.05

⊥ | ⌀ | 0 Ⓜ | A

6.4 MIN   A

SECTION A-A

MEANING

⌀0.08 POSN
TOL ZONE

AXIS PITCH
CYL THREAD

6.4 PROJ TOL

DATUM PLANE A

FUNCTIONAL GAGE (PRINCIPLES)

GO THREAD MEMBER GAGE PINS 4

⌀X.XX PLUS POS TOL AT
MMC (0.08) (4)

⌀X.XX GO MEMBER
PIN (4)

DATUM C SIM.
(ROTATION)

⌀38.2
(MMC) OF
SIM B DATUM (HOLE)
(LOCATION)

SIM DATUM A
(ORIENTATION)

6.4 (EQUAL TO PROJ TOL)

42.06
(2)

84.12
(2)

MMC RELATED TO MMC DATUM FEATURE

Since the threaded holes are to attach a mating part, we see that in assembly the critical position location of these holes will actually concern the inserted screws or the actual projection from these holes which must accommodate the mating part. Therefore, the location of the threaded holes is controlled by a *projected* position tolerance zone to represent the projected screw locations. As is seen in the callout and sectioned projected tolerance zone view, the tolerance zone is cylindrical and extends above, and perpendicular to, the datum plane A. The drawing specification and view using the thick chain line clearly shows the direction and extent of the projected tolerance zone.

Since in this case the threaded hole is a through hole, a drawing view is required to clearly show the direction from datum plane A that the tolerance zone projects. The length of extent of the tolerance zone is conveniently shown in this view as well. If projected tolerance zone is applied on a blind hole, the extent of the projected tolerance zone may be placed in the feature control frame as $\boxed{\oplus | \varnothing 0.08 \, \text{M} | \text{P} \, 6.4 | A | B \, \text{M} | C}$. A drawing sectional view may then be unnecessary.

The height of the projected position tolerance zone is determined by the application. It may be established by the thickness of the mating part through which the screws are to extend, or it may be determined by the thread hole depth when a thin mating part is involved, or any desired height which fulfills the part design requirements may be selected. When the projected tolerance zone is intended, it should be specified as shown with the symbol immediately following the stated tolerance and any material condition symbol in the feature control frame (see illustration). This system is also used on parts in which pins, studs, or other features are to be inserted and the critical assembled location is *above* the surface of the part.

The representative production gage shown on the preceding page also illustrates this principle. The gage is made up of the main gage body containing the datum B hole pin, the plate which contacts the datum surface A, a spring plate to provide orientation to datum C, four GO thread members per standard tolerances and, if desired, torquing flats to represent appropriate screw tightening and setting. The plate contains four holes which are 0.08 (the positional tolerance) larger than the four GO thread pins which are inserted into the threaded holes of the part through the gage plate. CMM simulation using computer soft-gaging techniques can be used in lieu of a hard-gage process.

The relationship of the GO thread member projections and the holes in the gage plate simulate the functional assembly requirement of the two mating parts and the four screws.

Gage-maker's tolerances would, of course, also apply but are not shown.

MMC RELATED TO MMC DATUM FEATURE

The illustration on the opposite page shows a circular pattern of 8 holes located by position tolerancing with respect to the center hole.

The requirement states "the eight holes are to be located at position within Ø0.05 at 10.25 MMC size of the holes with respect to datum A and datum hole B (at 26.04 MMC)."

As the actual size of the holes increases in production from MMC 10.25 to the high limit of 10.3, $\varnothing 0.05$ is added to the positional tolerance. As the datum hole B increases from MMC of 26.04 to 26.09, 0.05 is added to the hole pattern's positional tolerance as a group. Depending on the size increase of the holes, the total tolerance may be 0.15 instead of 0.05.

A production gage for this part is also shown. The pins in the pattern are spaced at 45° intervals on a 76.2 circle, with the pin size at 10.2 or at MMC size of the part hole 10.25 minus the positional tolerance of 0.05. The center pin is 26.04 (MMC size of the datum hole).

Gage-maker's tolerances also apply.

EXAMPLE

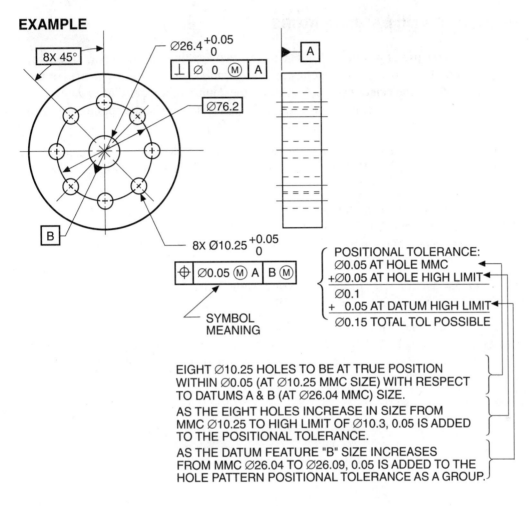

Ø26.4 +0.05 / 0

| ⊥ | Ø | 0 | Ⓜ | A |

A

Ø76.2

B

8X 45°

8X Ø10.25 +0.05 / 0

| ⊕ | Ø0.05 Ⓜ | A | B Ⓜ |

SYMBOL
MEANING

POSITIONAL TOLERANCE:
Ø0.05 AT HOLE MMC
+Ø0.05 AT HOLE HIGH LIMIT
Ø0.1
+ 0.05 AT DATUM HIGH LIMIT
Ø0.15 TOTAL TOL POSSIBLE

EIGHT Ø10.25 HOLES TO BE AT TRUE POSITION
WITHIN Ø0.05 (AT Ø10.25 MMC SIZE) WITH RESPECT
TO DATUMS A & B (AT Ø26.04 MMC) SIZE.

AS THE EIGHT HOLES INCREASE IN SIZE FROM
MMC Ø10.25 TO HIGH LIMIT OF Ø10.3, 0.05 IS ADDED
TO THE POSITIONAL TOLERANCE.

AS THE DATUM FEATURE "B" SIZE INCREASES
FROM MMC Ø26.04 TO Ø26.09, 0.05 IS ADDED TO THE
HOLE PATTERN POSITIONAL TOLERANCE AS A GROUP.

GAGE

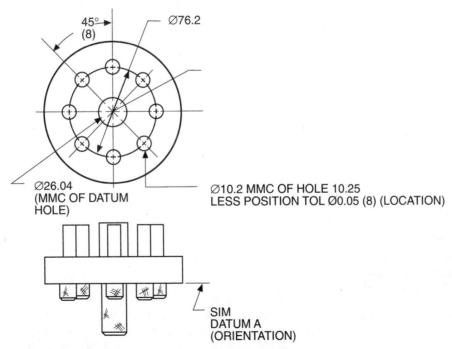

45°
(8)

Ø76.2

Ø26.04
(MMC OF DATUM
HOLE)

Ø10.2 MMC OF HOLE 10.25
LESS POSITION TOL Ø0.05 (8) (LOCATION)

SIM
DATUM A
(ORIENTATION)

MMC RELATED TO MMC DATUM FEATURE

This example shows the use of positional tolerancing on two locating pins on a part. The two pins are to take reference from the center hole which accommodates a shaft mechanism of the mating part (not shown). The center hole is selected as the datum reference. This example reminds us that positional tolerancing is not restricted to holes. It can also be applied to pins or to any feature on which a center (axis) is the basis for location.

The meaning of the position requirement is, "the two pins are to be located at true position within ⌀ 0.05 tolerance zone at 3.18 MMC size of the pins with respect to datums A, B (at MMC 12.5), and C."

Note that MMC of the pins is the high limit of size. As the two pins reduce in actual size from MMC of 3.18 to the low limit (LMC) of 3.1, 0.08 is added to the positional tolerance. As the datum feature B (the center hole) increases from MMC of 12.5 to 12.55 high limit, 0.05 is added to the two pins' positional tolerance as a group.

Depending on the production sizes of the bosses and datum hole, the total positional tolerance on the boss location may vary from ⌀ 0.05 to 0.18.

A representative gage to check the position location and the relationship with the datum hole is also shown. Note that the holes to check the position location of the pins are to the high limit of MMC of the pins *plus* the position diameter tolerance. The datum hole pin is to the MMC size of the datum hole of 12.5.

Gage-maker's tolerances also apply.

EXAMPLE

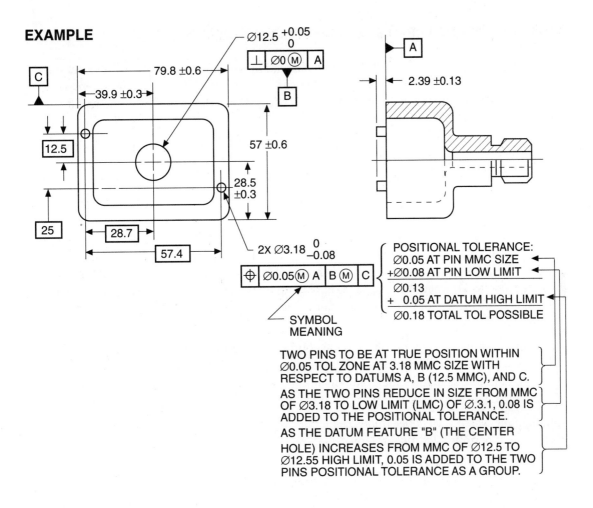

Ø12.5 +0.05 / 0

| ⊥ | Ø0 Ⓜ | A |

B

C

79.8 ±0.6

39.9 ±0.3

12.5

25

28.7

57.4

57 ±0.6

28.5 ±0.3

2X Ø3.18 0 / −0.08

| ⌖ | Ø0.05Ⓜ | A | B Ⓜ | C |

SYMBOL MEANING

A

2.39 ±0.13

POSITIONAL TOLERANCE:
Ø0.05 AT PIN MMC SIZE ◄
+Ø0.08 AT PIN LOW LIMIT ◄
Ø0.13
+ 0.05 AT DATUM HIGH LIMIT ◄
Ø0.18 TOTAL TOL POSSIBLE

TWO PINS TO BE AT TRUE POSITION WITHIN Ø0.05 TOL ZONE AT 3.18 MMC SIZE WITH RESPECT TO DATUMS A, B (12.5 MMC), AND C.

AS THE TWO PINS REDUCE IN SIZE FROM MMC OF Ø3.18 TO LOW LIMIT (LMC) OF Ø.3.1, 0.08 IS ADDED TO THE POSITIONAL TOLERANCE.

AS THE DATUM FEATURE "B" (THE CENTER HOLE) INCREASES FROM MMC OF Ø12.5 TO Ø12.55 HIGH LIMIT, 0.05 IS ADDED TO THE TWO PINS POSITIONAL TOLERANCE AS A GROUP.

GAGE

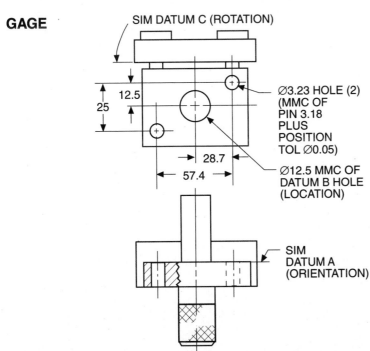

SIM DATUM C (ROTATION)

12.5

25

28.7

57.4

Ø3.23 HOLE (2)
(MMC OF
PIN 3.18
PLUS
POSITION
TOL Ø0.05)

Ø12.5 MMC OF
DATUM B HOLE
(LOCATION)

SIM
DATUM A
(ORIENTATION)

POSITION TOLERANCE—LEAST MATERIAL CONDITION (LMC)

Occasionally a method is required to control a situation which is essentially the reverse of the usual position relationship; that is, the stated position tolerance applies at the *least material condition,* LMC, of the feature or datum, instead of at MMC, and increases as the feature or datum *departs from* the least material condition.

Definition. Least Material condition (LMC) is the condition in which a feature of size contains the least amount of material within the stated limits of size: for example, maximum hole diameter, minimum shaft diameter. Least material condition is the condition opposite to MMC. For example, a shaft is at least material condition when it is at its *low* limit of size and a hole is at least material condition when it is at its *high* limit of size.

This method is applicable to special design requirements that will not permit MMC or that do not warrant the exacting requirements of RFS. It can be used to maintain critical wall thickness or critical center locations of features for which accuracy of location can be relaxed (position tolerance increased) when the feature leaves least material condition and approaches MMC. The amount of increase of positional tolerance permissible is equal to the feature actual size departure from least material condition.

The term "least material condition" and the abbreviation LMC have been used instead of "minimum material condition" (which is synonymous) to avoid confusion, since the abbreviation would be the same as that for maximum material condition. The symbol modifier Ⓛ is used to indicate the LMC requirement applicable to feature or datum.

Although the use of LMC does impose exacting requirements on both manufacturing and inspection, it permits additional tolerances.

Whenever least material condition (LMC) or Ⓛ is specified on a drawing, the position tolerance applies only when the feature is produced at its LMC size. See Fig. 1.

Additional positional tolerance is permissible but is dependent on, and equal to, the difference between the actually produced feature size (within its size tolerance) and LMC. See Fig. 2.

POSITION TOLERANCE—LEAST MATERIAL CONDITION (LMC)

EXAMPLE

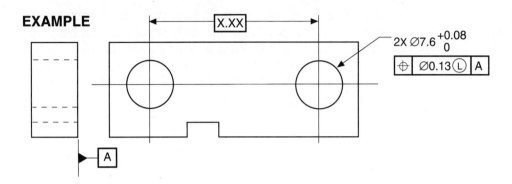

2X Ø7.6 $^{+0.08}_{0}$

| ⊕ | Ø0.13 Ⓛ | A |

A

MEANING

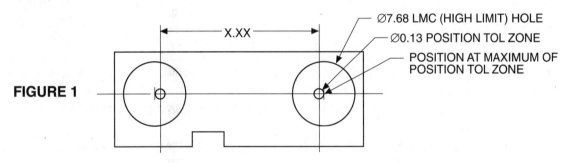

FIGURE 1

Ø7.68 LMC (HIGH LIMIT) HOLE
Ø0.13 POSITION TOL ZONE
POSITION AT MAXIMUM OF
POSITION TOL ZONE

X.XX

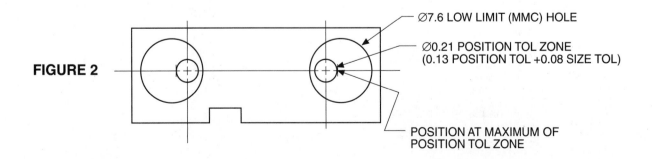

FIGURE 2

Ø7.6 LOW LIMIT (MMC) HOLE
Ø0.21 POSITION TOL ZONE
(0.13 POSITION TOL +0.08 SIZE TOL)

POSITION AT MAXIMUM OF
POSITION TOL ZONE

LMC RELATED TO RFS DATUM FEATURE

When a pattern of holes is dimensioned relative to the location of another hole, as shown on the following illustration, this hole is identified as a datum and the hole pattern is located dimensionally with respect to it. In this case, as contrasted with some earlier examples of similar parts using MMC, the holes are controlled on an LMC basis with a more critical relationship (RFS) to the datum center hole. In a situation of this kind the effect of the MMC principle could be detrimental to the necessary precision of the design requirement, so the design specifies the reverse principle, LMC. Such applications as gear centers, or other critical component interfaces, could require such considerations. LMC provides a compensating effect which states in essence that the hole location criterion is at its LMC size ("sloppiest") and can permit an increase in position tolerance as it (the hole) reduces in actual mating size (toward MMC) and "improves" the component/fit relationship. This also provides a dynamic alternative to specifying RFS to the four hole location.

In the example, the four holes are related to the center datum hole. As the position of the center datum hole shifts, the position of the four hole pattern itself must follow as dictated by the function of the part. Imagine that this part has a mating part with a shaft and four pins which must assemble with the five holes in the illustrated part.

Note that the 12.5 datum center hole is also located from datums (surfaces) A, B, and C. It is given a position tolerance of $\varnothing 0.3$ at MMC and a refined orientation, or perpendicularity, tolerance of $\varnothing 0.08$ at MMC. This properly controls the center hole relationship to the edge surfaces with a rather lenient position tolerance while the orientation to the primary datum A is maintained to closer tolerance. The center hole is identified as datum D so that the four 9.5 holes can be located with respect to it. Under MMC principles, the position and perpendicularity tolerances increase an amount equal to the actual mating hole size departure from MMC as shown in the illustration.

Wherever the datum hole D position varies in the design considerations, or in actual production, the four 9.5 hole pattern must follow. Note that the positional pattern dimensions originate at datum D to carry out this intent.

The four 9.5 holes are located by a position tolerance of $\varnothing 0.13$ at LMC with respect to datum A (for orientation), datum D regardless of feature size (RFS) (for location), and datum B (for rotation). Reference to the lower portion of the illustration under Meaning will assist in the understanding of the effect of the three datums and the importance of datum precedence.

The calculations of the $\varnothing 0.13$ positional tolerance for the $\varnothing 9.5$ holes proceeds as follows. Imagine the mating part pins as four $\varnothing 9.24$ MMC size pins. The fixed fastener method, as described in numerous places earlier in this text where MMC principles are applied is used; e.g.,

$$T = \frac{H - P}{2} \quad T = \frac{\varnothing 9.5 - \varnothing 9.24}{2} \quad T = \frac{0.26}{2} \quad T = \varnothing 0.13 \text{ on each part (or unequally divided).}$$

(Continued on page 154).

EXAMPLE

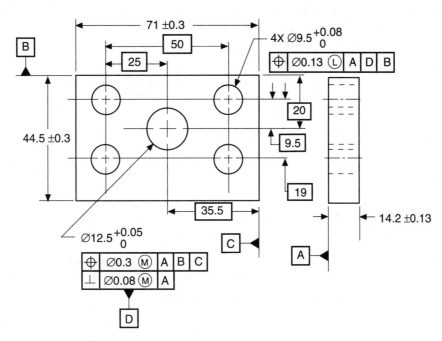

MEANING

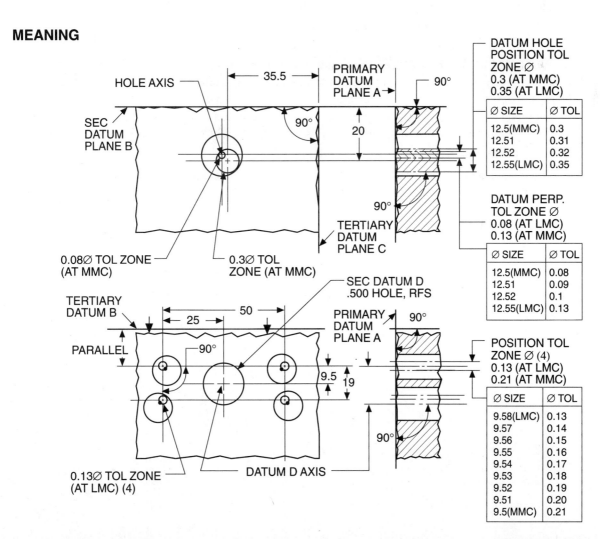

This will ensure that the part features will assemble as per routine positional tolerancing at MMC By then reversing the principle to LMC from MMC, the *same tolerance* (i.e., $\oslash$ 0.13 on each part) then applies at the LMC size limit of $\oslash$ 9.58. Thus, as the produced hole reduces in its actual mating size from LMC size, $\oslash$ 9.58, down to $\oslash$ 9.56, the positional tolerance is increased by $\oslash$ 0.02; and so on down to $\oslash$ 9.5 (MMC), whereby its positional tolerance is then $\oslash$ 0.21 as seen in the tabulation at the lower part of the illustration. What has then happened is that the criterion has simply been reversed and the reduction of the actual mating size of the hole permits its added displacement an equal amount. This compensating effect always ensures the mating feature's assembly essentially the same as if MMC had been specified. However, it permits the hole positional displacement to increase *only* "inward" (toward true position). In contrast, if MMC had been stated, as in earlier examples of this same part, the permitted positional displacement of the hole is always "outward." Thus, the compensating effect of LMC is evident as a valuable optional approach to protect wall thickness, strength, alignment, etc., and yet permit an equal amount of positional tolerance where MMC may have been specified instead.

The datum feature D is referenced on an RFS basis. This neutralizes the effect of size departure from MMC of the datum feature to the related hole pattern movement. It must be located from the axis of datum D and cannot move with respect to that axis. This could be in keeping with the precision of the part interface, the LMC requirement, and also could be to restrict the pattern movement as a concern for wall thickness between the outer hole surface limits relative to the outside surfaces. Where deemed necessary to design intent, LMC could also be applied to datum D in its established relationship to the small hole pattern. In which case, the four hole pattern, as a group, could shift an amount equal to the amount of such departure from LMC of the actual mating size of the datum hole D.

Functional gages and comparable techniques could be applied in the verification of the $\oslash$ 12.5 datum hole requirements as earlier illustrated on a similar part. However, functional gaging, of the physical variety, could *not* be performed on the four hole LMC requirement. It could be done with optical methods, graphic analysis, mathematically, etc., where the functional principles of LMC can be simulated as a gaging operation.

The advantages of LMC are (1) that it provides an often needed tool and (2) that it also provides an alternative other than RFS, where the MMC method is not desirable.

See also earlier examples in this section to determine the best choice of application on such pilot hole applications.

POSITION
SYMMETRY
OR PROFILE OF
NONCYLINDRICAL
FEATURES

(See also POSITION EXTENDED PRINCIPLES section pages 195-222)

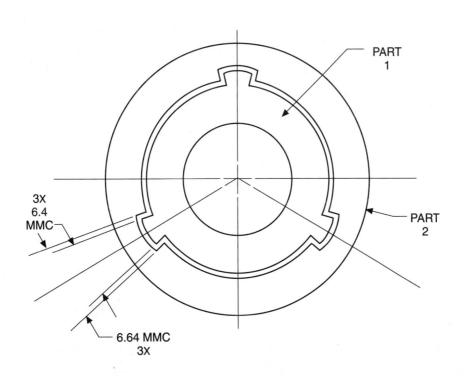

ASSEMBLY OF TWO PARTS WITH NONCYLINDRICAL FEATURES
(See following pages and pages 217–222)

155

NONCYLINDRICAL FEATURES—
SELECTION OF PROPER CONTROLS

There are three characteristics for controlling interrelated noncylindrical, symmetrical and unique shaped features:

1.   POSITION TOLERANCE (MMC, LMC, RFS)

2.   SYMMETRY TOLERANCE (RFS)

3.   PROFILE OF A SURFACE TOLERANCE

Any of the above methods provide effective control. However, it is important to select the most appropriate one to both meet the design requirements and provide the most economical manufacturing conditions. Below are recommendations to assist in selecting the proper control:

If the need is to control location of a noncylindrical or symmetrical size feature's *center plane of the actual mating envelope*, within a tolerance zone relative to datum planes, a center plane or axis (e.g., mating parts to assure proper assembly, function or interchangeability), use:

EXAMPLES:

POSITION

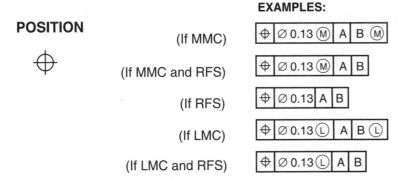

(If MMC)

(If MMC and RFS)

(If RFS)

(If LMC)

(If LMC and RFS)

If the need is to control location of a noncylindrical or symmetrical size feature, as based upon *a tolerance zone at the boundary periphery* or about the true position centerplanes, use:

POSITION

(If MMC, RFS, or LMC, select as appropriate above)

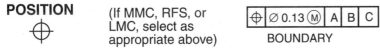

BOUNDARY

If the need is to control the symmetry of a noncylindrical or symmetrical size feature's *derived median line*, as established by all median points of opposed or correspondingly located surface elements, within a tolerance zone relative to a datum center plane, e.g., to control balance, center of mass, function, etc.), use:

SYMMETRY   (Always RFS)

If the need is to control form, orientation, profile and location of a noncylindrical unique shaped size feature, *within a tolerance zone of that shape*, relative to datum planes, use:

PROFILE OF A SURFACE   (Surface only)

POSITION ⊕
MMC WITH RESPECT TO A CENTER PLANE AND RELATED TO A DATUM FEATURE

Position tolerance relationships are more often associated with round holes or features and establish a cylindrical tolerance zone around theoretically exact axes. The cylindrical tolerance zone is not applicable to slots, dial markings, tabs, etc., for which non-cumulative tolerance and MMC aspects of positional tolerancing may also be desired.

Such features may be allowed to vary with respect to a center plane rather than an axis. The position tolerance zone is a total wide zone with one-half the total tolerance assigned to each side of the center plane.

In this example, as based upon an assembly relationship shown on this section introduction page, two mating parts illustrate the calculations and relationships. The top part could be either a thin metal part or a type of drive shaft with three tab projections. The mating part below might be a sleeve or collar which must fit the upper part. To simplify initial explanation, side views and primary (orientation) datums are not shown on either part.

Both parts have corresponding datum reference diameters which are related, in turn, to the controlled features of each individual part. The datums are identified by the letter A in the datum feature symbol. The feature control frame for the top part (Example 1) reads, "these features (3 tabs) must be at true position within 0.12 total wide zone with the feature at MMC size and with respect to datum A at MMC." Although the symbol used is the same as that for cylindrical zones, there is no confusion, since the drawing always clearly shows the feature being dimensioned and the ⌀ symbol is not used, thus designating the tolerance zone as a noncylindrical width zone.

The feature control frame for the bottom part (Example 2) reads, "these features (3 slots) must be at true position within 0.12 total wide zone with the features at MMC size and with respect to datum A at MMC."

Note that the tolerance zones are *not* cylindrical but are total widths equally disposed about, and parallel to, the center plane as established by the 120° basic angles and extending the full depth and length of the produced feature.

When designated as shown, the width of the tolerance zone is always total and is equally disposed on either side of the true position center plane. In this case, the total wide zone is 0.12, with 0.06 on each side of the basic center plane.

The calculations of the positional tolerance zones for mating parts of this type are shown at the upper right. They are based on the same reasoning previously discussed for "fixed fasteners" using cylindrical features. The tolerance zones in this case are, however, not cylindrical but total width.

As in any positional calculations, the MMC sizes of the two corresponding mating part features are used to determine their individual positional tolerances. The MMC width of the tab, 6.4, is subtracted from the MMC width of the slot, 6.64, giving a combined clearance of 0.24. This is divided by the fixed factor 2 to give the total tolerance zone for each mating part feature at MMC. As previously discussed on round feature positional tolerance calculations, the total combined tolerance (in this case 0.24) may be divided as desired in other combinations (e.g., 0.16 and 0.08, 0.14 and 0.1, etc.). See page 167 for further detail on calculation methods.

NONCYLINDRICAL MATING PART FEATURES

EXAMPLE 1   **MEANING**

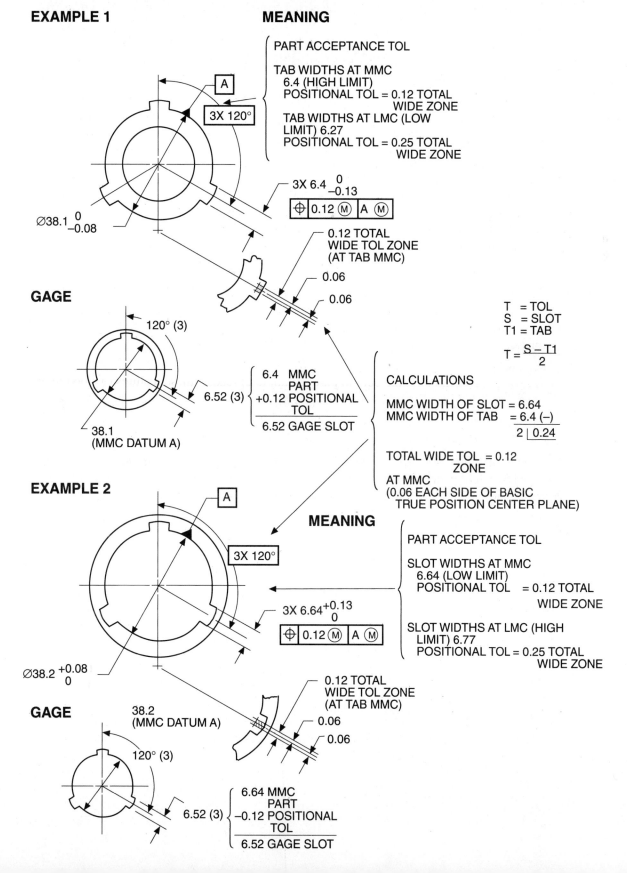

PART ACCEPTANCE TOL

TAB WIDTHS AT MMC
  6.4 (HIGH LIMIT)
    POSITIONAL TOL = 0.12 TOTAL
                              WIDE ZONE
  TAB WIDTHS AT LMC (LOW
  LIMIT) 6.27
    POSITIONAL TOL = 0.25 TOTAL
                              WIDE ZONE

A

3X 120°

∅38.1 $^{0}_{-0.08}$

3X 6.4 $^{0}_{-0.13}$

⊕ | 0.12 Ⓜ | A | Ⓜ

0.12 TOTAL
WIDE TOL ZONE
(AT TAB MMC)

0.06

0.06

GAGE

120° (3)

6.52 (3)

38.1
(MMC DATUM A)

6.4   MMC
      PART
+0.12 POSITIONAL
      TOL
─────────────
6.52 GAGE SLOT

T = TOL
S = SLOT
T1 = TAB

$$T = \frac{S - T1}{2}$$

CALCULATIONS

MMC WIDTH OF SLOT = 6.64
MMC WIDTH OF TAB = 6.4 (−)
                    ───────
                    2 | 0.24

TOTAL WIDE TOL = 0.12
              ZONE
AT MMC
(0.06 EACH SIDE OF BASIC
  TRUE POSITION CENTER PLANE)

EXAMPLE 2

A

3X 120°

MEANING

PART ACCEPTANCE TOL

SLOT WIDTHS AT MMC
  6.64 (LOW LIMIT)
    POSITIONAL TOL   = 0.12 TOTAL
                              WIDE ZONE

SLOT WIDTHS AT LMC (HIGH
    LIMIT) 6.77
    POSITIONAL TOL = 0.25 TOTAL
                              WIDE ZONE

3X 6.64 $^{+0.13}_{0}$

⊕ | 0.12 Ⓜ | A | Ⓜ

∅38.2 $^{+0.08}_{0}$

38.2
(MMC DATUM A)

0.12 TOTAL
WIDE TOL ZONE
(AT TAB MMC)

0.06

0.06

GAGE

120° (3)

6.52 (3)

6.64 MMC
     PART
−0.12 POSITIONAL
     TOL
─────────────
6.52 GAGE SLOT

In Example 1, the notation "Part Acceptance Tolerance" indicates that the total positional tolerance zone increases from 0.12 to 0.25 as the actual size of the produced tab width reduces from MMC of 6.4 to 6.27.

The same is true for Example 2. The slot width positional tolerance increases from 0.12 to 0.25 as the slot actual size increases to high limit size of 6.77. Simulated gages are also shown.

Note that the 6.4 MMC tab width in Example 1 is accommodated by a 6.52 gage slot determined by adding the 0.12 positional tolerance to the MMC size of the part tab. The gage for Example 2 shows the reverse, with the positional tolerance of 0.12 being subtracted from the 6.64 MMC size of the part slot to establish the gage tab size of 6.52 These are the virtual condition sizes, respectively of the two mating features.

Functionally gaging these parts will permit additional positional tolerance for the tabs and slots as a group as the datum feature sizes depart from MMC. However, because of the unique geometry, the actual amount of this effect (due to the datum diameter relationship to the slot and tab flats) may not be conveniently predicted or calculated.

MMC WITH RESPECT TO A LINE (DIAL) AND RELATED DATUM FEATURE

Positional tolerancing may also be applied effectively to parts which do not mate. Since position tolerancing is noncumulative, it is ideally suited to dial markings, etc.

The example illustrates a dial with graduated markings which have critical angular location with respect to one another, and with the center hole identified as the datum reference. The length of the markings are relatively unimportant. In cases such as this, it is usually important that the angular distances between the markings not accumulate tolerance. The use of basic angle locations with position tolerance zones straddling these locations is ideal.

Referring to Meaning, note that the tolerance zones for the groove center lines are parallel to the 15° basic center planes which are taken from the actual center of the 4.8 datum hole B. The tolerance zones are total *wide* tolerance zones with half the zone on each side of these basic center planes. For each marking, the tolerance zone is 0.3 total when the marking grooves are at MMC size of 0.8. The tolerance zones can increase to as much as 0.43 if the dial markings are produced to their least material condition (high limit) of 0.93.

The center hole, datum feature B, is specified at RFS. The RFS datum is used to ensure that the size variation of the center datum hole will not affect the position pattern location with respect to the center hole axis, B; the part is "squared-up" (stabilized) to primary datum plane A.

USE OF "LEAST MATERIAL CONDITION" APPLICATION[*]

When greatest accuracy of the slot angular position is required while the slot marking is at its largest width 0.93 (high limit size), the principles of "least material condition" (LMC) may be desired (see bottom of next page).

---

[*]See pages 150-154.

MEANING

These markings are located at true position within 0.3 total wide zone at marking LMC size of 0.93 with respect to datum A and B (RFS). As the marking groove actual size gets smaller, or departs from LMC (approaches MMC), the position tolerance is increased an equal amount; e.g., at the marking groove size of 0.8 width, the position tolerance is 0.43 total wide zone.

EXAMPLE

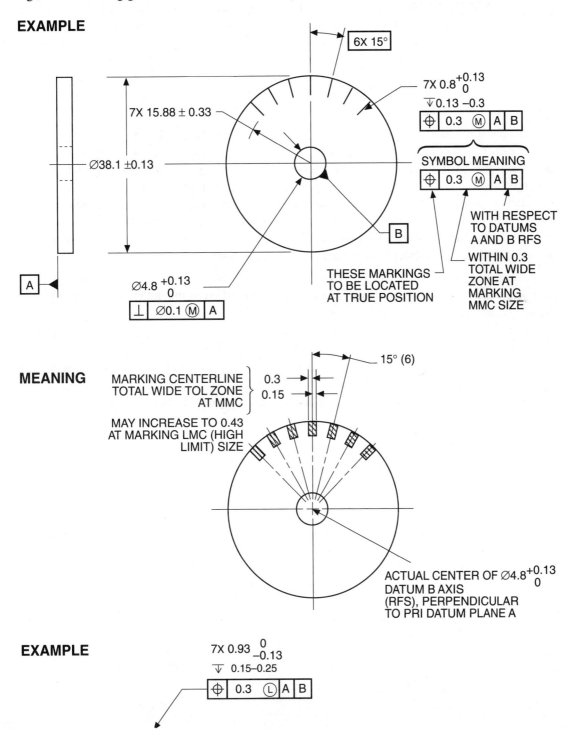

MEANING

MARKING CENTERLINE
TOTAL WIDE TOL ZONE } 0.3
AT MMC } 0.15

MAY INCREASE TO 0.43
AT MARKING LMC (HIGH
LIMIT) SIZE

15° (6)

ACTUAL CENTER OF $\varnothing 4.8^{+0.13}_{0}$
DATUM B AXIS
(RFS), PERPENDICULAR
TO PRI DATUM PLANE A

EXAMPLE

7X 0.93 $^{0}_{-0.13}$
⩗ 0.15–0.25

⊕ | 0.3 | Ⓛ | A | B

POSITION (MMC) WITH RESPECT TO CENTER PLANES (MATING PARTS), BIDIRECTIONAL TOLERANCING

This illustration presents a unique mating part relationship. Position tolerancing principles are extended to utilize a combination of total wide position requirements. The horizontal slots of the upper part are to mate with screw clearance holes of the lower part.

The illustration clearly shows that assembly requirements demand closer control of the hole location in the vertical direction than in the horizontal direction since the slots can compensate for considerable horizontal variation. Therefore the conventional cylindrical zone position method does not seem appropriate but total wide zone methods appear to satisfy the requirement. Using this approach, we can specify separate positional tolerance zones of different values for vertical and horizontal displacements of the slot and holes, i.e., bidirectional tolerancing.

In Example 1, it is quite apparent from the elongated slots that a width type positional tolerancing rather than diametral positional tolerancing is desired. However, in Example 1 as well as in Example 2, the absence of the diameter symbol $\oslash$ and the direction of the arrows is the key indicator that the total wide tolerance zone is to apply. The vertical and horizontal extension lines from the features indicate the directions of the tolerance zones, and the datums and tolerance of the feature control frame state the orientation and total width of the tolerance zone. Note that the tolerance zone consists of the total width equally disposed over the centerplanes; it extends the full depth of the feature perpendicular to the primary datum.

One may ask why the positional tolerancing system was used here, since the resulting tolerance zone seems to be the same as that which would have been obtained from coordinate dimensioning and tolerancing. The answer is that the positional MMC system recognizes that the *actual size* to which these features are produced affects their position relationships. As the feature actual sizes depart from MMC, that is, as the slots or holes get larger in their size tolerance range, additional positional tolerance is permitted. For instance, in Example 1, the 0.3 total wide tolerance zone on the slots could be increased to as much as 0.43, and the 0.3 and 0.8 total wide zones on the holes in Example 2 could be increased to as much as 0.43 and 0.93, respectively, depending on the actual sizes to which these features are produced.

Functional gaging techniques could be used on this part.

See also Extended Use of Position Tolerancing for similar applications.

MMC—NONCYLINDRICAL MATING PART FEATURES

EXAMPLE 1

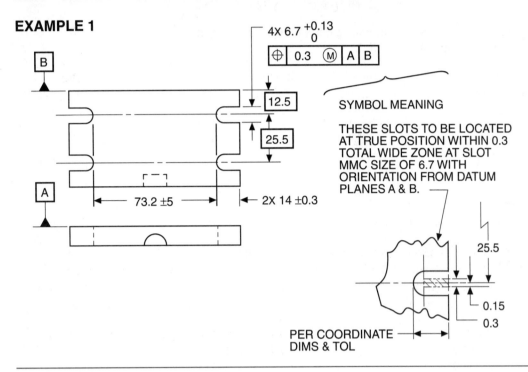

4X 6.7 +0.13 / 0

⊕ | 0.3 | Ⓜ | A | B

12.5

25.5

73.2 ±5   2X 14 ±0.3

SYMBOL MEANING

THESE SLOTS TO BE LOCATED
AT TRUE POSITION WITHIN 0.3
TOTAL WIDE ZONE AT SLOT
MMC SIZE OF 6.7 WITH
ORIENTATION FROM DATUM
PLANES A & B.

25.5

0.15

0.3

PER COORDINATE
DIMS & TOL

EXAMPLE 2

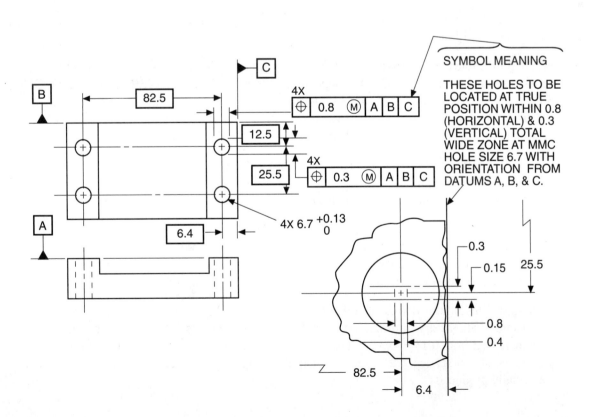

82.5

4X
⊕ | 0.8 | Ⓜ | A | B | C

12.5

25.5

4X
⊕ | 0.3 | Ⓜ | A | B | C

4X 6.7 +0.13 / 0

6.4

SYMBOL MEANING

THESE HOLES TO BE
LOCATED AT TRUE
POSITION WITHIN 0.8
(HORIZONTAL) & 0.3
(VERTICAL) TOTAL
WIDE ZONE AT MMC
HOLE SIZE 6.7 WITH
ORIENTATION FROM
DATUMS A, B, & C.

0.3

0.15

25.5

0.8

0.4

82.5

6.4

MMC—NONCYLINDRICAL MATING PART FEATURES

This illustration shows a pair of mating parts involving noncylindrical features. Part 1 is to fit within the opening of part 2.

Part 1 has a width of $25.46_{-0.15}^{0}$ which is to fit within the $25.6_{0}^{+0.08}$ opening width on part 2. Simultaneously, the $12.72_{0}^{+0.1}$ slot on part 1 is to fit onto the $12.6_{-0.08}^{0}$ projection on part 2.

The $12.72_{0}^{+0.1}$ slot on part 1 has a position feature control symbol which states, "this feature is to be at true position within 0.13 at MMC size of the feature with respect to datums A and B at MMC size." The face surface of the part is established as primary datum A and the width as secondary datum B.

Part 2 has an identical position feature control frame on the 12.6 dimension, and the 25.6 opening is established as datum B and the face surface as primary datum plane A.

Figure 1(a) shows the relationship of these two parts as they would appear if both parts were produced perfectly at the feature MMC sizes. Note the common center or median planes established on both parts. The parts are assembled in Fig. 1(b). All details shown in Figures 1, 2 and 3 relate in perpendicular orientation to the primary datum plane A and central to secondary datum B width.

Figure 2(a) illustrates the slot feature on part 1 offset the maximum permissible amount (0.065) at the extreme of the 0.13 total tolerance zone when the part is at MMC size. Also, the mating projection of part 2(b) is shown offset in the opposite direction the maximum permissible amount (0.065) at the extreme of the 0.13 total tolerance zone when the part is at MMC size.

Figure 2(b) shows the assembly of the two parts. They still assemble satisfactorily. Figure 2 also emphasizes that the 0.13 total tolerance zone, as stated in the symbol boxes on parts 1 and 2, applies at the MMC size of the features and is the maximum tolerance permissible under this condition.

Figure 3 illustrates the increase in the permissible total position tolerance zone as the feature actual sizes *depart* from MMC to the opposite extreme of LEAST MATERIAL CONDITION. For part 1 (Fig. 3a), with the slot at its *high* limit MMC size of 12.82 and the datum width at its *low* limit LMC size of 25.31, the permissible position tolerance zone becomes 0.38 total or a 0.19 offset off the median plane of the slot with respect to the datum median plane.

For part 2 in Fig. 3(b), with the projection at its low limit LMC size of 12.52 and the datum opening width at its high limit LMC size of 25.68, the position tolerance zone becomes 0.29 total or a 0.145 offset off the median plane of the projection with respect to the datum median plane.

Figure 3(b) shows the assembly of the two parts under these conditions. They still assemble satisfactorily with considerably more clearance as a result of the feature actual size variation to size limits opposite MMC, or their LEAST MATERIAL CONDITION.

From these illustrations it is evident that positional MMC applications permit greater tolerance and ensure a satisfactory fit of mating parts. For example, the possible tolerance on part 1 has been increased from 0.13 to 0.38 and on part 2 from 0.13 to 0.29. The actual tolerance to be realized is, of course, dependent upon the actual mating sizes to which the concerned features are actually produced.

MAXIMUM MATERIAL CONDITION (MMC)
EXAMPLES

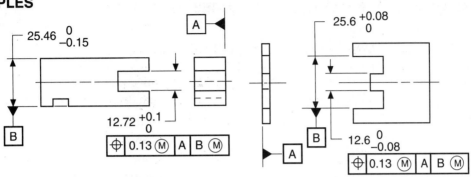

PERFECT POSITION AT MMC
MEANING

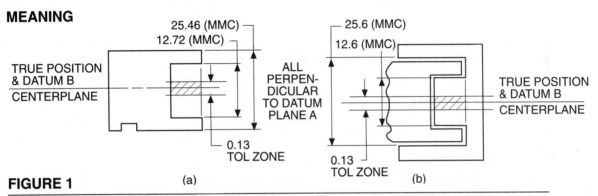

FIGURE 1 (a)   (b)

POSITION TOLERANCE ZONES AT MMC

POSITION TOLERANCE ZONES AT LMC

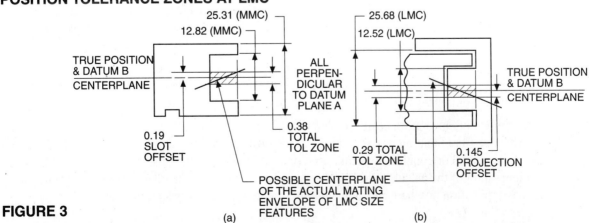

FIGURE 2 (a)   (b)

FIGURE 3 (a)   (b)

NONCYLINDRICAL MATING PART FEATURES

MMC CALCULATIONS TO DETERMINE TOLERANCE

In this example we present the calculations required to determine the positional tolerance for the mating parts shown in the previous example. Note the expanded "fixed fastener" equation which can be used in such situations.

Since one part is to fit within the other, the first step is to determine the clearance of features and which feature is to receive the position tolerance. In this case, it seems more functional to control the position of the slot in part 1 and the position of the projection in part 2. The clearance of the two mating part features is to be 0.12 minimum. The projection on part 2 is 12.6 and the slot on part 1 is larger at 12.72. These are MMC sizes, or the largest projection possible on part 2 and the smallest slot possible on part 1.

The width features on both parts are given 0.14 clearance at MMC size of the features and are selected as the datum features for each part.

Under the subheading, POSITION TOLERANCE CALCULATIONS, the 12.6 MMC size of the projection on part 2 is subtracted from the 12.72 MMC size of the slot on part 1. This results in a difference of 0.12. Next the 25.46 MMC datum projection feature of part 1 is subtracted from the 25.6 MMC datum slot of part 2, resulting in a difference of 0.14.

The 0.12 result of the first calculation and the 0.14 result of the second calculation are added to give the 0.26 total combined positional tolerance for both parts and their interrelated features. This total tolerance is then divided to establish the required position tolerance on *each* individual part. How we allocate the total tolerance is optional, so long as it totals the calculated combined tolerance, in this case 0.26.

For the purposes of this example, the 0.26 total tolerance was divided evenly, with 0.13 selected as the position tolerance for both the 12.72 slot on part 1 and the 12.6 projection on part 2. These two figures, 0.13 plus 0.13, total 0.26 and comply with the 0.26 allowable total combined positional tolerance calculated.

Once the position tolerance is established for both mating part features based on their relationship to each other and to common datum axes, possible extra position tolerance for each part may be determined as shown in the lower half of the figure.

To do these calculations, we must first determine the relationship of one mating part feature to another and, then, we must consider each of these part features individually with respect to the actual size variations which could occur within their size tolerances. As has been shown, the size of features affects their positional tolerances, and it is this fact that makes positional tolerancing advantageous, since it permits economical production with greater tolerances and ensures assembly of the mating parts.

On part 1, the permissible tolerance may be increased from 0.13 to 0.38 and on part 2 the permissible tolerance may be increased from 0.13 to 0.29. The *actual* tolerance permissible in each case is, of course, dependent on the *actual* sizes of the features as produced.

This method of calculating position tolerances assumes the possibility of zero clearance—zero interference fits of mating part features if all features are at extreme tolerance limits. It also assumes parallel orientation or permissible float of one part to the other at assembly. Additional compensation of the calculated tolerance values should be considered as necessary for any particular application or where additional datum orientation may restrict this float.

The "fixed fastener" equation is again used in this case. Since there is only *one feature* relative to its datums on each part, the effect of the datum feature departure from MMC size can be immediately added into the consideration by expansion of the equation to include the MMC advantage together (feature and its datum). If more than one feature is related to the datum, each size feature would need to be calculated individually relative to its corresponding mating part feature. Thus, any departure from MMC of the datum feature would be added float of the pattern as a group.

EXAMPLES

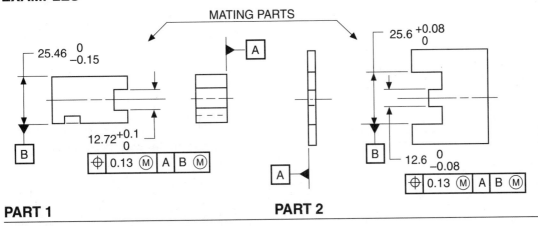

PART 1   **PART 2**

POSITION TOLERANCE CALCULATIONS

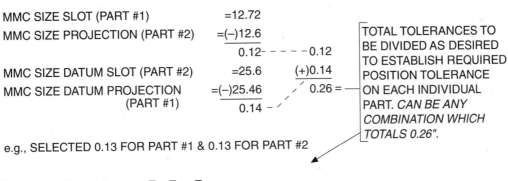

MMC SIZE SLOT (PART #1)   =12.72
MMC SIZE PROJECTION (PART #2)   =(−)12.6
                                            0.12 − − − − 0.12

MMC SIZE DATUM SLOT (PART #2)   =25.6   (+)0.14
MMC SIZE DATUM PROJECTION   =(−)25.46   0.26 =
         (PART #1)   0.14

TOTAL TOLERANCES TO BE DIVIDED AS DESIRED TO ESTABLISH REQUIRED POSITION TOLERANCE ON EACH INDIVIDUAL PART. *CAN BE ANY COMBINATION WHICH TOTALS 0.26".*

e.g., SELECTED 0.13 FOR PART #1 & 0.13 FOR PART #2

$$T = \frac{(S - P) + (D_1 - D_2)}{2} \qquad 2T = T_1 + T_2$$

S = SLOT   PART 1
P = PROJ   PART 2
D_1 = DATUM SLOT PART 2
D_2 = DATUM PROJ PART 1

EXTRA TOLERANCE FOR EACH PART

PERMISSIBLE SLOT POSITION TOL AS FEATURE SIZES DEPART FROM MMC:

STATED POSITION TOL WITH SLOT AT 12.72 MMC   =0.13
PLUS TOTAL 12.72 SLOT SIZE TOL   +0.1

POSITION TOL WITH DATUM WIDTH AT 25.46 MMC   =0.23
PLUS TOTAL 25.46 DATUM WIDTH SIZE TOL   +0.15

TOTAL POSITION TOL WITH SLOT & DATUM WIDTH AT   =0.38
LEAST MAT'L CONDITION (LARGEST SLOT, SMALLEST
DATUM WIDTH)

PART 1

(Continued on next page).

PERMISSIBLE PROJECTION POSITION TOL AS FEATURE SIZES
DEPART FROM MMC:

STATED POSITION TOL WITH PROJECTION AT .495 MMC	=0.13
PLUS TOTAL .495 PROJECTION SIZE TOL	+0.08
POSITION TOL WITH DATUM WIDTH AT 1.005 MMC	=0.21
PLUS TOTAL 1.005 DATUM SLOT SIZE TOL	+0.08
TOTAL POSITION TOL WITH PROJECTION & DATUM OPENING AT LEAST MAT'L CONDITION (SMALLEST PROJ., LARGEST DATUM OPENING)	=0.29

PART 2

NONCYLINDRICAL MATING PART FEATURE GAGES

Functional gages may be utilized on noncylindrical parts when position tolerancing is used.

Part 1 shown at the right is the same part used in the preceding examples. Immediately below part 1 is a representative functional gage used to check position on this part. Gage-maker's tolerances would, of course, be included in the actual construction of this gage.

From the previous explanations and this illustration, we see that as the part feature sizes depart from MMC within their size tolerances, they become equally acceptable to the gage while permitting greater position tolerance. The gage essentially simulates the mating part situation, and therefore a part which passes the gage will assemble with its mating part.

Note, however, that the high and low limits of size are determined by other gages or measurements. The illustrated gage checks only the position requirement.

Part 2 is the mating part used in the previous examples. Below it is a representative functional gage used to check the position on this part. Gage-maker's tolerances would, of course, also be included in the actual construction of this gage.

As in the mating part, above, we see that as the part feature sizes of part 2 depart from MMC within their size tolerances, they become equally acceptable to the gage while permitting greater position tolerance. The gage essentially simulates the mating part situation, and therefore a part which passes the gage will assemble with its mating part.

As in part 1, the high and low limits of size are determined by other gages or measurements. The illustrated gage checks the position requirement only.

It should be noted that these gages would require a stabilizing plate (i.e. datum simulator) to establish the primary simulated datum feature A surface. The gage portions for the controlled features may need to be moveable if a hard gage is used. Optical or CMM processes could adequately stabilize to datum A prior to measurement.

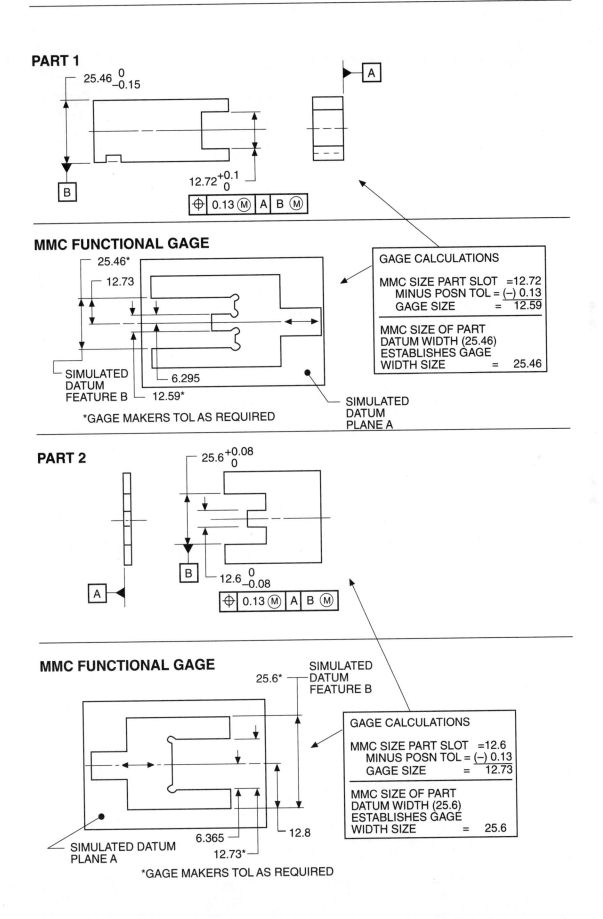

PART 1

25.46 $^{0}_{-0.15}$

12.72 $^{+0.1}_{0}$

| ⊕ | 0.13 Ⓜ | A | B Ⓜ |

A

B

MMC FUNCTIONAL GAGE

25.46*

12.73

6.295

12.59*

SIMULATED DATUM FEATURE B

*GAGE MAKERS TOL AS REQUIRED

SIMULATED DATUM PLANE A

GAGE CALCULATIONS

MMC SIZE PART SLOT = 12.72
MINUS POSN TOL = (−) 0.13
GAGE SIZE   =   12.59

MMC SIZE OF PART DATUM WIDTH (25.46) ESTABLISHES GAGE WIDTH SIZE   =   25.46

PART 2

25.6 $^{+0.08}_{0}$

12.6 $^{0}_{-0.08}$

| ⊕ | 0.13 Ⓜ | A | B Ⓜ |

B

A

MMC FUNCTIONAL GAGE

25.6*

SIMULATED DATUM FEATURE B

6.365

12.73*

12.8

SIMULATED DATUM PLANE A

*GAGE MAKERS TOL AS REQUIRED

GAGE CALCULATIONS

MMC SIZE PART SLOT = 12.6
MINUS POSN TOL = (−) 0.13
GAGE SIZE   =   12.73

MMC SIZE OF PART DATUM WIDTH (25.6) ESTABLISHES GAGE WIDTH SIZE   =   25.6

MMC—NONCYLINDRICAL FEATURES

The example below illustrates control of a flat feature in a position relationship to a cylindrical datum. This situation is typical of many parts for which position functional techniques will assure agreement with design intent and also permit functional or receiver gaging.

The part is shown with a representative functional gage. Note that the possible position tolerance increases from 0.25 to 0.46 maximum as the feature actual sizes depart from MMC toward the opposite or least material limits. The actual position tolerance depends on the actual sizes to which the features are produced with the amount of increase equal to the departure from MMC size.

EXAMPLE

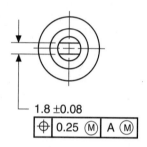

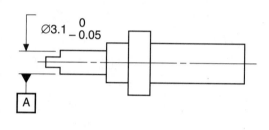

GAGE

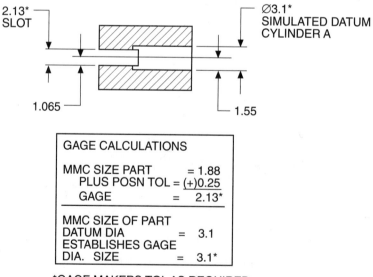

GAGE CALCULATIONS	
MMC SIZE PART	= 1.88
PLUS POSN TOL	= (+)0.25
GAGE	= 2.13*
MMC SIZE OF PART DATUM DIA ESTABLISHES GAGE DIA. SIZE	= 3.1 = 3.1*

*GAGE MAKERS TOL AS REQUIRED

RFS—NONCYLINDRICAL FEATURES

Noncylindrical feature application may require RFS precision. In the case of this example, the part is also a symmetrical one where the features must be controlled with respect to their location relative to one another. Imagine that due to part function, the dynamics of the MMC principles cannot be allowed on this part.

The example shows a part using position tolerancing on the slot; the part lower surface and width are established as the datum features. The functional requirement is to relate the slot location central to the outside width of the part while stabilized to the lower surface. To simplify explanation, size dimensions and tolerances of the slot and the datum width have been omitted in the lower example. In order to establish some basis for the relationship of the slot feature to the width datum feature, we must have a common reference plane. This is the center plane of the datum B feature, and thus it is also the true position center plane. The resulting center plane of the actual mating envelope of the tolerance feature (the slot) is determined which is then compared to the datum B center plane which is simultaneously the true position center plane. The results of this comparison determines whether or not the tolerance is met.

The meaning below shows the relationship and establishment of the positional tolerance zone with respect to the datum B centerplane. Note that the centerplane of datum B establishes the middle (true position) of the 0.13 wide tolerance zone simultaneously. The centerplane of the actual mating envelope of the produced slot (RFS) must lie between the two confining planes of the 0.13 wide tolerance zone. Note that the resulting centerplane of the slot may fall at an angle anywhere within the tolerance zone. (See also the following example and illustrations, next two pages, for more detail.)

EXAMPLE

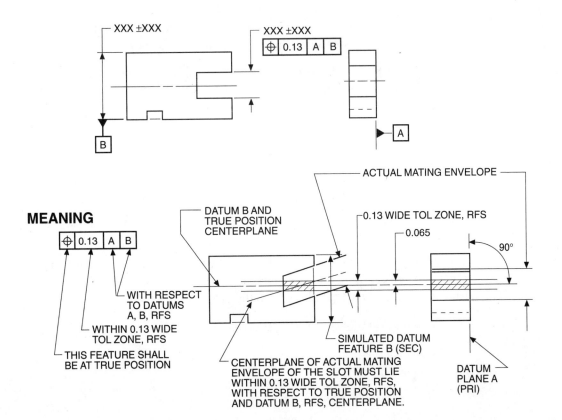

RFS—NONCYLINDRICAL FEATURES

In this example, a part similar to that in the previous illustration has been used. Size dimensions and tolerances have also been added to the slot and datum features. This example, as the previous one, is also an RFS application.

The meaning below the example in Fig. 1a illustrates the establishment of the 0.13 tolerance zone about the center plane of the datum feature. Note that the tolerance zone remains the same, 0.13, when the datum feature actual size and the slot actual size vary anywhere in their size tolerance range. See Fig. 1b.

In either case, and also at any place between these sizes, the centerplane is established by the datum B size but the 0.13 tolerance zone remains the maximum permissible tolerance.

The total tolerance zone for the slot is therefore 0.13 regardless of the actual size of the datum feature or the slot.

In Figures 2(a) and (2b) we show the manner in which the centerplane of the slot, regardless of its size, may vary within the position tolerance zone. In 2(a) the slot is shown at 12.5 MMC and in 2(b) at 12.6 LMC. To be acceptable, the centerplanes of either of these extremes of the slot size must fall within the 0.13 tolerance zone. Note that the slot centerplanes in either example fall within the tolerance zone so these parts are acceptable.

Figure 3 shows that the slot centerplane of its actual mating envelope may fall anywhere within this tolerance zone.

In Figure 4 we show one method of verifying the position tolerance on this part in order to clarify how theoretical centerplanes presented on drawings are converted to actual measurements. The slot has been offset to the limit of the 0.13 tolerance zone. A 12.5 wide slot and a 25 wide datum feature have been assumed for this example. The part is rested against datum plane A (primary) and on one side of the 25 datum feature B, or its width. A maximum measurement is taken from the measuring surface to the top underside of the slot.

The part is then rotated 180°, or turned over, to the opposite side of the 25 datum feature B and against datum plane A (primary) and a similar maximum measurement is taken. The difference between these two measurements, as may be seen in the illustration, actually determines the total variation of the slot center plane with respect to the datum center plane and the tolerance zone.

A preferred (and actually more correct) method, although usually requiring more verification equipment (such as a CMM), is to establish the datum center plane simultaneously from the part width. See the DATUM section for more detail.

Comparison between the resulting value and the tolerance requirement determines the acceptance of the slot position.

As seen in this example, the resulting 0.13 total variation of the measurements is equal to the 0.13 required tolerance. Therefore, the slot is positional, or central, to datum B within the required tolerance and is also square (perpendicular) to the primary datum A.

EXAMPLE

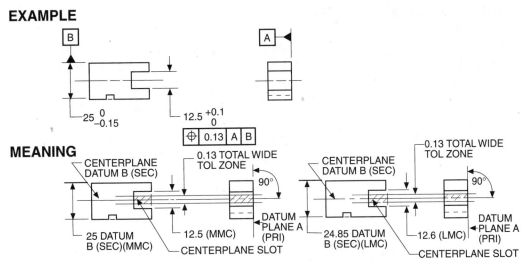

MEANING

TOTAL TOL ZONE FOR SLOT IS 0.13 REGARDLESS OF SIZE OF DATUM OR SLOT

FIGURE 1(a)   **FIGURE 1(b)**

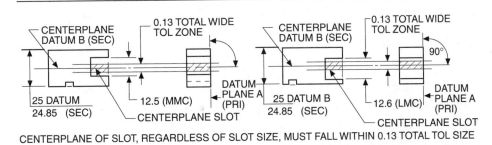

CENTERPLANE OF SLOT, REGARDLESS OF SLOT SIZE, MUST FALL WITHIN 0.13 TOTAL TOL SIZE

FIGURE 2 (a)   **FIGURE 2 (b)**

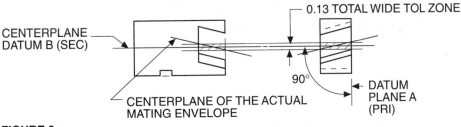

FIGURE 3

VERIFICATION

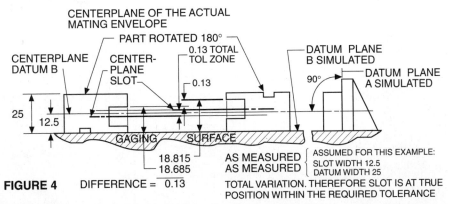

FIGURE 4   DIFFERENCE = 0.13

ASSUMED FOR THIS EXAMPLE:
SLOT WIDTH 12.5
DATUM WIDTH 25

TOTAL VARIATION. THEREFORE SLOT IS AT TRUE
POSITION WITHIN THE REQUIRED TOLERANCE

LMC—NONCYLINDRICAL FEATURES

Occasionally a method is required to control a situation which is essentially the reverse of the usual position relationship; that is, the stated position tolerance applies at the *least material condition*, LMC, of the feature or datum, instead of at MMC, and increases as the feature or datum *departs from* the least material condition.

Definition. Least material condition (LMC) is the condition in which a feature of size contains the least amount of material within the stated limits of size: for example, maximum hole diameter, minimum shaft diameter. Least material condition is the condition opposite to MMC. For example, a shaft is at least material condition when it is at its *low* limit of size and a hole is at least material condition when it is at its *high* limit of size.

This method is applicable to special design requirements that will not permit MMC or that do not warrant the exacting requirements of RFS. It can be used to maintain critical wall thickness or critical center locations of features for which accuracy of location can be relaxed (position tolerance increased) when the feature leaves least material condition and approaches MMC. The amount of increase of positional tolerance permissible is equal to the feature actual mating size departure from least material condition.

The term "least material condition" and the abbreviation LMC have been used instead of "minimum material condition" (which is synonymous) to avoid confusion, since the abbreviation would be the same as that for maximum material condition. The symbol modifier Ⓛ is used to indicate the LMC requirement applicable to feature or datum.

Although the use of LMC does impose exacting requirements on both manufacturing inspection, it does permit additional tolerances.

Whenever least material condition (LMC) or Ⓛ is specified on a drawing, the position tolerance applies only when the feature is produced at its LMC size.

Additional positional tolerance is permissible but is dependent on, and equal to, the difference between the produced feature actual mating envelope size (within its size tolerance) and LMC.

The illustration gives an example of LMC applied to noncylindrical features. As shown under MEANING, the tolerance zone increases as the feature actual mating envelope size departs from LMC toward MMC. In this case because of part function and required precision relative to the center hole, datum B, the slot relationships are relative to datum B, RFS.

Author advisory: LMC could be applied to the datum feature as well if desired. However, it may be well to try to avoid such use due to somewhat complicated details that then arise. RFS gives precision and has a more readily understood meaning.

EXAMPLE

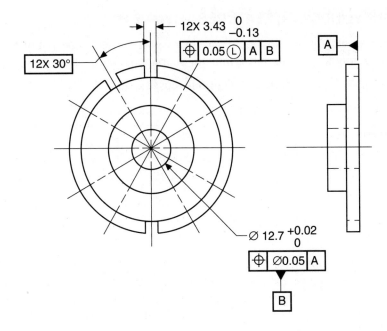

12X 3.43 $^{0}_{-0.13}$

| ⊕ | 0.05 Ⓛ | A | B |

12X 30°

A

Ø 12.7 $^{+0.02}_{0}$

| ⊕ | Ø0.05 | A |

B

MEANING

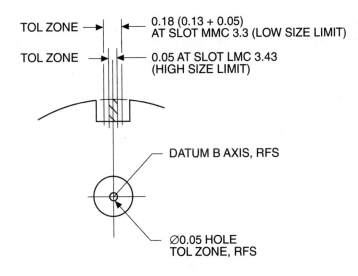

TOL ZONE → 0.18 (0.13 + 0.05)
AT SLOT MMC 3.3 (LOW SIZE LIMIT)

TOL ZONE → 0.05 AT SLOT LMC 3.43
(HIGH SIZE LIMIT)

DATUM B AXIS, RFS

Ø0.05 HOLE
TOL ZONE, RFS

RFS—NONCYLINDRICAL FEATURES—SYMMETRY ⌯

Where a feature, such as the slot on the part below, is required to be symmetrical to the datum center plane of another feature, the symmetry characteristic may be used.

To specify symmetry the following conditions must exist:

1. Noncylindrical features (slots, tabs, projections, etc.) only are to be controlled.

2. The material condition RFS only is to apply.

3. The feature and its datum must be symmetrically configured to each other.

4. The datum feature is usually noncylindrical but may be cylindrical if appropriate to the part.

Symmetry-Symmetry is that condition where the median points of all opposed or correspondingly-located elements of two or more feature surfaces are congruent with the axis or center plane of a datum feature.

Symmetry finds its primary use where centrality of the part's mass, equal distribution of wall thickness, strength, balance, etc., is of concern. Its basis is to require that all opposed elements of opposed surfaces establish a series of median points. These points develop a derived median line which must be contained within the symmetry tolerance zone about the datum center plane; see the illustration below.

Where possible, position tolerance can be considered as another option on symmetrical parts and noncylindrical features, permitting the symmetry characteristic to be specified on requirements as referenced above.

Similar to concentricity, symmetry may require a detailed and time-consuming analysis to derive data for the verification process. A CMM process may be required to comply with such requirements.

EXAMPLE

MEANING

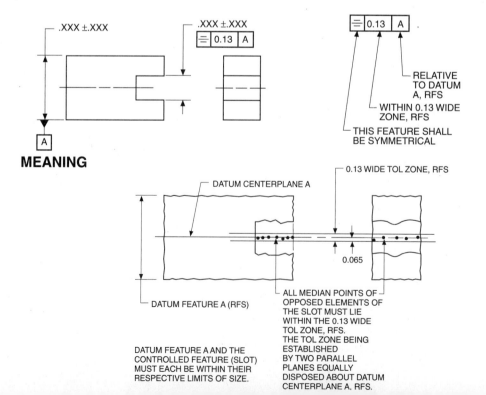

NONCYLINDRICAL FEATURES—PROFILE—COMPOSITE TOLERANCE ⌒

Profile tolerance can be used for controlling size, form, orientation, and location of features. Basic dimensions are required to define and describe the tolerance zone within which the feature surface or surfaces must lie. Previous sections of this text have described profile of a surface and profile of a line details on other applications. Profile of a surface is commonly used upon, but not restricted to, the more unusual shaped features. It is also used to control conicity and coplanarity. This section of the text extends the use of profile of a surface to also locate a feature or features of the noncylindrical variety and introduce the composite tolerance method on profile tolerance.

The example shown on page 178 is a specially shaped feature (aperture, vent, fit). Its shape, size, form, orientation and location is controlled by all-around profile of a surface. The upper segment of the composite feature control frame controls the feature orientation and location relative to the datum reference frame planes A, B, C within the 0.8 tolerance zone. The lower segment controls the feature size, form, and orientation (not location) within the 0.13 tolerance zone. Note the smaller tolerance zone (0.13) is free to move about within the larger zone while maintaining orientation (perpendicularity) to datum plane A. The Meaning figures illustrate the foregoing description.

Page 179 illustrates further expansion and use of profile of a surface to single segment profile control, composite profile with the addition of a secondary datum in the lower segment (for orientation only application) and a profile and position control where the MMC principle may be used. Each illustration contains an example and brief description of meaning.

PROFILE—COMPOSITE TOLERANCE ◠

EXAMPLE

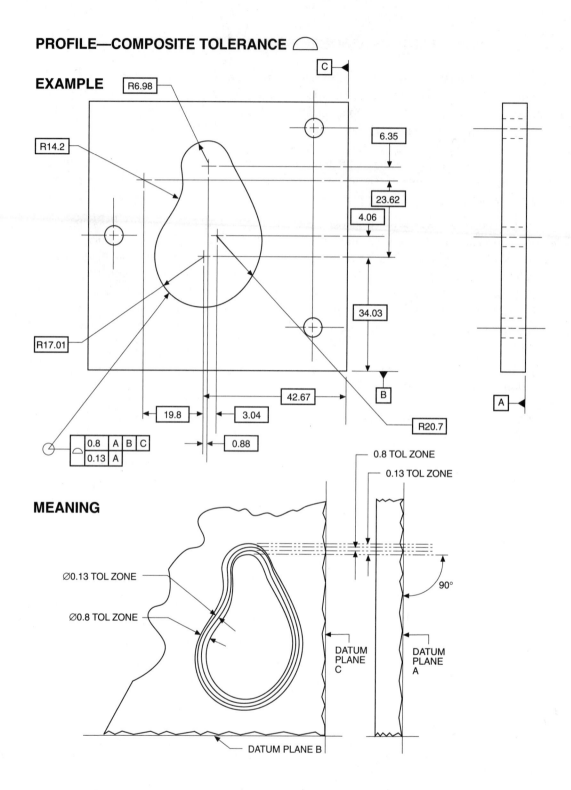

MEANING

PROFILE—SINGLE ENTRY—COMPOSITE—PROFILE/POSITION TOLERANCE

EXAMPLES

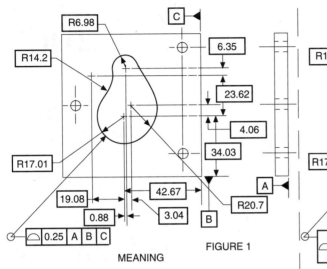

FIGURE 1

MEANING

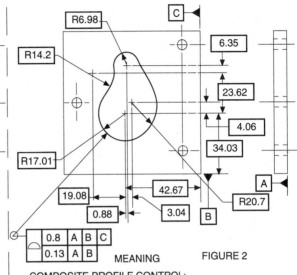

FIGURE 2

MEANING

PROFILE CONTROL:
ONE REQUIREMENT CONTROLS FEATURE
SIZE, FORM, ORIENTATION AND LOCATION.
TOLERANCE ZONE 0.25 (SIMILAR TO LARGER
(0.8) TOL ZONE OF ILLUSTRATION ON
PREVIOUS PAGE), A SINGLE TOLERANCE
ZONE, RELATIVE TO DATUM PLANES A, B, C.

COMPOSITE PROFILE CONTROL:
ONE REQUIREMENT CONTROLS FEATURE
LOCATIONWITHIN 0.8 TOL ZONE
TO DATUMS A, B, C. THE *2ND REQM'T*, 0.13,
CONTROLS FEATURE SIZE, FORM AND
ORIENTATION RELATIVE TO DATUMS A & B.
THE 0.13 ZONE (SIMILAR TO THE PREVIOUS
PAGE ILLUSTRATION) MAY FLOAT WITHIN
THE LARGER ZONE, 0.8, *BUT MUST
MAINTAIN PARALLEL ORIENTATION (NOT
LOCATION)* RELATIVE TO DATUM PLANE B.

MEANING

PROFILE *AND* POSITION CONTROL:
PROFILE REQUIREMENT CONTROLS SIZE,
FORM, AND ORIENTATION OF FEATURE
WITHIN 0.15 TOL ZONE RELATIVE
TO DATUM PLANE A. *POSITION REQM'T*
CONTROLS ORIENTATION AND LOCATION
OF FEATURE WITHIN 0.25 BOUNDARY ZONE,
i.e., THE MMC SIZE OF THE FEATURE MINUS
0.125 PER SIDE, OR TO ITS VIRTUAL
CONDITION. AS THE FEATURE SIZE DEPARTS
FROM MMC, THE BOUNDARY ZONE SIZE
INCREASES THAT AMOUNT.

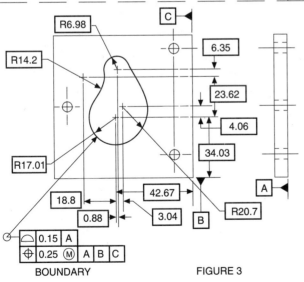

BOUNDARY

FIGURE 3

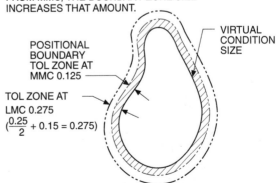

POSITIONAL
BOUNDARY
TOL ZONE AT
MMC 0.125

TOL ZONE AT
LMC 0.275
$(\frac{0.25}{2} + 0.15 = 0.275)$

VIRTUAL
CONDITION
SIZE

POSITION
OF
COAXIAL
FEATURES

COAXIAL FEATURES—
SELECTION OF PROPER CONTROL

There are four characteristics for controlling interrelated coaxial features:

1.   RUNOUT TOLERANCE (circular or total)   (RFS)

2.   POSITION TOLERANCE   (MMC or RFS)

3.   CONCENTRICITY TOLERANCE   (RFS)

4.   PROFILE OF A SURFACE   (RFS DATUM)

Any of the above methods provides effective control. However, it is important to select the *most appropriate* one to both meet the design requirements and provide the most economical manufacturing conditions. (See also details of preceding and following sections.)

Below are recommendations to assist in selecting the proper control:

If the need is to control only CIRCULAR cross-sectional elements in a composite relationship to the datum axis, RFS, e.g., multidiameters on a shaft, use:

CIRCULAR RUNOUT ⟋   **EXAMPLE** | ⟋ | 0.05 | A – B |

(This method controls any composite error effect of circularity, concentricity, and circular cross-sectional profile variations.)

If the need is to control the TOTAL cylindrical or profile surface in composite relative to the datum axis RFS, e.g., multi-diameters on a shaft, bearing mounting diameters, etc., use:

(This method controls any composite error effect of circularity, cylindricity, straightness, coaxiality, angularity, and parallelism.)

NOTE   Runout is always implied as an RFS application. It cannot be applied on an MMC basis, since an MMC situation involves functional interchangeability or assemblability (probably of mating parts), in which case POSITION tolerance would be used. See below.

If the need is to control the total cylindrical or profile surface and its actual mating envelope axis relative to the datum axis on an MMC or RFS basis, e.g., on mating parts to assure interchangeability or assemblability, use:

POSITION ⊕   **(IF MMC)**   **EXAMPLE** | ⊕ | ⌀0.13 Ⓜ | A Ⓜ |

   (IF RFS)   **EXAMPLE** | ⊕ | ⌀0.13 | A |

   OR RFS DATUM | ⊕ | ⌀0.13 Ⓜ | A |

If the need is to control such as the *axis* of one or more features in composite relative to a *datum axis,* RFS, e.g., to control such as balance of a rotating part, use:

CONCENTRICITY ◎   **EXAMPLE** | ◎ | ⌀0.08 | A – B |

NOTE   Concentricity is always implied as an RFS application. Variations in size (departure from MMC size, out-of-circularity, out-of-cylindricity, etc.) do not in themselves conclude *axis* error.

If the need is to control the total cylindrical or profile surface simultaneously with the size dimension(s) (using basic dimensions for both), relative to a datum axis, e.g., precise fit, multi-diameters, etc., use:

PROFILE OF A SURFACE ⌒   **EXAMPLE** | ⌒ | 0.13 | A |

POSITION TOLERANCING (MMC) OF COAXIAL FEATURES ⊕

This illustration on page 183 shows a common application of position tolerancing of coaxial features. A functional MMC relationship is desired.

A functional datum is selected and the feature is specified to be "at true position (coaxial) within ⌀0.04 tolerance zone with feature at MMC with respect to datum A at MMC."

A functional gage (or equivalent technique) may be used on this part. The interpretation and the illustration gage show how the part remains functionally acceptable as the concerned features depart from their MMC (worst condition) size within their size tolerance range. Further, the advantages of position tolerancing are realized since greater position tolerance is permitted as the feature and datum actual mating envelope sizes depart from MMC. Functionally good parts are always accepted on this basis; as in all types of MMC position tolerance, size tolerance *and* position tolerance are considered together.

Positional tolerance controls axis or center plane displacement. This part illustrates the manner in which position of coaxial features incorporates the same considerations as conventional positional tolerancing of hole patterns. The axes provide the common denominator needed to relate displacement of one feature to another and to allow calculations and gage determinations based on functional design requirements. The functional gage will accommodate permissible axis displacement in terms of the surface configuration, including both form and position errors. The functional gage checks *position* only. The hole high size limit and low size limit (if necessary) are checked separately.

Zero tolerancing on position of coaxial features may occasionally be useful (see page 206).

POSITION TOLERANCING (RFS) OF COAXIAL FEATURES ⊕

Where the *axis of the actual mating size,* on an RFS basis is required, and no additional positional tolerance can be permitted in the feature being controlled, the callout:

| ⊕ | ⌀ 0.04 | A |

(RFS per Rule #2)

would be substituted in the illustration on page 183.

This will permit lobing, flats, surface irregularities, etc., so long as all actual local size requirements are met and the axis is contained within stated positional tolerance, RFS.

Functional gaging principles could not be used here and sophisticated verification equipment would be required. Although this method differs from a concentricity requirement it is similar in complexity.

COAXIAL FEATURES

EXAMPLE

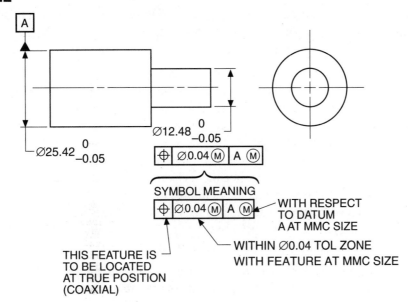

SYMBOL MEANING

WITH RESPECT
TO DATUM
A AT MMC SIZE

WITHIN Ø0.04 TOL ZONE
WITH FEATURE AT MMC SIZE

THIS FEATURE IS
TO BE LOCATED
AT TRUE POSITION
(COAXIAL)

MEANING

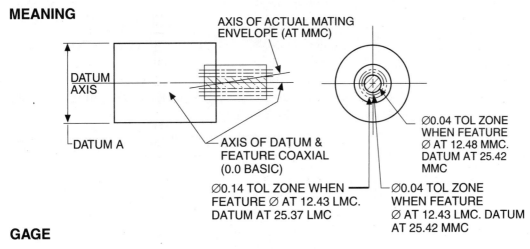

AXIS OF ACTUAL MATING
ENVELOPE (AT MMC)

DATUM
AXIS

DATUM A

AXIS OF DATUM &
FEATURE COAXIAL
(0.0 BASIC)

Ø0.14 TOL ZONE WHEN
FEATURE Ø AT 12.43 LMC.
DATUM AT 25.37 LMC

Ø0.04 TOL ZONE
WHEN FEATURE
Ø AT 12.48 MMC.
DATUM AT 25.42
MMC

Ø0.04 TOL ZONE
WHEN FEATURE
Ø AT 12.43 LMC. DATUM
AT 25.42 MMC

GAGE

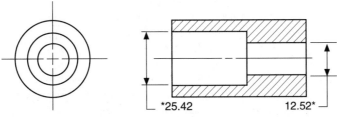

25.42   12.52

*GAGE MAKERS TOL APPLY

COMPARISON BETWEEN COAXIAL FEATURES SPECIFIED RFS (RUNOUT) AND COAXIAL FEATURES SPECIFIED MMC (POSITION)

In a mating part situation, positional tolerancing of coaxial features can be used to good advantage. In this illustration we compare a part with coaxial features dimensioned and toleranced using total runout (RFS) with the same part dimensioned and toleranced using position (MMC).

The example at the top shows a part similar to that of the previous example except that total runout (RFS) tolerancing is used. Note that the 12.48 diameter in the top example has a feature control frame which states "this feature is to be within 0.04 total runout when mounted on datum A." This is an RFS application where the stated runout tolerance of 0.04 is the total allowable tolerance FIM (TIR, FIR) regardless of the size to which the 12.48 diameter and 25.42 datum diameter are produced. The meaning is shown at the right; the probable method of checking is also illustrated. It is seen that the dial indicator reading of the rotated 12.48 diameter with respect to the 25.42 datum diameter and its axis will entirely determine the runout reading and thus the acceptance or rejection of the part. The dial indicator reading must be 0.04 or less for part acceptance, and this tolerance is maximum.

A closer study of this part and its function may reveal that it has a *mating part* into which it must fit. Knowing this, we can determine the control of the dimensions and tolerances of the *two* parts as they are related to each other, to ensure proper fit and economic production. Where this problem of assemblability or functional requirement exists, position tolerancing and its associated advantages should be considered.

Note that the same part is illustrated under the MMC (SPECIFIED AS POSITION) subheading. The dimensions are identical. The position characteristic symbol is, however, indicated with a $\oslash$ 0.04 tolerance requirement and is specified as an MMC application. At the right is the mating part containing the holes which must fit the shaft at the left. It also utilizes position control with a tolerance requirement of $\oslash$ 0.06 at MMC.

Figures 1, 2, and 3 illustrate the mating part positional relationships and the manner in which additional tolerance is achieved dependent on actual part size variations.

In Fig. 1(a), the 0.04 tolerance zone permits a 0.02 axis displacement of the 12.48 diameter on the shaft with respect to the 25.42 datum diameter A axis. On the mating part (Fig. 1b), the 0.06 tolerance zone on the hole permits 0.03 axis displacement of the 12.5 hole with respect to the 25.5 datum hole A axis. This displacement includes all the errors of form and position of the feature which have an effect. Actually, in the use of MMC position, the reference to axes is primarily theoretical as a common denominator for comparison and calculation. In reality, the functional gage (or technique) which represents each mating part and its fit determines acceptance of the part on the basis of surface contact as related to the axis displacement.

The sectional view in Fig. 1(b), showing the shaft inserted into the hole, illustrates the fit of the two parts at MMC size with the 12.48 shaft diameter and the 12.5 hole diameter displaced in opposite directions to the full extent of the position tolerance. The parts assemble satisfactorily.

(Continued on page 186).

RFS (SPECIFIED AS TOTAL RUNOUT)   MEANING

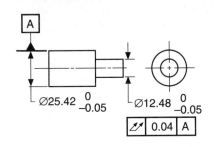
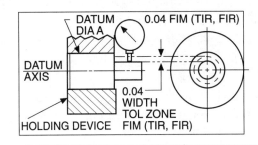

MMC (SPECIFIED AS POSITION)   MATING PART

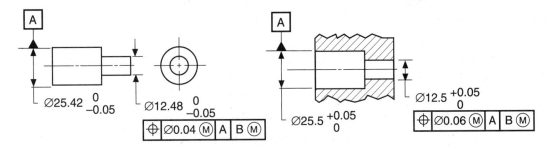

POSITION TOLERANCE ZONES AT MMC

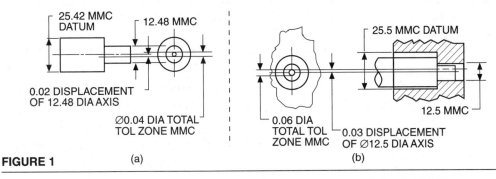

FIGURE 1   (a)   (b)

POSITION TOLERANCE ZONES AT LEAST MATERIAL CONDITION

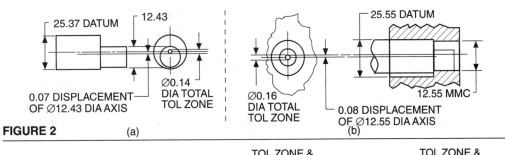

FIGURE 2   (a)   (b)

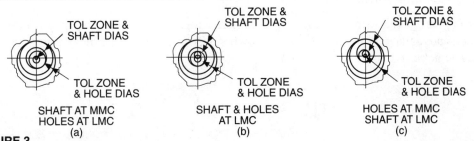

FIGURE 3

In this situation the part shaft and hole sizes are at MMC; that is, we have the largest permissible shaft size and the smallest permissible hole size. If the parts assemble under these conditions, it is reasonable to assume that as the shaft gets smaller and the hole larger, the tolerance could be increased while continuing to permit assembly.

Figure 2(a) illustrates how the position tolerance zone increases as the part and hole actual sizes depart from MMC to the opposite size extreme (LEAST MATERIAL CONDITION). Note that the 12.48 shaft diameter is now 12.43 and the 25.42 datum is now 25.37, or the low limit of the shaft diameter sizes. In Fig. 2(b) the 12.5 hole diameter is now 12.55 and the 25.5 datum diameter is now 25.55, or the high limit of the hole sizes.

It may now be seen that the reduction in shaft actual sizes and increase in hole actual sizes permit a greater displacement of the mating features, resulting in a permissible ⌀0.14 total position tolerance zone on the stepshaft at the left and a ⌀0.16 total position tolerance zone on the mating part at the right.

The sectional view at far right showing the shaft inserted into the hole illustrates the fit. The parts still assemble satisfactorily.

With the use of position, the permissible tolerance has been increased from ⌀0.04 to ⌀0.14 on the shaft at the left and from ⌀0.06 to ⌀0.16 on the mating part at the right.

Referring to the part examples under the MMC (POSITION) subheading, note that a rule-of-thumb method of determining the maximum possible position tolerance is to *add* the stated tolerance found in the feature control frame and the total size tolerance of the features involved. For example, on the shaft at left, the position tolerance of ⌀0.04 found in the feature control frame plus the 0.05 total size tolerance of the 12.48 shaft diameter, plus the 0.05 total size tolerance of the shaft 25.42 datum diameter (or 0.04, plus 0.05, plus 0.05) equals 0.14 total. On the mating part at the right, the position tolerance of ⌀0.06 found in the feature control frame, plus the 0.05 total size tolerance of the 12.5 hole, plus the 0.05 total size tolerance of the 25.5 datum hole diameter (or 0.06, plus 0.05, plus 0.05) equals 0.16 total.

Remember, however, that the resulting tolerance is dependent on the actual size to which the features are produced. Thus, the tolerance under an MMC application is usually somewhere between the stated tolerance and the opposite limit as determined by actual feature sizes.

Figure 3 illustrates the assembly relationship of the shaft and holes at various extremes. Figure 3(a) shows the shaft at MMC or its largest size, with the mating part holes at LEAST MATERIAL CONDITION, or their *largest* size. Figure 3(b) shows both parts at LEAST MATERIAL CONDITION or at the size limits opposite to the calculated MMC sizes; that is, the shaft is at its smallest allowable size and the holes are at their largest allowable size. This is the condition illustrated in Fig. 2; it allows the greatest amount of tolerance on both parts. Figure 3(c) shows the holes at MMC or their smallest size and the shaft at LEAST MATERIAL CONDITION or its smallest size. Note that the parts assemble in each case.

Numerous other combinations are, of course, also possible and acceptable, so long as the established size and location tolerances are held.

These examples again illustrate that in position tolerancing, size tolerances affect position tolerances. It is advantageous to consider this fact whenever possible so that maximum tolerance yield may be achieved.

RFS (SPECIFIED AS TOTAL RUNOUT)   MEANING

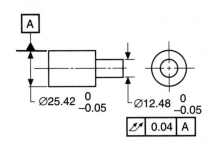

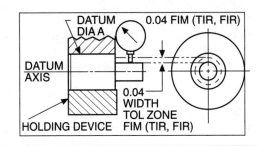

MMC (SPECIFIED AS POSITION)   MATING PART

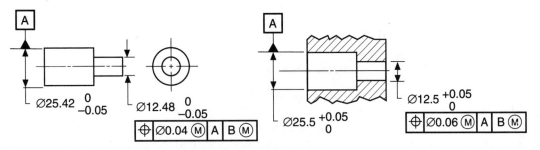

POSITION TOLERANCE ZONES AT MMC

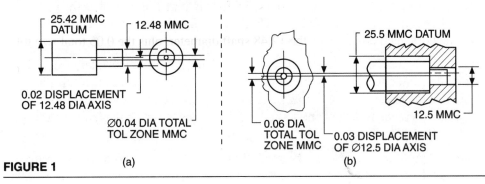

FIGURE 1   (a)   (b)

POSITION TOLERANCE ZONES AT LEAST MATERIAL CONDITION

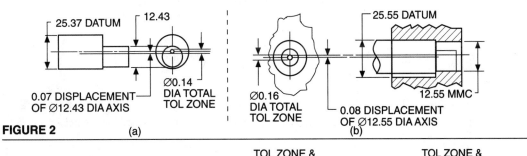

FIGURE 2   (a)   (b)

FIGURE 3

MMC CALCULATIONS TO DETERMINE TOLERANCE

The actual calculations necessary to determine position tolerances of coaxial features of mating parts are illustrated in this example. The mating parts are the same as those in the preceding example. The method shown here is valid only for simple two coaxial feature parts as illustrated. It is a variation of the "fixed fastener" method.

Note that we have selected functional datums: the diameters of mating features, the 25.42 shaft diameter, and the 25.5 hole diameter. This establishes a common datum axis and the relationship between the diameters of the two parts. This information is necessary for calculation. Selected *clearances* of the holes to the mating shaft diameters are established by the design requirements. These clearances in turn establish the position tolerances to be stated for these features.

Note that the first step illustrated under the subheading POSITION TOLERANCE CALCULATIONS shows the 12.48 MMC shaft size of part 2 subtracted from the 12.5 MMC hole size of part 1, resulting in a difference of 0.02. Also, the 25.42 MMC datum shaft size of part 2 is subtracted from the 25.5 MMC datum hole size of part 1, resulting in a difference of 0.08.

The 0.02 result of the first calculation and the 0.08 result of the second calculation are added to give the 0.1 total combined position tolerance for both parts and for their interrelated features. This total tolerance is then divided to establish the required position tolerance on each individual part. The division can be made in any combination, so long as the individual part tolerances total no more than the calculated combined tolerance, in this case, 0.1. For this example, 0.06 has been selected as the position tolerance for the 12.5 hole in part 1, and 0.04 as the position tolerance for the 12.48 diameter shaft of part 2. These two figures total 0.1 and comply with the allowable total combined position tolerance.

Dependent on feature actual size variation, we may realize an extra or bonus tolerance. This possible additional position tolerance as the feature actual sizes depart from MMC size is calculated as shown on the lower portion of the illustration. Remember, however, that this increase in position tolerance is dependent upon the *actual* sizes to which the features are produced within their size tolerances.

Since most parts are produced somewhere between the high and low size tolerance extremes, the actual position tolerance permissible on part number 1 would be somewhere between 0.06 and 0.16, say approximately 0.12. It could, however, be 0.16 and still provide the proper fit to the mating part. The actual position tolerance permissible on part 2 would probably be somewhere between 0.04 and 0.14, say approximately 0.1. It could, however, be 0.14 and still provide the proper fit to the mating part.

The advantages of using position tolerances on coaxial features are now apparent.

1. Greater tolerance is permitted.

2. The fit of the mating part is guaranteed when the dimension and the tolerance requirement are calculated simultaneously and consideration is given to the possible variables which can occur in the assembly relationship.

The method of calculating position tolerances described above assumes the possibility of a zero interference-zero clearance condition of the mating part features at extreme tolerance limits. Additional compensation of the calculated tolerance values should be considered as necessary relative to a particular application. Where more than two coaxial features are involved, the calculation method would use the conventional "fixed fastener" method where each feature is calculated individually for its appropriate position tolerance. See pages 116–119 for principles.

POSITION TOLERANCE—COAXIAL FEATURES—MATING PARTS

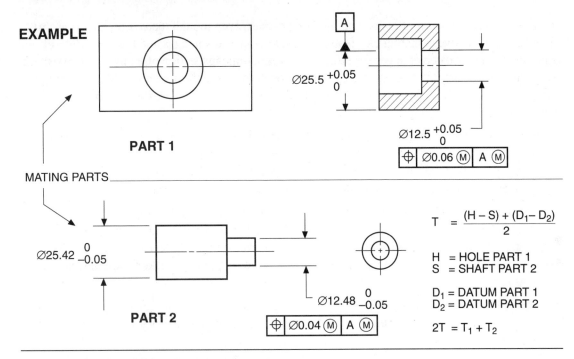

EXAMPLE

PART 1

MATING PARTS

PART 2

$$T = \frac{(H - S) + (D_1 - D_2)}{2}$$

H = HOLE PART 1
S = SHAFT PART 2

D_1 = DATUM PART 1
D_2 = DATUM PART 2

$2T = T_1 + T_2$

POSITION TOLERANCE CALCULATIONS

MMC SIZE SLOT (PART #1)   = 12.5
MMC SIZE PROJECTION (PART #2)   = (−)12.48
                                                             ‾‾‾‾‾‾‾‾
                                                               0.02 ‑ ‑ ‑ ‑ 0.02

MMC SIZE DATUM SLOT (PART #2)   = 25.5   (+)0.08
MMC SIZE DATUM PROJECTION   = (−)25.42   0.1 =
        (PART #1)   ‾‾‾‾‾‾‾
                                                              0.08

e.g., SELECTED 0.06 FOR PART #1 & 0.04 FOR PART #2

> TOTAL TOLERANCES TO BE DIVIDED AS DESIRED TO ESTABLISH REQUIRED POSITION TOLERANCE ON EACH INDIVIDUAL PART. *CAN BE ANY COMBINATION WHICH TOTALS 0.1".*

EXTRA TOLERANCE FOR EACH PART

PERMISSIBLE HOLE POSITION TOL AS HOLE ACTUAL SIZES DEPART (GET LARGER) FROM MMC:

STATED POSN TOL WITH HOLE AT 12.5 MMC	= 0.06
PLUS TOTAL 12.5 HOLE TOL	+ 0.05
POSN TOL WITH DATUM HOLE 25.5 AT MMC	= 0.11
PLUS TOTAL 25.5 DATUM HOLE TOL	+ 0.05

PART 1   TOTAL POSN WITH BOTH HOLES AT LEAST MAT'L CONDITION   = 0.16

PERMISSIBLE SHAFT POSITION TOL AS SHAFT ACTUAL SIZES DEPART (GET SMALLER) FROM MMC:

STATED POSN TOL WITH SHAFT AT 12.48 MMC	= 0.04
PLUS TOTAL 12.48 DIA TOL	+ 0.05
POSN TOL WITH DATUM SHAFT DIA 25.42 AT MMC	= 0.09
PLUS TOTAL 25.42 DATUM DIA TOL	+ 0.05

PART 2 TOTAL POSN TOL WITH BOTH SHAFT DIAS AT LEAST MAT'L CONDITION   = 0.14

MMC GAGES

Positional tolerances may, of course, be used on any part, even though there is no direct assembly relationship between mating part features. The tolerance advantages and the possibility of using functional or receiver gages for effective and economic inspection often makes the position technique very desirable under such conditions. However, where the assemblability of mating part features is involved, position tolerancing is always recommended. Position tolerancing of coaxial features of mating parts is an ideal example of this use.

Functional or receiver gages for coaxial features of mating parts simulate the fit of the actual part features in a manner similar to the hole pattern functional gages. As in the hole pattern functional gage, part feature actual mating size departure from MMC permits greater part acceptance based on the functional interrelationship of size and position variations.

The illustration shows the two mating parts previously discussed and examples of functional gages to check the position requirements of each part. The upper illustration, part 1, is shown with a gage for functionally checking the position tolerance of the holes. The simple calculations to determine the gage dimensions are shown at the right. Gage-maker's tolerances also apply as required.

The lower illustration shows part 2 and a functional gage for checking the position of the shaft diameters of this part. The calculations are shown at right. Gage-maker's tolerances also apply as required.

The illustrated gages will check only the position tolerances of the part. Sizes of the associated features require a separate size check.

PART 1

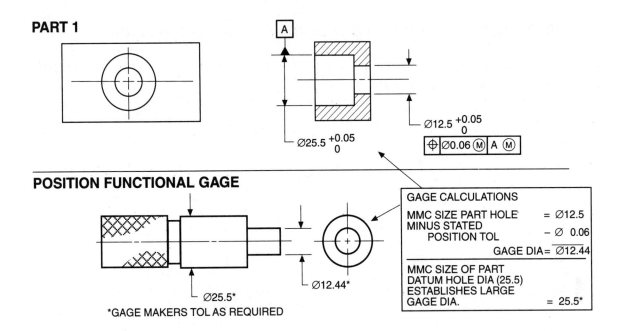

$\emptyset 25.5$ $^{+0.05}_{0}$

$\emptyset 12.5$ $^{+0.05}_{0}$

⊕ | $\emptyset 0.06$ Ⓜ | A | Ⓜ

POSITION FUNCTIONAL GAGE

$\emptyset 25.5^*$

$\emptyset 12.44^*$

*GAGE MAKERS TOL AS REQUIRED

GAGE CALCULATIONS

MMC SIZE PART HOLE	= $\emptyset 12.5$
MINUS STATED POSITION TOL	− $\emptyset\ 0.06$
GAGE DIA=	$\emptyset 12.44$

MMC SIZE OF PART
DATUM HOLE DIA (25.5)
ESTABLISHES LARGE
GAGE DIA.   = 25.5*

PART 2

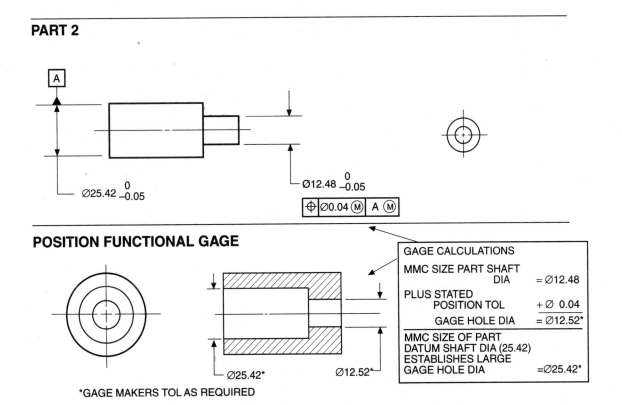

$\emptyset 25.42$ $^{0}_{-0.05}$

$\emptyset 12.48$ $^{0}_{-0.05}$

⊕ | $\emptyset 0.04$ Ⓜ | A | Ⓜ

POSITION FUNCTIONAL GAGE

$\emptyset 25.42^*$

$\emptyset 12.52^*$

*GAGE MAKERS TOL AS REQUIRED

GAGE CALCULATIONS

MMC SIZE PART SHAFT DIA	= $\emptyset 12.48$
PLUS STATED POSITION TOL	+ $\emptyset\ 0.04$
GAGE HOLE DIA	= $\emptyset 12.52^*$

MMC SIZE OF PART
DATUM SHAFT DIA (25.42)
ESTABLISHES LARGE
GAGE HOLE DIA   =$\emptyset 25.42^*$

RELATION TO DATUM SURFACES USING A RADIAL HOLE PATTERN

Radial hole patterns may be controlled by position tolerancing in a manner similar to that used on more conventional hole patterns.

Since the four hole pattern in the example at the right requires location as well as feature-to-feature relationship (in-line), composite positional tolerancing may be used to advantage.

Datums are established relative to the part function, the face surface A and the 28.2 diameter, datum B. Imagine this part mating with another at assembly on the surface A and located and centered on B so pins will pass into the holes. MMC is desired so the hole and mating part pin sizes are used to determine the hole-to-hole position tolerances.

In this example, the holes are desired to be within one plane (coplanar), in-line (colinear), and spaced 90° apart. Therefore, as a pattern they may vary from that basic configuration only to the extent of permissible tolerances. In this case, the pattern as an entity (established from the holes themselves) may vary from position (and parallel to A) within 0.8 diameter at MMC. In the pattern, the holes may vary from their basic colinear alignment within 0.13 diameter at MMC. The axes of the four holes must lie within both the pattern locating tolerance zones and the feature relating tolerance zones.

In terms of gaging this part, two separate gages or gaging operations would be required: one to orient datums A and B and evaluate the 0.8 diameter tolerance, and another to evaluate the 0.13 diameter tolerance. The latter could be accomplished by passing gage pins of 3.07 diameter through the pairs of two holes in-line simultaneously. The hole size requirements must, of course, also be met.

As the actual sizes of the four holes depart from MMC, more position tolerance is acquired (as has been previously described in other examples). In this case, although the example is more complex, the principles are the same.

EXAMPLE

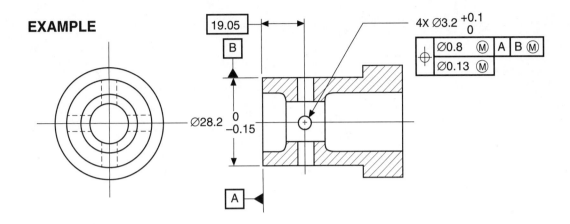

4X ⌀3.2 +0.1 / 0

⊕	⌀0.8 Ⓜ	A	B Ⓜ
	⌀0.13 Ⓜ		

19.05

B

⌀28.2

0 / −0.15

A

MEANING

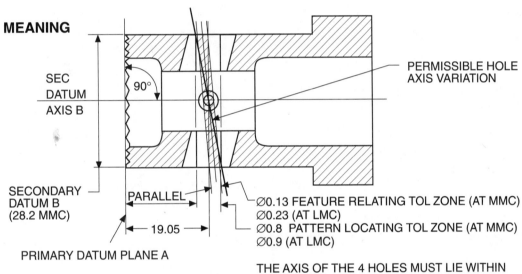

SEC DATUM AXIS B

90°

PERMISSIBLE HOLE AXIS VARIATION

SECONDARY DATUM B (28.2 MMC)

PARALLEL

19.05

⌀0.13 FEATURE RELATING TOL ZONE (AT MMC) ⌀0.23 (AT LMC)
⌀0.8 PATTERN LOCATING TOL ZONE (AT MMC) ⌀0.9 (AT LMC)

PRIMARY DATUM PLANE A

THE AXIS OF THE 4 HOLES MUST LIE WITHIN BOTH THE PATTERN LOCATING TOL ZONES (⌀0.8 AT MMC) AND THE FEATURE RELATING TOL ZONES (⌀0.13 AT MMC)

POSITION
EXTENDED
PRINCIPLES

SEPARATE PATTERNS OF HOLES

Patterns of holes (or features) on a part which are either shown by common basic or untoleranced dimensions or are related to common datums are normally assumed to be related and therefore considered a single composite pattern. Occasionally, convenience in the placement of dimensions or the need to show common dimension origination from related datum features on the drawing can unintentionally impose restrictions on production or inspection.

If certain holes (or features) in a BASIC dimensioned pattern or patterns are related to one another functionally but not to others which are also shown in the same pattern, or if we wish to produce or gage certain portions of the pattern separately, we may indicate this by the notation SEP REQT or SEPARATE REQUIREMENT beneath or adjacent to the concerned feature control frame or frames. Coding the appropriate holes ⌖ on the drawing is helpful in clarifying the drawing intent.

Hole (or feature) patterns on a part separated by rectangular coordinates, related to *different* datums, or having no direct tie by dimensional arrangement, are considered to be separate patterns; no further notation is required in this case.

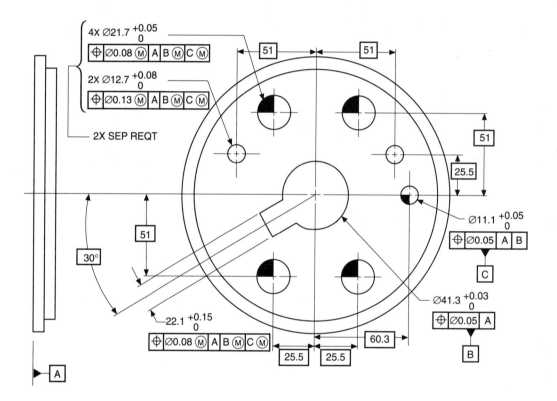

PERPENDICULARITY (SQUARENESS) CONTROL WITHIN POSITION TOLERANCING

Where out-of-perpendicularity of threaded holes could cause inserted screws, bolts, studs, pins, or dowels to interfere with mating parts or result in cocked seating of screw or bolt heads, closer control or perpendicularity *within* the position tolerance may be required.

Perpendicularity control may be specified on the drawing, using one of the methods illustrated below.

PERPENDICULARITY SPECIFICATION ON DRAWING

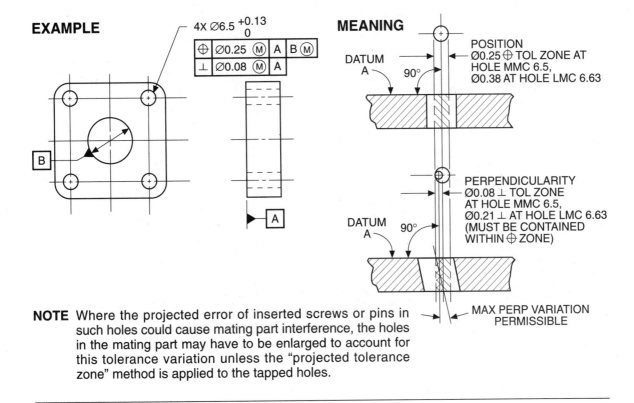

EXAMPLE

4X Ø6.5 $^{+0.13}_{0}$

| $\oplus$ | Ø0.25 Ⓜ | A | B Ⓜ |
| $\perp$ | Ø0.08 Ⓜ | A | |

MEANING

DATUM A   90°

POSITION
Ø0.25 $\oplus$ TOL ZONE AT HOLE MMC 6.5, Ø0.38 AT HOLE LMC 6.63

PERPENDICULARITY
Ø0.08 $\perp$ TOL ZONE AT HOLE MMC 6.5, Ø0.21 $\perp$ AT HOLE LMC 6.63 (MUST BE CONTAINED WITHIN $\oplus$ ZONE)

DATUM A   90°

MAX PERP VARIATION PERMISSIBLE

NOTE Where the projected error of inserted screws or pins in such holes could cause mating part interference, the holes in the mating part may have to be enlarged to account for this tolerance variation unless the "projected tolerance zone" method is applied to the tapped holes.

TO CALCULATE

CALCULATION TO COMPENSATE POSITIONAL TOLERANCE FOR OUT-OF-PERPENDICULARITY OR WHERE PROJECTED TOLERANCE ZONE NOT USED:
WHERE: D = MINIMUM DEPTH OF HOLE OR PART THICKNESS FOR FIXED FASTENER (OR PIN)
      P = MAXIMUM THICKNESS OF PART WITH CLEARANCE HOLE OR PROJECTION OF FASTENER (OR PIN)
TO CALCULATE POSITION TOLERANCE:

$$T = \frac{H - F}{\left(1 + \frac{2P}{D}\right)}$$

TO CALCULATE CLEARANCE HOLE MMC SIZE:

$$H = F + T_1 + T_2\left(1 + \frac{2P}{D}\right)$$

PROJECTED TOLERANCE ZONE

The projected tolerance zone method prevents the condition shown in Fig. 1, where interference could possibly exist with conventional positional tolerancing. The variation from perpendicularity of the portion of the bolt passing through the mating part is of concern. Therefore the location and perpendicularity of the tapped hole is of importance insofar as it affects this extended portion of the bolt. The projected tolerance zone method (Fig. 2) eliminates this interference.

With this method, we can use conventional "fixed fastener" calculations to determine the position tolerance. Further, specifying by this method means that gaging techniques will simulate the mating part relationship, and the projected perpendicularity error will be accounted for in the tolerance and in the gaging.

FIGURE 1

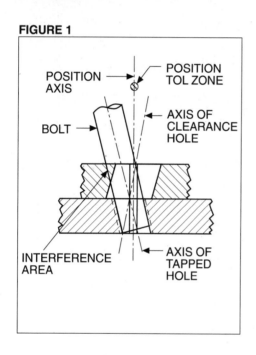

FIGURE 2

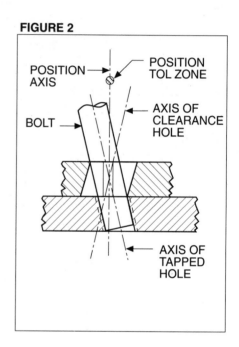

EXAMPLE   **MEANING**

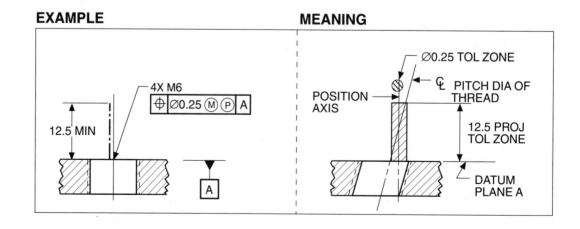

ROUND HOLES–0 TOLERANCE (PERFECT POSITION AT MMC)

"Zero" (0) position tolerancing is a technique adaptable to situations requiring functional inter-changeability and maximum tolerance advantage in the feature size, form, orientation, and position interrelationships. Where mating parts and features are simply to mate up or "GO" and tangent contact of the mating features could occur, zero tolerancing is technically acceptable.

However, under some conditions, zero position tolerancing is *not* appropriate. For example, where specific running clearances, fit, or similar special mating feature conditions are required, zero position tolerancing will not, in general, be technically applicable. There are other consid-erations, also, which require evaluation to determine whether or not zero position tolerancing is applicable.

For most position applications and examples shown elsewhere in this text, we could have used the zero position tolerancing method. It is an optional method for stating many common posi-tion mating part requirements.

As foregoing sections of this text have emphasized, position tolerances are usually established on the basis of MMC size relationships of mating part features. The feature sizes are the crite-ria with which the process of developing the position tolerances starts. The designed clearance between the mating components is the basis for the position tolerances which are stated on the drawing and applied in manufacture. When the features specified by position tolerances are actually produced, any actual mating size departure from the MMC size (e.g., enlarging size of a hole) adds to the permissible position tolerance.

In zero position tolerancing the same principles apply, except that the position tolerance stated is always a fixed "zero," with all the tolerance placed on the size dimension. This, of course, assumes that the actually produced feature will show some deviation from MMC, which is then added to the "zero" tolerance to give a working position/orientation tolerance.

In either conventional or zero position tolerancing, size, form, orientation, and position variations are considered simultaneously as a composite value. This is really the fundamental principle (along with the MMC principle) on which functional position tolerancing is based. The reason for this is that related mating part features perform their function in the space limitations provided, regardless of whether that space is derived from size, form, orientation, or position variation.

There is some controversy about zero position tolerancing, with both proponents and oppo-nents contributing valuable comments. However, a good understanding of the principles involved, their proper application, and an awareness of their limitations will go a long way toward ensuring proper use of this effective geometric tolerancing tool.

A forceful argument for the use of zero tolerancing arises in situations where a produced part with a positional hole pattern might be acceptable to a functional gage yet be rejectable on the basis of a low limit "GO" size violation, with the result that functionally good parts might be scrapped.

At this point, we wish to emphasize that in conventional position tolerancing, the stated size tolerance can be used for size, form, orientation, and position variations as the feature size departs from MMC, whereas a stated position tolerance may be used only for orientation and position variations. Actual size tolerance variations of the feature from MMC size can thus add to the position tolerance; but, according to standard practices, unused position variations can-not be added to size tolerances.

The above principle is best described by referring to the CONVENTIONAL POSITION TOLERANCE APPLICATION example which follows on page 204. The notation at the bottom of the illustration states that if the hole is produced in perfect location, its size will be permitted to exceed the low limit 6.5 (MMC) size down to the virtual condition size of 6.3. The virtual condition size is developed from the MMC size of the hole, 6.5, minus the stated position tolerance, 0.2. This is, of course, also the functional gage pin size, and represents the mating part feature at its extreme condition of assembly.

Obviously, not many parts will be produced in perfect position. However, the point is that when the position tolerance, or any portion of it, has *not* been used up, the *size* of the feature should reap the benefit (e.g., the hole should be allowed to get smaller an equal amount and yet be accepted as functional).

Note that the virtual condition size developed in the ZERO POSITION TOLERANCE example on page 205 is also 6.3. In this method the virtual condition size and the MMC size are the *same*. The 0.2 position tolerance of the "conventional" example has been shifted to the size tolerance in the "0" method. The result for both parts, insofar as resulting position gaging is concerned, is the same; but the above-mentioned "rejectable" part (low limit size violation) will be acceptable under the zero method.

The 6.3 diameter, or virtual condition size, on both examples represents the most extreme condition of assembly at MMC, or the most extreme mating condition of the hole which would accept the mating part feature. Thus, the zero method recognizes, accounts for, and uses the full tolerance advantage. Size tolerances, in particular, are fully utilized with no loss of form, or position tolerance.

Further analysis of zero tolerancing, however, provokes questions that tend to temper some of its advantages. First, for the less experienced or uninitiated user, zero tolerancing presents a psychological barrier: the zero may give a false impression of the "perfection" expected. Second, the designer may feel that he is relinquishing excessively broad discretion to the production departments, thus abdicating design responsibilities in favor of production conveniences such as larger size tolerances. Other comments heard from those hesitant to make full use of zero tolerancing are: it will be misunderstood, it costs more, it is only for gaging convenience, it may result in line-to-line fits.

These objections are primarily due to lack of understanding and experience. Improved understanding and repeated experience with practical applications will readily convince the user of the advantages of zero tolerancing.

Zero form tolerancing, occasionally used on perpendicularity, is not discussed here. However, examples of it are given in appropriate sections of the text. The reasoning underlying zero *orientation* tolerancing is the same as that on which zero position tolerancing is based except for the added consideration of position relationships. A position tolerance, of course, contains orientation tolerance. An understanding of the zero orientation control method, however, might contribute to better acceptance of the zero position principles.

ZERO POSITION

To clarify the reasons underlying the use of the "zero" position method versus the "conventional" position method, further discussion is presented in Figs. 1 and 2.

In Fig. 1, one of the holes illustrated in the 0 method example is shown with reference to the gage pin (or simulated mating part component). It is seen that the zero (0) position specification requires a perfect part (perfect form, orientation, and perfect position) when at MMC, or virtual condition size.

FIGURE 1

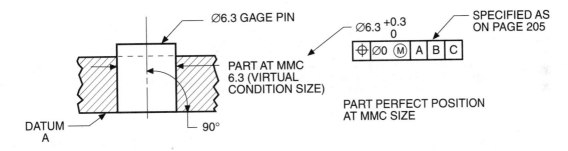

FIGURE 2

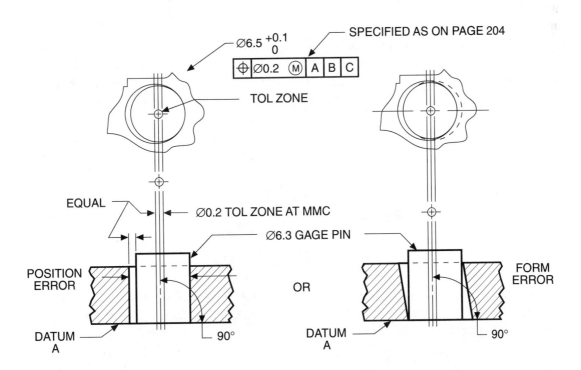

Since there must be *some* clearance between the hole and the inserted mating component or they will not assemble, there is an immediate deviation from the perfect "zero clearance-zero interference" situation and some tolerance is acquired.

Further, since additional tolerance latitude is usually required to assure assembly, we may find it useful to establish a fixed position tolerance. Thus we calculate an acceptable tolerance on the basis of MMC sizes and use the "conventional" position method.

Figure 2 on page 201 illustrates the "conventional" method and the established position tolerance. The tolerance of 0.2 will permit either position or orientation error (or a combination of both) to this extent, when the feature is at MMC. With the same size gage pin as in Fig. 1 on page 201, we see that the position tolerance of 0.2, plus the size tolerance of 0.1, is equivalent to the 0.3 size tolerance obtained by the zero method in Fig. 1.

An application which will demonstrate the effectiveness of zero tolerancing, is a very critical relationship of locating dowels relative to locating holes on a mating part.

As an example, imagine $6.35\,{}^{0}_{-0.038}$ on the locating dowels and $6.4\,{}^{+0.038}_{0}$ on the locating holes. Using the conventional position "fixed fasteners" method, the calculations are:

MMC size hole   6.4
MMC size dowel   6.35
$\overline{}$
            2/0.05
$\overline{}$

=0.025----Position Tolerance on both hole and dowel

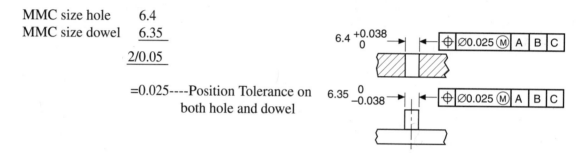

The actual position tolerance in production on both parts would be somewhere between 0.025 and 0.063 (increase due to MMC departure). A functional gage pin size to check the holes would be 6.375 (hole MMC 6.4, MINUS POSN TOL 0.025 = 6.375). Since the gage pin represents the worst condition (virtual condition size) of the mating dowel at 6.375, the hole size could be acceptable functionally at 6.375; yet this exceeds the stated hole size low limit.

The dowel size, too, could be functional at 6.375 (dowel MMC 6.35, PLUS 0.025 = 6.375) which represents the mating part hole at the extreme condition (virtual condition size). This exceeds the stated dowel size high limit.

However, the zero position method can provide more total tolerance and yet guarantee proper control if stated as:

$$6.375\,{}^{0}_{-0.063}\ \ \text{(dowel)}\quad\text{and}\quad 6.375\,{}^{+0.063}_{0}\ \ \text{(hole).}$$

A comparison of the two methods in terms of the full tolerance range difference between the hole and dowel which determines usable size, orientation, and position tolerance is shown on the following page.

FEATURE SIZE RANGE	USABLE SIZE TOL	USABLE POSN TOL

Conventional Method

⌀ Hole   6.4   to   6.438   0.126   (Hole) up to 0.063 DIA

⌀ Dowel   6.35   to   6.312   −0.05   (Dowel) up to 0.063 DIA

              0.05   to   0.126   ⌀0.076 USABLE

                     LOSE .002 SIZE TOL:

Zero Method

⌀ Hole   6.375 to   6.438   ⌀0.126 USABLE   (Hole) up to 0.063 DIA

⌀ Dowel   6.375 to   6.312   (Dowel) up to 0.063 DIA

              0   to   0.126

              FULL UTILIZATION OF SIZE TOL   SAME POSN TOL RESULTS

The foregoing example assumes that there is only a remote probability that both hole and dowel will actually be produced at exactly 6.375; hence some clearance will be present. However, if this should cause concern and some compensation is desired, a slight adjustment of specified size limits (and virtual condition size) can be made to eliminate the problem and ensure that no line-to-line condition will result.

As an alternative to the zero position tolerancing method, some companies have established in-house standards which permit the use of a functional gage as a hole "GO" size gage, thus allowing, for example, the low limit size of a hole to deviate simulating the zero method and the mating part relationship. However, transfer of work to outside sources would then require special documentation to assure that specifications are properly interpreted.

The illustrations on the following pages further illustrate the principles described on page 201–203. Note that the "conventional" method on page 204 establishes a fixed position tolerance which presents the possibility of a part being *acceptable* to a functional gage but *rejectable* due to exceeding the low limit (MMC) size. The zero method on page 205, however, eliminates that problem. It does, however, retain the inherent considerations previously discussed. Use of zero position tolerancing should be based upon the best balance of design/production objectives.

CONVENTIONAL POSITION TOLERANCE APPLICATION FOR COMPARISON WITH ZERO POSITION TOLERANCE

POSITION TOLERANCE (CONVENTIONAL)

EXAMPLE

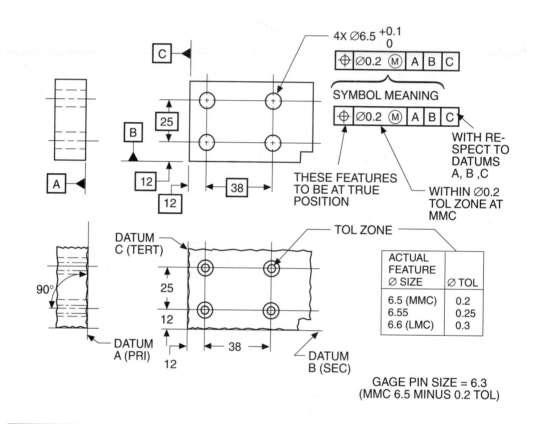

4X ⌀6.5 +0.1 / 0

⊕ ⌀0.2 Ⓜ A B C

SYMBOL MEANING

⊕ ⌀0.2 Ⓜ A B C

WITH RE-SPECT TO DATUMS A, B ,C

THESE FEATURES TO BE AT TRUE POSITION

WITHIN ⌀0.2 TOL ZONE AT MMC

DATUM C (TERT)

TOL ZONE

DATUM A (PRI)

DATUM B (SEC)

ACTUAL FEATURE ⌀ SIZE	⌀ TOL
6.5 (MMC)	0.2
6.55	0.25
6.6 (LMC)	0.3

GAGE PIN SIZE = 6.3
(MMC 6.5 MINUS 0.2 TOL)

MEANING

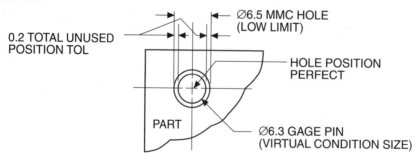

0.2 TOTAL UNUSED POSITION TOL

⌀6.5 MMC HOLE (LOW LIMIT)

HOLE POSITION PERFECT

PART

⌀6.3 GAGE PIN (VIRTUAL CONDITION SIZE)

ASSUMING THAT THE GAGE PIN REPRESENTS THE MOST EXTREME MATING CONDITION, AS POSITION APPROACHES PERFECT, IT IS EVIDENT THAT THE HOLE SIZE COULD GO DOWN TO 6.3 (0.2 BELOW 6.5 LOW LIMIT OF HOLE) AND STILL PASS THE GAGE PINS, HOWEVER, PARTS BELOW THE LOW (MMC) LIMIT HOLE SIZE OF 6.5 WOULD BE REJECTED ON SIZE YET ARE GOOD PARTS.

EXAMPLE

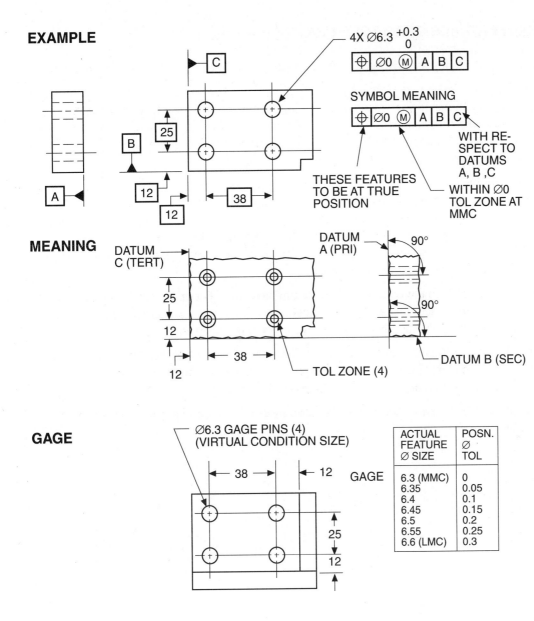

4X ⌀6.3 +0.3 / 0

| ⊕ | ⌀0 Ⓜ | A | B | C |

SYMBOL MEANING

| ⊕ | ⌀0 Ⓜ | A | B | C |

WITH RE-SPECT TO DATUMS A, B ,C

WITHIN ⌀0 TOL ZONE AT MMC

THESE FEATURES TO BE AT TRUE POSITION

MEANING

DATUM C (TERT)

DATUM A (PRI)

90°

90°

25

12

38

12

TOL ZONE (4)

DATUM B (SEC)

GAGE

⌀6.3 GAGE PINS (4) (VIRTUAL CONDITION SIZE)

38

12

GAGE

25

12

ACTUAL FEATURE ⌀ SIZE	POSN. ⌀ TOL
6.3 (MMC)	0
6.35	0.05
6.4	0.1
6.45	0.15
6.5	0.2
6.55	0.25
6.6 (LMC)	0.3

ADVANTAGES OF ZERO POSITIONAL TOLERANCE

1. *Hole "GO" plug gage not needed* if using "GO" functional gage which checks positional location of pattern and hole low limit size simultaneously.

2. *No unused positional tolerance* when using zero positional tolerance method. As locations approach perfection under a conventional (specified pos tol e.g., 0.2) position application, the unused positional tolerance *cannot* be added to the *size* tolerance. Therefore, under some conditions good parts are rejected as they exceed the low limit size.

COAXIAL FEATURES–ZERO TOLERANCE

Zero position tolerancing as illustrated in the preceding examples may also be applied to coaxial features.

Where position relationships between coaxial features must be very exact, zero tolerancing may provide the necessary control. As in other position applications, it is appropriate for mating parts. It has the advantage of fully utilizing *size* tolerances as these interrelate with form, orientation, and position relationships.

The example on the next page shows zero tolerancing applied to a simple part. Position tolerance is acquired as the feature actual sizes (feature and datum feature) depart from MMC. When the part is everywhere at its MMC size, it must be perfectly located. However, in normal applications, the sizes produced are somewhere within the size tolerance range and thus develop allowable positional tolerance. The actually produced sizes establish the position tolerance to the amount of size departure from MMC.

Note that the sizes of both the feature *and* the datum have an effect on the total relationship and the developed tolerance. Just as in all applications of position tolerancing, the part must be produced and its size established before the position tolerance is fixed.

A functional or receiver gage is also illustrated. The principles involved are identical to those in the previous examples. However, with the zero method, the high limit "GO" size of the diameter is also checked, along with the position requirement.

For this application, the use of maximum size tolerances and functional principles has been facilitated by the zero method.

EXAMPLE

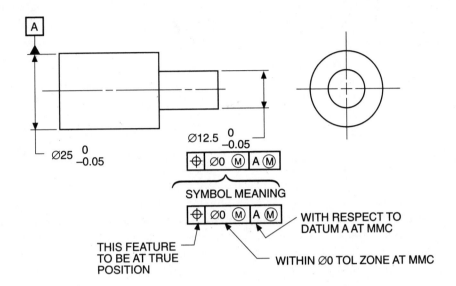

SYMBOL MEANING

THIS FEATURE
TO BE AT TRUE
POSITION

WITHIN Ø0 TOL ZONE AT MMC

WITH RESPECT TO
DATUM A AT MMC

MEANING

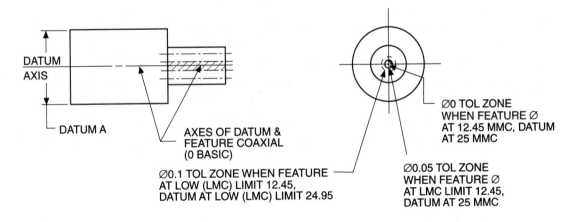

DATUM
AXIS

DATUM A

AXES OF DATUM &
FEATURE COAXIAL
(0 BASIC)

Ø0.1 TOL ZONE WHEN FEATURE
AT LOW (LMC) LIMIT 12.45,
DATUM AT LOW (LMC) LIMIT 24.95

Ø0 TOL ZONE
WHEN FEATURE Ø
AT 12.45 MMC, DATUM
AT 25 MMC

Ø0.05 TOL ZONE
WHEN FEATURE Ø
AT LMC LIMIT 12.45,
DATUM AT 25 MMC

GAGE

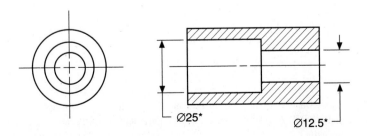

Ø25*

Ø12.5*

*GAGE MAKERS TOL APPLY

POSITION TOLERANCE–PRINCIPLE OF RECIPROCITY [*]

The principle of reciprocity is perhaps the most advantageous method of specifying position tolerances to mating holes or pins which recognizes, and utilizes, the interplay and interdependence of allowable locational and size tolerances. The example at right figure (a) illustrates a part using this principle.

When pins and holes interrelate in assembly, the space filled or cleared in the relationship between these features is a composite value of their surface-to-surface contact as limited by their *size* clearances and the *location* of the axes of these feature surfaces. This principle permits both size and locational tolerances to interact with one another; whereas in the conventional method if called out as shown in the figure (b), can only take advantage of added tolerance as derived from the actual mating envelope size of the feature as it departs from MMC. Any *unused* position tolerance is lost. The use of the symbol ® specifies the "principle of reciprocity" as an extension of the maximum material condition principle. As the feature actual size departs from MMC, the bonus or extra positional tolerance is acquired, as with usual MMC methods. However, with the reciprocity principle, as the calculated positional tolerance as stated on the drawing becomes excess (i.e., all of the positional tolerance is not used), the unused positional tolerance can be used to deviate the actual mating envelope size; e.g., in the case of the hole, reduce its actual mating envelope size an amount equal to the unused positional tolerance stated on the drawing. The columns headed Ⓜ and Ⓜ ® in figure (d) can be compared to see the resulting difference between these two methods.

Where zero positional tolerancing is used, see fig (c) and right hand column of fig (d), note that the size nominal value is reduced and specified at ⌀ 7.49 instead of ⌀ 7.62. This size reduction transfers the ⌀ 0.13 position tolerance to smaller size and makes the positional tolerance zero.

Each of these options as shown under the three columns of fig (d) have a distinct use as dependent upon the designer's choice as to which best suits the design requirement. Earlier discussion in this text gives detailed explanation of the option between conventional and zero positional tolerancing. Those differences are summarized here under the first and third columns of fig (d) for comparison to the reciprocity method (in the middle column).

The reciprocity method would typically be used where the designer desires this reciprocal interaction between size and location seeking the dynamic results as described. With such an application, size deviation is permitted to be a variable, widening the size tolerance range only as necessary.

Note that if a functional position tolerance gage were used, the exact same gage could be used on any of the three methods; the virtual condition ⌀ 7.49 would be consistent to each. If CMM options were used, soft-gaging computer simulation of the reciprocal method could be readily achieved with the actual mating envelope size and locational data derived. The "hard" functional gage shown represents the physical embodiment of how size and position variation can occur with their collective variations permitted in whatever manner and amount, so long as there is no violation of ⌀ 7.49, virtual condition, and the hole LMC has also been respected.

The principle of reciprocity can also be applied on an LMC basis (i.e., Ⓛ ®) although such use would be somewhat unusual and be rather complex. If done, the methodology shown at right would be reversed in the tabulated data and the LMC symbology appropriately stated.

---

[*] Author advisory: The principle of reciprocity method is included here as advisory and for information only. It is not included in the USA standards. At the time of this writing, this concept is proposed for addition to ISO standard ISO 2692, Maximum Material Requirement. This concept was earlier found in older texts by the author under the title "Maximum True Position Principle" in texts dated 1966. It was presented as a USA paper in ISO deliberations around 1971 in Zurich, Switzerland. At that time it was considered too advanced for inclusion in ISO standards. Mr. Matsumoto of Japan and Mr. Eberle of Switzerland also put forth this idea around that time. Mr. Matsumoto described the method the "Extreme Positional Method" and suggested the symbol Ⓔ.

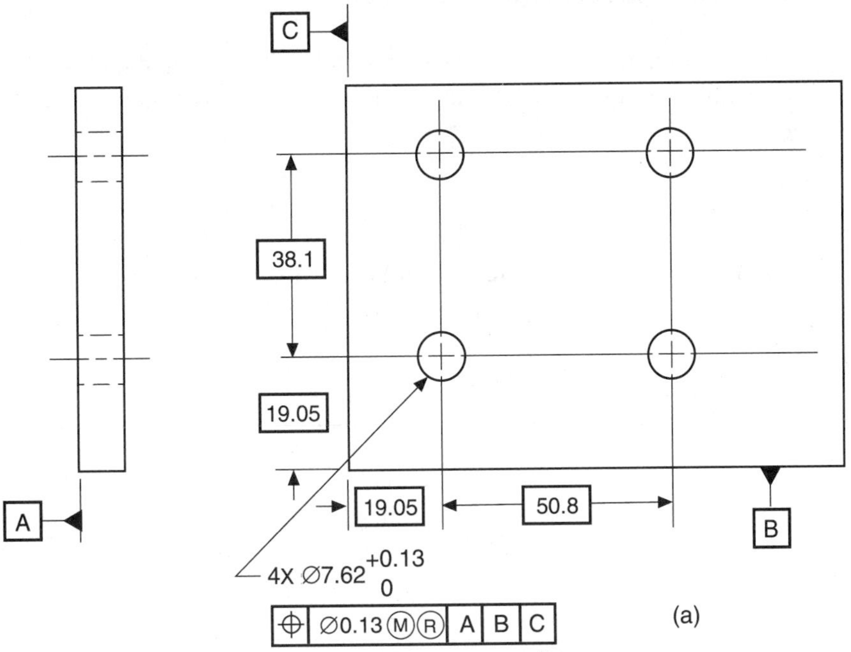

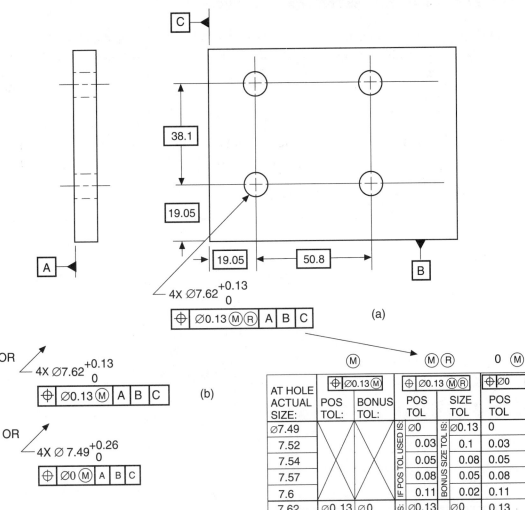

(a)

OR

4X Ø7.62 +0.13 0

| ⊕ | Ø0.13 Ⓜ | A | B | C |

(b)

OR

4X Ø 7.49 +0.26 0

| ⊕ | Ø0 Ⓜ | A | B | C |

	Ⓜ		Ⓜ Ⓡ		0 Ⓜ
	⊕ Ø0.13 Ⓜ		⊕ Ø0.13 Ⓜ Ⓡ		⊕ Ø0 Ⓜ
AT HOLE ACTUAL SIZE:	POS TOL:	BONUS TOL:	POS TOL	SIZE TOL	POS TOL
			IF POS TOL USED IS:	BONUS SIZE TOL IS:	
Ø7.49			Ø0	Ø0.13	0
7.52			0.03	0.1	0.03
7.54			0.05	0.08	0.05
7.57			0.08	0.05	0.08
7.6			0.11	0.02	0.11
			IF POS TOL ALLOWED IS:	BONUS POS TOL IS:	
7.62	Ø0.13	Ø0	Ø0.13	Ø0	0.13
7.64	0.15	0.02	0.15	0.02	0.15
7.67	0.18	0.05	0.18	0.05	0.18
7.7	0.21	0.08	0.21	0.08	0.21
7.72	0.23	0.1	0.23	0.1	0.23
7.75	0.26	0.13	0.26	0.13	0.26

(d)

GAGE

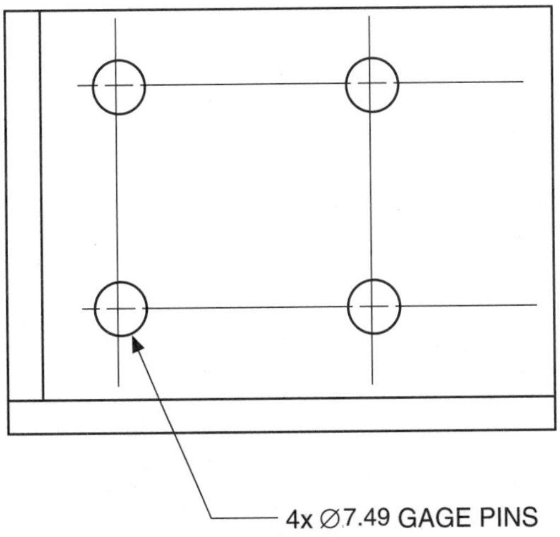

(e)

NOTE: SAME POSITIONAL TOL FUNCTIONAL GAGE USED FOR ALL THREE APPLICATIONS. THE EFFECT AND USE OF THE SIZE TOLERANCE DIFFERS. FOR RECIPROCITY THE FUNCTIONAL GAGE WOULD CHECK ONLY A WORST CASE LIMIT. SINCE THE TOLERANCES ARE DYNAMIC VARIABLES, DEPENDENT UPON ONE ANOTHER AS THE ACTUAL MATING SIZE PRODUCED AND THE AMOUNT OF UNUSED POSITIONAL TOLERANCE INTERMIX, ONLY A CMM, OR EQUIVALENT ANALYSIS, COULD TRULY AND COMPLETELY VERIFY RECIPROCITY REQUIREMENTS.

(f)

MATING PARTS, FIXED FASTENERS

The following series of figures further illustrates mating part situations and the use of the fixed fastener method of calculation. It covers the typical applications listed below.

MATING PARTS, FIXED FASTENERS

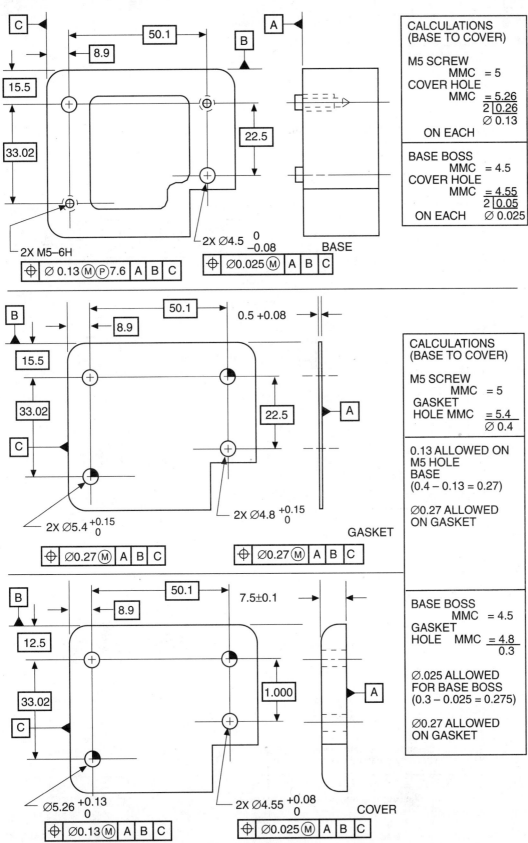

CALCULATIONS
(BASE TO COVER)

M5 SCREW
MMC = 5
COVER HOLE
MMC = 5.26
2 | 0.26
⌀ 0.13
ON EACH

BASE BOSS
MMC = 4.5
COVER HOLE
MMC = 4.55
2 | 0.05
ON EACH ⌀ 0.025

CALCULATIONS
(BASE TO COVER)

M5 SCREW
MMC = 5
GASKET
HOLE MMC = 5.4
⌀ 0.4

0.13 ALLOWED ON
M5 HOLE
BASE
(0.4 − 0.13 = 0.27)

⌀0.27 ALLOWED
ON GASKET

BASE BOSS
MMC = 4.5
GASKET
HOLE MMC = 4.8
0.3

⌀.025 ALLOWED
FOR BASE BOSS
(0.3 − 0.025 = 0.275)

⌀0.27 ALLOWED
ON GASKET

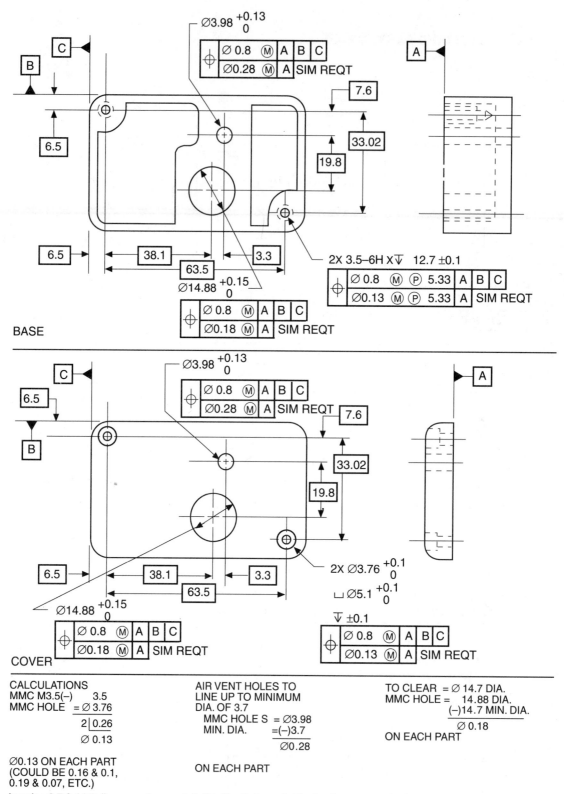

BASE

COVER

CALCULATIONS
MMC M3.5(–)   3.5
MMC HOLE = Ø 3.76
                    2 | 0.26
                    Ø 0.13

Ø0.13 ON EACH PART
(COULD BE 0.16 & 0.1,
0.19 & 0.07, ETC.)

AIR VENT HOLES TO
LINE UP TO MINIMUM
DIA. OF 3.7
   MMC HOLE S = Ø3.98
   MIN. DIA.   =(–)3.7
                      Ø0.28

ON EACH PART

TO CLEAR = Ø 14.7 DIA.
MMC HOLE =   14.88 DIA.
                    (–)14.7 MIN. DIA.
                    Ø 0.18
ON EACH PART

Imagine 3.7 & 14.7 dias as columns of air (floating fasteners). Floating fastener method may be used since overlap of edges is permissible; one dia need not clear the other so long as the 3.7 & 14.7 dias will pass.

NOTE: SIM REQT= Simultaneous requirement for all indicated feature relating tolerances as single pattern.

MATING PARTS AND FIXED FASTENERS

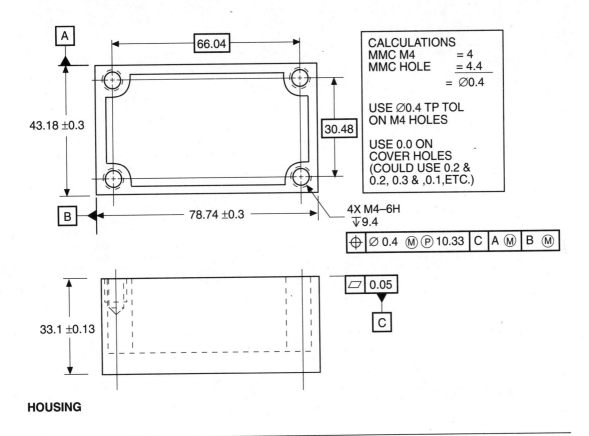

CALCULATIONS
MMC M4   = 4
MMC HOLE = 4.4
          = ⌀0.4

USE ⌀0.4 TP TOL
ON M4 HOLES

USE 0.0 ON
COVER HOLES
(COULD USE 0.2 &
0.2, 0.3 & ,0.1,ETC.)

HOUSING

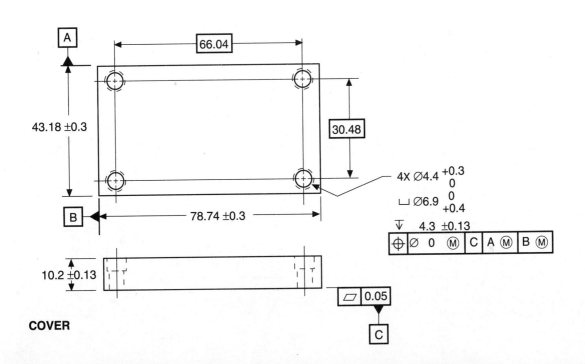

COVER

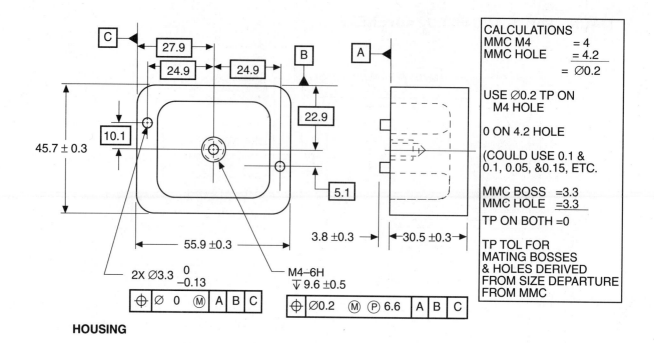

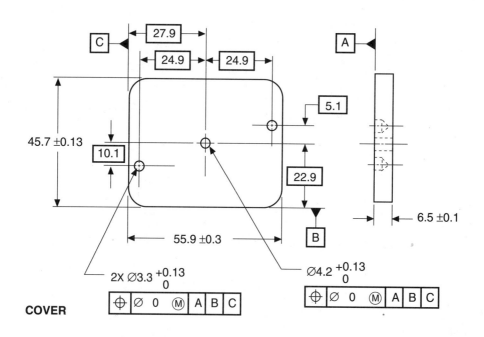

HOUSING

COVER

COMBINATION AND UNIQUE APPLICATIONS

Position tolerancing may be applied in numerous ways. The principles are extended in the following illustrations to unique combinations of DIA and TOTAL tolerances, square holes, etc. Note that the established principles can be adapted to many different situations.

ROUND HOLES, GREATER POSITIONAL TOLERANCE IN ONE DIRECTION THAN IN THE OTHER—BIDIRECTIONAL

EXAMPLE

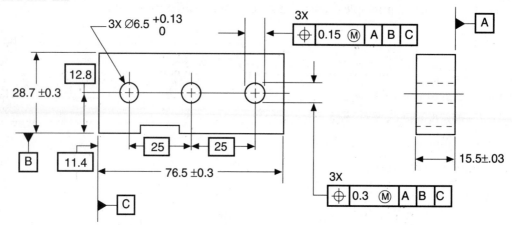

MEANING

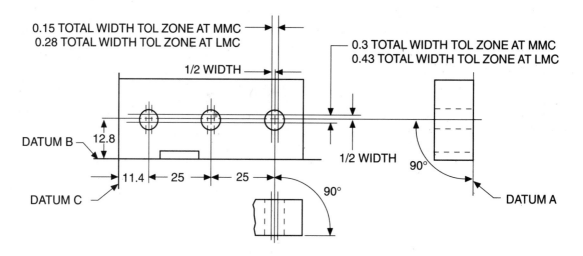

GAGE

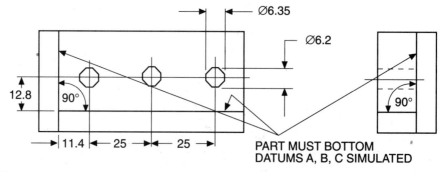

NOTE: Gage makers tol apply

DIAMETER AND TOTAL WIDE TOLERANCE ZONES, ROUND HOLES AND ELONGATED HOLES—BIDIRECTIONAL

EXAMPLE

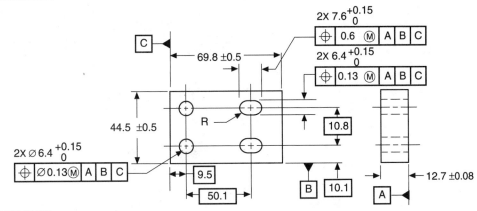

MEANING

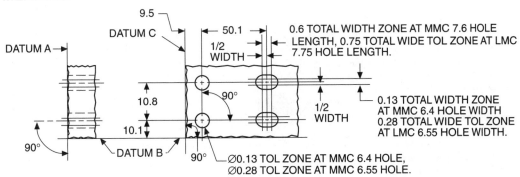

NOTE: Remaining part profile tolerance zones are determined by the overall size limits

GAGE

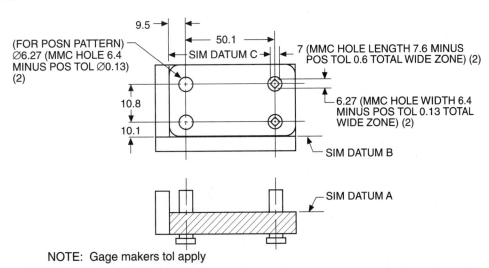

NOTE: Gage makers tol apply

ELONGATED HOLES, BOUNDARY TOLERANCE ZONES

EXAMPLE

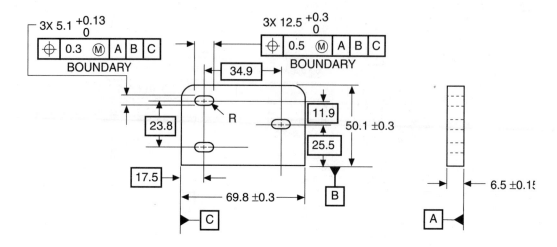

MEANING

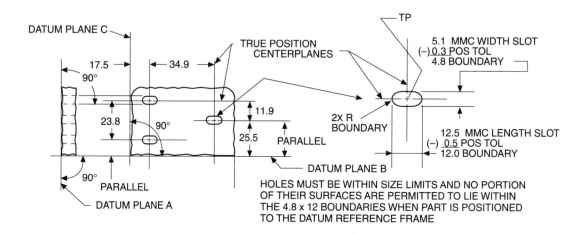

HOLES MUST BE WITHIN SIZE LIMITS AND NO PORTION
OF THEIR SURFACES ARE PERMITTED TO LIE WITHIN
THE 4.8 x 12 BOUNDARIES WHEN PART IS POSITIONED
TO THE DATUM REFERENCE FRAME

GAGE

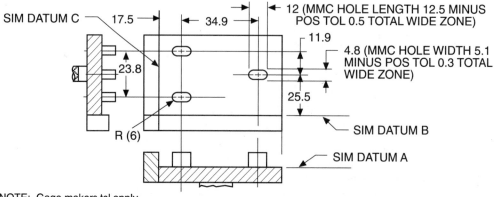

NOTE: Gage makers tol apply

PATTERN AND FEATURE ORIENTATION FROM DATUM SURFACES

USING PART PROFILE AS DATUM

EXAMPLE

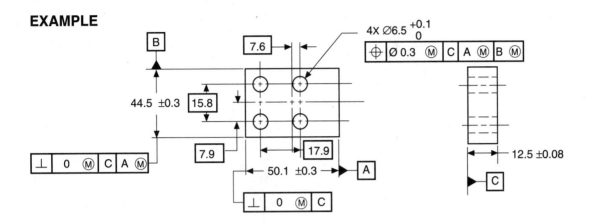

MEANING

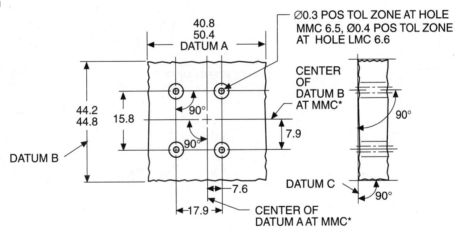

*Centers may shift within a zone equal to the departure of datum features A & B sizes from MMC.

GAGE

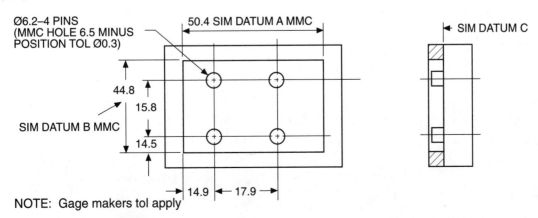

NOTE: Gage makers tol apply

TOTAL WIDE TOLERANCE ZONES, FEATURES TO BE SYMMETRICALLY LOCATED—BIDIRECTIONAL

EXAMPLE

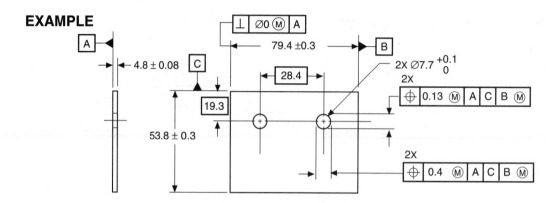

MEANING

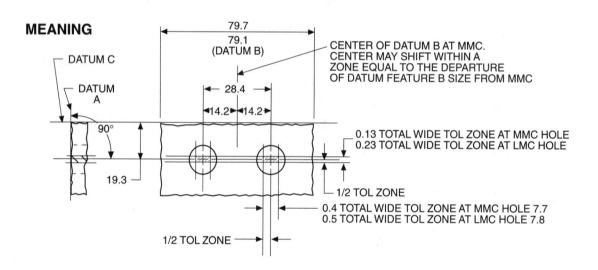

GAGE

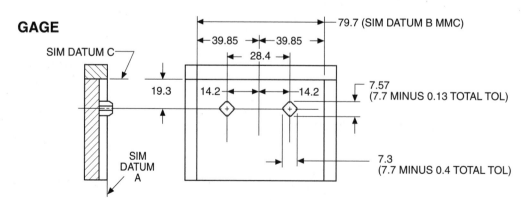

NOTE: Gage makers tol apply

SQUARE HOLES, TOTAL WIDE ZONE POSITIONAL TOLERANCE

EXAMPLE

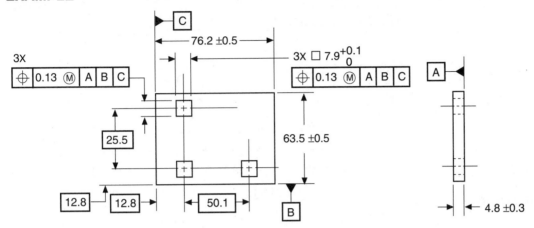

MEANING

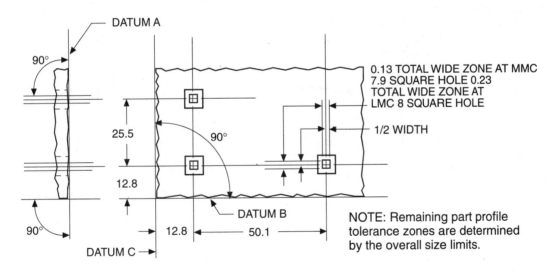

0.13 TOTAL WIDE ZONE AT MMC 7.9 SQUARE HOLE 0.23 TOTAL WIDE ZONE AT LMC 8 SQUARE HOLE

1/2 WIDTH

NOTE: Remaining part profile tolerance zones are determined by the overall size limits.

GAGE

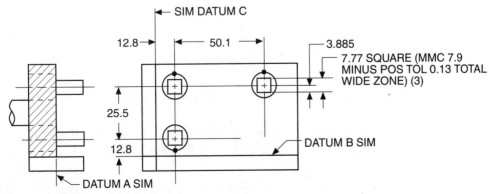

NOTE: Gage makers tol apply

SQUARE HOLES, GREATER POSITIONAL TOLERANCE ONE DIRECTION THAN ANOTHER—BIDIRECTIONAL

EXAMPLE

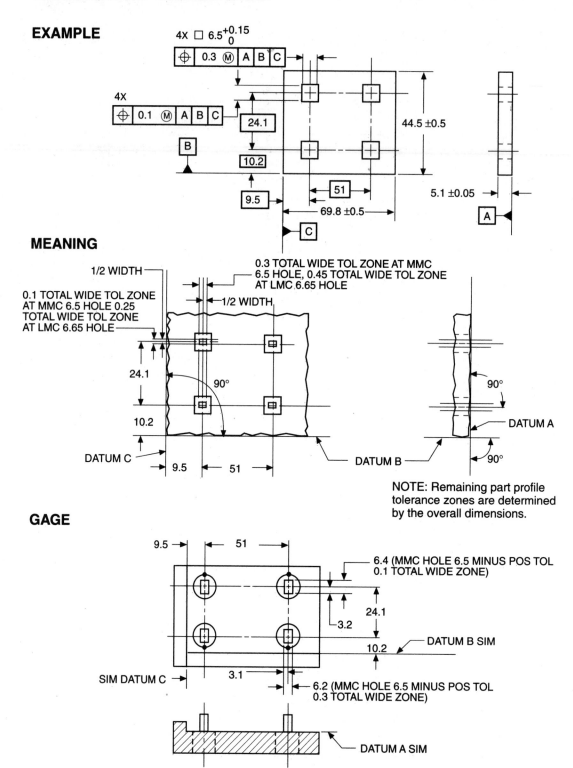

MEANING

GAGE

NOTE: Remaining part profile tolerance zones are determined by the overall dimensions.

NOTE: Gage makers tol apply

DATUMS

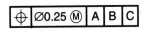

DATUMS

In the previous sections we have made continual reference to datums and have implied the importance of their use. Intelligent and effective application of the principles of geometric dimensioning and tolerancing depends to a great extent on a good understanding of the kinds of datums used, their definitions, proper selections, and the interpretations implied by the various uses.

The first portion of this section on datums covers the various kinds of datums, their definitions, establishment, relationship to each other and to measuring planes, and their meaning.

The last portion of this section will cover applications, selection, and further sample part examples.

DATUMS

A datum is a theoretically exact point, axis, or plane derived from the true geometric counterpart of a specified datum feature. A datum is the origin from which the location of geometric characteristics of features of a part are established. Datums are established by specified features or surfaces. Where orientation or position relationships are specified from a datum, the features involved are located with respect to this datum and not with respect to one another.

Every feature on a part can be considered a possible datum. That is, every feature shown on a drawing depicts a theoretically exact geometric shape as specified by the design requirements. However, a feature normally has no practical meaning as a datum unless it is actually used for some functional relationship between features. Thus a datum appearing on an engineering drawing can be considered to have a dual nature: it is (1) a "construction" datum, which is the geometrically exact representation of any part feature, and (2) a "relationship" datum, which is any feature used as a basis for a functional relationship with other features on the part. Since the datum concept is used to establish relationships, the "relationship" datum is the only type used on engineering drawings.

By the above definition, a datum on an engineering drawing is always assumed to be "perfect." However, since perfect parts cannot be produced, a datum on a physically produced part is assumed to exist in the contact of the actual feature surface with precise manufacturing or inspection equipment such as machine tables, surface plates, gage pins, etc. These are called datum simulators which create simulated datum planes, axes, etc., and, while not perfectly true, are usually of such high quality that they adequately *simulate* true references. This contact of the actual feature with precise equipment is also assumed to simulate functional contact with a mating part surface.

DATUM FEATURE (SURFACE HOLE, SLOT, DIAMETER, ETC.)

A *datum feature* is an actual feature of a part that is used to establish a *datum*. A *feature* is the general term applied to a physical portion of a part, such as a surface, hole, or slot.

By the definition of a DATUM, which is assumed to be theoretically exact, a "datum surface" or a "datum feature" is technically *not* the datum. These terms are, however, used in referring to the actual part feature or surface *from which* a datum is established when the feature or surface is used as a reference or contact with a tool, surface plate, or other checking device (the datum simulator) to establish simulated datum planes or axes. A "datum surface" or "datum feature" refers to the actual part feature or surface used to *establish* the datum. Since "datum surfaces" and "datum features" are actual things, they cannot be perfect, and thus they include all the real irregularities and inaccuracies of the surface or feature. (See examples under DATUM PLANE and further explanations following.)

DATUM POINT

A datum point is that which has position but no extent, such as the apex of a pyramid or cone, the center point of a sphere, or a reference point on a surface for functional, tooling, or gaging purposes.

The "apex of a pyramid or cone," "the center point of a sphere," or random points, lines or areas (extremities) on a feature surface are considered *construction* datums, which are used to construct the geometry of the feature. "A reference point on a surface for functional, tooling, or gaging purposes" is considered a functional *relationship* datum and has meaning in the drawing specification. Such datums must be specified.

An example of a construction datum point and a specified datum point are shown following. Construction datum points are usually the random extremities or high points of an actual specified datum feature, which may be used to establish primary, secondary, and tertiary datum planes of orientation for the part. They occur at random locations as a part of the actually produced datum surfaces. Using the principles of geometry, we establish a "primary" datum plane by three (minimum) datum points, a "secondary" datum plane by two (minimum) datum points, and a "tertiary," or third, datum plane by one (minimum) datum point.

Where datum planes are to be established for specific locations on a part surface, be it for design, or for manufacturing and inspection repeatability, they may be *specified* as datum target points, using datum target symbols, as shown. Location dimensions are required with specified datum target points.

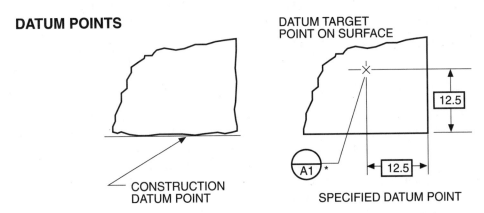

DATUM POINTS

DATUM TARGET
POINT ON SURFACE

12.5

A1 *

12.5

CONSTRUCTION
DATUM POINT

SPECIFIED DATUM POINT

*SPECIFIED DATUM TARGET, SEE FOLLOWING PAGES FOR EXPLANATION.

DATUM LINE, DATUM AXIS

A datum line is that which has length but not breadth or depth, such as the intersection line of two planes, the center line or axis of holes or cylinders, or a reference line for functional, tooling, or gaging purposes.

Examples showing a *construction* datum line and a *specified* datum line shown on the following page. Note that a hole (part a) contains a datum center line or axis (as every round hole does) which is derived from the intersection of two imaginary planes at an angle of 90°. The datum center line is also an axis as derived from the geometric center of a cylinder. These are construction datums inherent in the geometry.

However, unless the hole is *specified* as a datum feature (part c) with a relationship to other features, it has no meaning as a datum. When stated as datums, the same construction datums that are shown in part (a) become part of the requirement as established from the actual datum hole surface. (See also DATUM AXIS in following text.)

Where a specific datum line is to be established on a surface for design function, or manufacturing and inspection repeatability, it may be *specified* as a datum target line, as shown in part (b).

DATUM LINE, DATUM AXIS (continued)

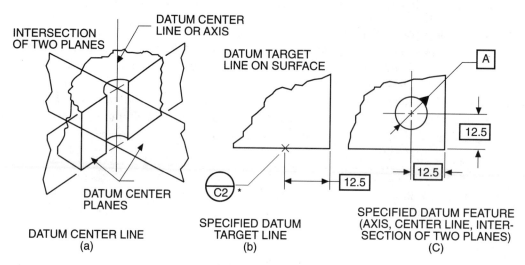

DATUM CENTER LINE
(a)

SPECIFIED DATUM
TARGET LINE
(b)

SPECIFIED DATUM FEATURE
(AXIS, CENTER LINE, INTER-
SECTION OF TWO PLANES)
(C)

*SPECIFIED DATUM TARGET, SEE FOLLOWING PAGES FOR EXPLANATION.

DATUM AXIS

A datum axis is a theoretically exact axis derived from the true geometric counterpart of a specified datum feature (i.e., a cylinder). A datum axis is the origin from which the relationship of geometric characteristics of features of a part are established. The simulated datum axis is the center of the simulated datum cylinder as established by extremities or contacting points of the datum feature simulator cylindrical surface. For an external feature, the datum is the axis of the smallest circumscribed cylinder which contacts the feature surface. For an internal feature, the datum is the axis of the largest inscribed cylinder which contacts the feature surface.

Since measurements or reference cannot be made from *theoretical* cylinders, etc., they are therefore assumed to exist, not in the part itself, but in the contact of the part with more precise manufacturing or inspection equipment (the datum simulators). Fixture or gage pins, or gage cylinders, are *not* true cylinders, but they are usually of such high quality that they adequately simulate true cylinders, etc., and therefore are considered "true." These references are called *simulated* datum cylinders and axes. See following further illustrations and explanation.

Author advisory: Any physical differences between the "true geometric counterpart of the datum feature," the "simulated datum feature," and the "simulated datum plane" or "simulated datum axis" represent a necessary compromise between unpredictable part, tool and gage surface precision variables versus practical application within human capabilities to approach such finite exactness. To attempt to bridge this gap between the theory and the realities involved, the "simulated datum feature," the "simulated datum plane," and the "true geometric counterpart of the datum feature" are, for all practical purposes, normally considered consolidated into one composite result when applied under typical every day conditions. This means that when the "simulated datum feature (via the datum feature simulator)" and the part "datum feature" are brought into contact, "the datum (point, axis, or plane)" is considered to be, thus, established about as well as it can be done under normal operating conditions and constraints. Where sophisticated computerized electronic verification or evaluation methods are undertaken, it may be possible and/or necessary to detect and respect such minute differences. Under these conditions, results are based upon recognition, detection, and distinction of these values. Because of the above referenced electronic options and more sophisticated requirements arising in industry, the standards must recognize this more detailed nomenclature to adequately describe the datum concepts where necessary.

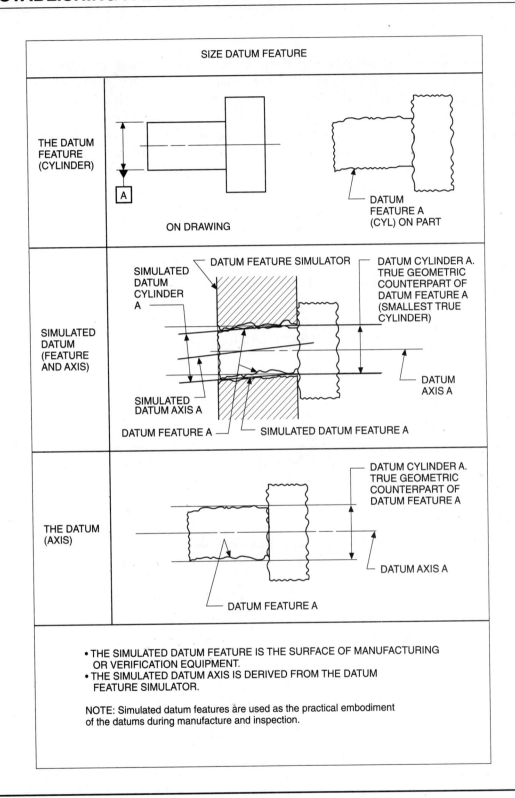

SIZE DATUM FEATURE	
THE DATUM FEATURE (CYLINDER)	ON DRAWING — DATUM FEATURE A (CYL) ON PART
SIMULATED DATUM (FEATURE AND AXIS)	SIMULATED DATUM CYLINDER A — DATUM FEATURE SIMULATOR — DATUM CYLINDER A. TRUE GEOMETRIC COUNTERPART OF DATUM FEATURE A (SMALLEST TRUE CYLINDER) — DATUM AXIS A — SIMULATED DATUM AXIS A — DATUM FEATURE A — SIMULATED DATUM FEATURE A
THE DATUM (AXIS)	DATUM CYLINDER A. TRUE GEOMETRIC COUNTERPART OF DATUM FEATURE A — DATUM AXIS A — DATUM FEATURE A

- THE SIMULATED DATUM FEATURE IS THE SURFACE OF MANUFACTURING OR VERIFICATION EQUIPMENT.
- THE SIMULATED DATUM AXIS IS DERIVED FROM THE DATUM FEATURE SIMULATOR.

NOTE: Simulated datum features are used as the practical embodiment of the datums during manufacture and inspection.

DATUM PLANE

A datum plane, when developed from a primary datum feature which is nominally flat (a "plane" surface) is a theoretically exact plane established by the extremities or contacting points of the datum feature surface with a *simulated* datum plane. The simulated datum plane is constructed from the datum simulator (surface plate or other checking device).

229

Since measurements or references cannot be made from *theoretical* planes, the planes are therefore assumed to exist, not in the part itself, but in the contact of the part with more precise manufacturing or inspection equipment. Machine tables, surface plates, fixture surfaces, etc., are *not* true planes, but they are usually of such high quality that they adequately simulate true planes and are therefore considered true references. These references are called datum simulators and construct the *simulated* datum planes. Such reference planes can also be developed by coordinate movements of tooling or gaging equipment, such as on a coordinate measuring machine.

In the example below, the desired surface is shown as a datum feature and a datum plane. Also shown is the establishment of the datum plane from the actual part feature surface through contact of the surface extremities with the simulated datum plane. A relationship of this datum plane to other features on the part would be shown on the drawing.

DATUM PLANE ESTABLISHED FROM A DATUM FEATURE (SURFACE)

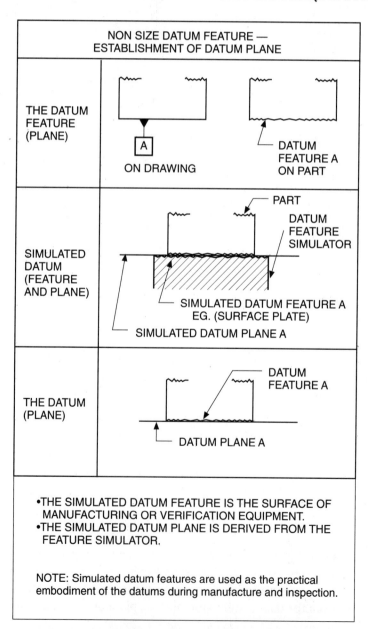

POSITION TOLERANCE TO DATUM CENTER PLANE, RFS

DATUM CENTER PLANE

A datum center plane is the theoretically exact center of the true geometric counterpart of the specified datum feature as established via the simulated datum. For an external width feature, the datum is the center plane between two parallel planes which, at *minimum* separation, contact the corresponding surfaces of the feature.

EXAMPLE

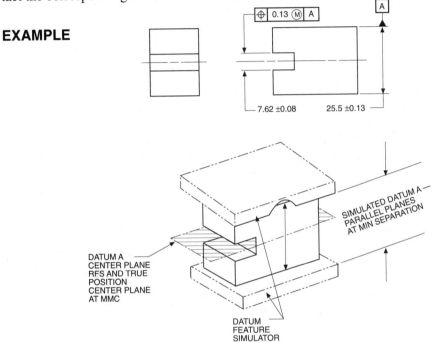

For an internal feature, the datum is the center plane between two parallel planes which, at *maximum* separation, contact the corresponding surfaces of the feature.

EXAMPLE

MEANING

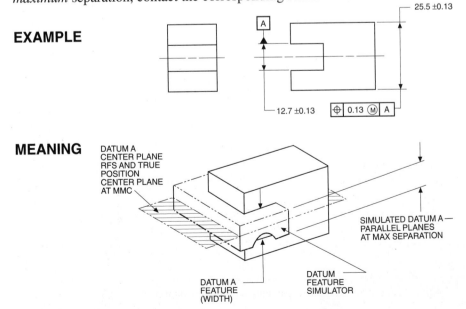

ESTABLISHING DATUM PLANES FROM DATUM SURFACES— DATUM REFERENCE FRAME

Datum planes are theoretically perfect reference planes. Datum planes are always established by, or are relative to, actual or physical features. The most common datum plane is the type established from a datum surface.

The example under "DATUM PLANE" illustrated the conventional establishment of a primary datum plane from a primary datum surface. In establishing datum planes in relation to defining or measuring a part, at least two and usually three datum and measuring planes are considered in locating features. In part configuration, other than cylindrical, there are usually three planes of orientation. These three planes conform to the relationship of the conventional geometric X, Y, and Z axes and resulting planes of orientation.

The three planes are referred to as a datum reference frame and are composed of a primary, secondary, and tertiary (third) datum plane as established from the appropriate actual datum surfaces. Unless otherwise specified or controlled, the largest or most important surface is usually selected as the primary datum, the next largest or most important as the secondary datum, and the remaining surface as the tertiary plane. Design functional requirements should be the first criterion for the establishment of datum priorities. Where the datums are to be specified, they are identified with appropriate letters as previously discussed.

All datum planes contained within a datum reference frame are 90° BASIC, or perpendicular, to each other by interpretation. The dimension lines or center lines related to and/or shown perpendicular to these datum planes are implied as 90° BASIC, or perpendicular, to the datum planes. The resulting measuring planes are therefore considered to be mutually perpendicular to one another.

The illustration shows the establishment, orientation, and relationship of three datum planes from the part surfaces to determine the datum reference frame. The relationship of the measuring planes to the datum planes is also illustrated.

The PRIMARY datum surface A establishes the first relationship of the part for proper orientation. The three (minimum) extremities or contact points of the surface establish the primary datum *plane* A.

The SECONDARY datum surface B establishes a further relationship of the part for proper orientation. The two (minimum) extremities or contact points of the surface establish the secondary datum *plane* B at 90° BASIC to datum plane A.

The TERTIARY (third) datum surface C completes the part orientation. One (at least) extremity or contact point of the surface establishes the third datum *plane* C at 90° BASIC to both datum planes A and B.

Auxiliary datums, specified with respect to feature relationships not related to or functional with the part main datum reference frame, would not be considered 90° BASIC to the main datum reference frame, but rather as 90° BASIC to the auxiliary datum reference frame.

ESTABLISHING DATUM PLANES FROM DATUM SURFACES/FEATURES—THREE PLANE CONCEPT—DATUM REFERENCE FRAME

EXAMPLE

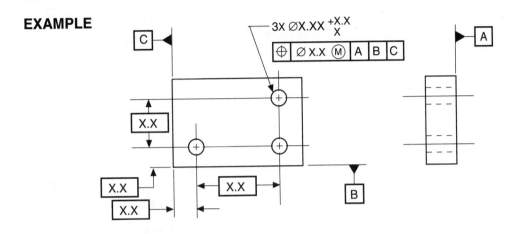

ESTABLISHING THE DATUM PLANES

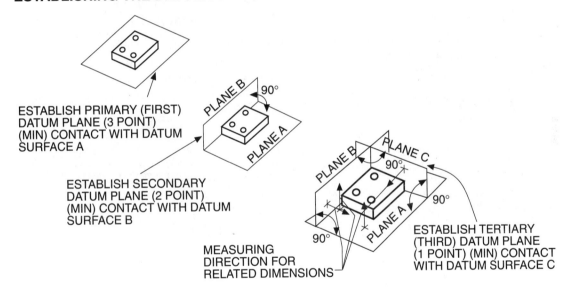

ESTABLISH PRIMARY (FIRST) DATUM PLANE (3 POINT) (MIN) CONTACT WITH DATUM SURFACE A

ESTABLISH SECONDARY DATUM PLANE (2 POINT) (MIN) CONTACT WITH DATUM SURFACE B

MEASURING DIRECTION FOR RELATED DIMENSIONS

ESTABLISH TERTIARY (THIRD) DATUM PLANE (1 POINT) (MIN) CONTACT WITH DATUM SURFACE C

ESTABLISHING DATUM PLANES FROM DATUM POINTS, LINES, OR AREAS (PARTIAL DATUM SURFACES) USING DATUM TARGETS

Where datum orientation is required on parts of irregular contour, such as castings, forgings, sheet metal, etc., datum targets provide a valuable tool. Specified datum targets which serve as means of constructing special datum planes of orientation can be of three types: points, lines, or areas. Datum targets establish the necessary datum reference frame and, in addition, ensure repeatable part location for manufacturing and inspection operations.

Datum targets are also used to indicate special, or more critical, design requirements where functional part feature relationships are to be indicated from specific points, lines, or areas on the part surface.

Datum points and lines have been previously defined. A datum area is a datum established from a partial datum surface. On a drawing, a datum area is outlined with phantom lines and identified by diagonal slash lines. It may be of any shape.

The previously mentioned geometric rules of establishing the datum planes of orientation (primary, secondary, tertiary) must be observed. However, where datum targets are used, particularly on irregular castings or forgings, the number of targets may vary from the conventional geometrical 3–2–1 point orientation as determined by the part configuration.

The locations and/or sizes of datum points, lines, or areas are controlled by BASIC or untoleranced dimensions and imply exactness within standard tooling, gaging, or shop tolerances. Toleranced locations or sizes may be used with datum target symbols if such practice is preferred.

The drawings below illustrate the use of datum target symbols to establish datum planes and part orientation.

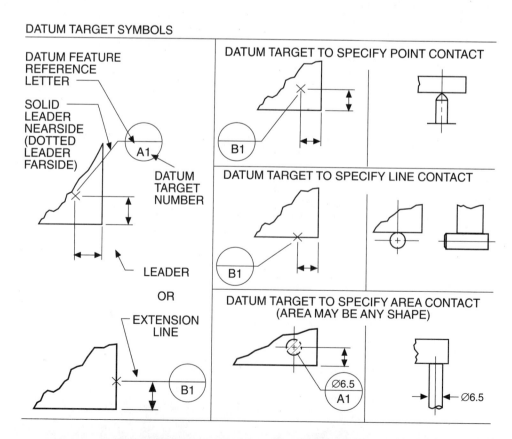

DATUM TARGET SYMBOLS

DATUM FEATURE REFERENCE LETTER

SOLID LEADER NEARSIDE (DOTTED LEADER FARSIDE)

A1

DATUM TARGET NUMBER

LEADER

OR

EXTENSION LINE

B1

DATUM TARGET TO SPECIFY POINT CONTACT

B1

DATUM TARGET TO SPECIFY LINE CONTACT

B1

DATUM TARGET TO SPECIFY AREA CONTACT (AREA MAY BE ANY SHAPE)

Ø6.5

A1

Ø6.5

DATUM TARGETS

PART ORIENTATION TO DATUM PLANES AS ESTABLISHED BY POINTS, LINES, OR AREAS

DATUM PLANES ESTABLISHED FROM DATUM POINTS AND AREAS ON NOMINALLY FLAT DATUM SURFACES:

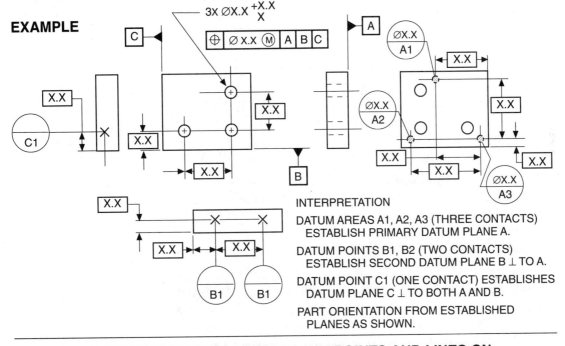

INTERPRETATION

DATUM AREAS A1, A2, A3 (THREE CONTACTS)
  ESTABLISH PRIMARY DATUM PLANE A.

DATUM POINTS B1, B2 (TWO CONTACTS)
  ESTABLISH SECOND DATUM PLANE B ⊥ TO A.

DATUM POINT C1 (ONE CONTACT) ESTABLISHES
  DATUM PLANE C ⊥ TO BOTH A AND B.

PART ORIENTATION FROM ESTABLISHED
  PLANES AS SHOWN.

DATUM PLANES ESTABLISHED FROM DATUM POINTS AND LINES ON IRREGULAR CONTOUR SURFACES:

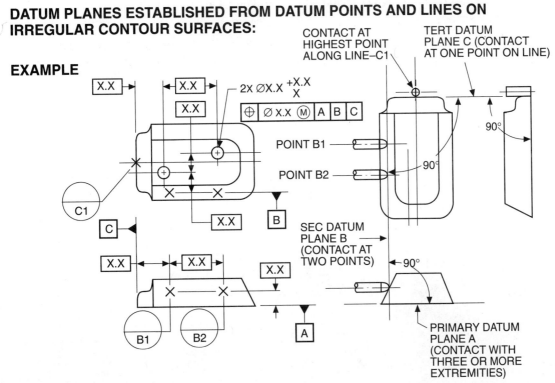

ESTABLISHING A DATUM AXIS FROM DATUM FEATURES

For a rectangular part, the three reference planes are readily visualized since the flat surfaces of the part resemble datum planes, and orientation to the three geometric planes X, Y, and Z is obvious. However, this is not true for cylindrical part surfaces since they bear no resemblance to datum planes and their relationship to the X, Y, and Z planes of orientation is not so obvious. Due to the geometry of a cylinder, a cylindrical part may be assumed to be simultaneously located in two planes of orientation. These two planes may be visualized as center planes intersecting 90° BASIC to each other at the datum axis. As discussed in the section *Datum Axis*, on outside cylinders the datum axis is established from the minimum circumscribed simulated datum cylinder contacting the actual cylindrical surface (RFS). On inside cylinders, it is established from the maximum inscribed simulated datum cylinder contacting the actual cylindrical surface (RFS).

Where the MMC principle is to be applied, the datum axis of a cylindrical part is established by the axis of the datum surface cylinder at MMC size. Under this condition, the actually produced feature axis may vary within a tolerance zone equal to the departure from MMC size of the feature. This situation is illustrated in the example, PRIMARY DATUM FEATURE MMC.

The illustrations following expand upon the application of datums to cylindrical parts. The examples and the illustrated interpretation of each should be self-explanatory. Note that a cylindrical feature, due to its geometry, establishes orientation in two directions simultaneously: the two imaginary datum construction planes intersect at the datum axis. However, only one datum letter identification is used to establish the datum feature (cylinder).

The examples also illustrate the use of a secondary datum plane with the primary datum feature (cylinder) and the use of the feature as *secondary* datum. In the latter instance, note that the secondary datum feature (cylinder) (perpendicular, 90° BASIC to the primary datum plane) may have its axis at variance with the *actual* feature axis. This possibility, however, is recognized and is a part of the design criterion whenever datums are used in this sequence on a part of cylindrical shape.

Where rotational orientation of the cylindrical part relative to features is required, a tertiary datum is necessary. The example, TERTIARY (THIRD) DATUM FEATURES RFS on pages 239 and 275, illustrates this type of requirement.

ESTABLISHING A DATUM AXIS FROM DATUM FEATURES

PRIMARY DATUM FEATURE RFS

PART

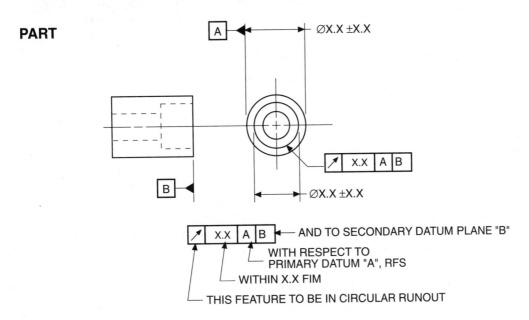

| ↗ | X.X | A | B | ← AND TO SECONDARY DATUM PLANE "B"

WITH RESPECT TO
PRIMARY DATUM "A", RFS

WITHIN X.X FIM

THIS FEATURE TO BE IN CIRCULAR RUNOUT

MEANING

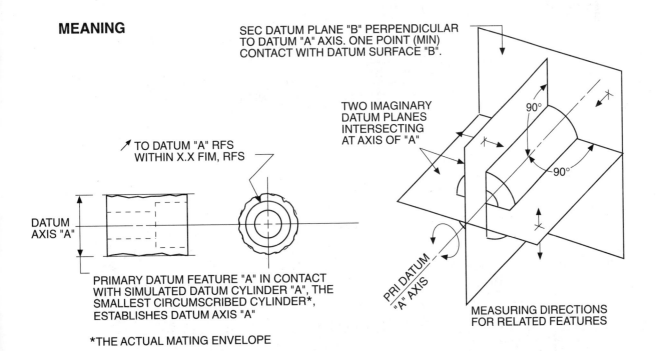

SEC DATUM PLANE "B" PERPENDICULAR
TO DATUM "A" AXIS. ONE POINT (MIN)
CONTACT WITH DATUM SURFACE "B".

TWO IMAGINARY
DATUM PLANES
INTERSECTING
AT AXIS OF "A"

90°

90°

PRI DATUM
"A" AXIS

MEASURING DIRECTIONS
FOR RELATED FEATURES

↗ TO DATUM "A" RFS
WITHIN X.X FIM, RFS

DATUM
AXIS "A"

PRIMARY DATUM FEATURE "A" IN CONTACT
WITH SIMULATED DATUM CYLINDER "A", THE
SMALLEST CIRCUMSCRIBED CYLINDER*,
ESTABLISHES DATUM AXIS "A"

*THE ACTUAL MATING ENVELOPE

PRIMARY DATUM FEATURE MMC

PART

A

Ø X.XX ± X.X

X.X

⊕ │ Ø X.X Ⓜ │ A Ⓜ │ B ◄── AND SECONDARY DATUM PLANE "B"

WITH RESPECT TO PRIMARY DATUM "A" AT MMC

WITHIN ØX.X AT MMC

THIS FEATURE TO BE LOCATED AT TRUE POSITION

4X ØX.XX ±X.X

⊕ │ Ø X.X Ⓜ │ A Ⓜ │ B

X.XX ±X.X

B

ØX.XX ±X.X

⊕ │ Ø X.X Ⓜ │ A Ⓜ │ B

DATUM PLANE "B" ⊥ TO DATUM "A" FEATURE AXIS AT MMC ONE POINT (MIN) CONTACT WITH DATUM SURFACE "B".

MEANING

ØX.X ⊕ TOL ZONE AT FEATURE MMC

POSSIBLE AXIS OF DATUM "A" FEATURE AWAY FROM MMC.

⊕ TO DATUM "A" AT MMC WITHIN .XXX AT MMC SIZE OF FEATURE

DATUM AXIS

TWO IMAGINARY DATUM PLANES INTERSECTING AT AXIS OF "A" AT MMC SIZE

90°

90°

AXIS ZONE EQUAL IN SIZE TO DIFFERENCE BETWEEN ACTUAL SIZE AND MMC SIZE OF DATUM "A" FEATURE

PRIMARY DATUM FEATURE "A" TRUE CYLINDER ESTABLISHED BY MMC SIZE OF PRIMARY DATUM FEATURE

DATUM "A" FEATURE FINISHED AWAY FROM MMC (ACTUAL MATING ENVELOPE).

DATUM "A" AXIS

POSSIBLE AXIS OF DATUM FEATURE WHEN AWAY FROM MMC.

MEASURING DIRECTIONS FOR ANY RELATED FEATURES

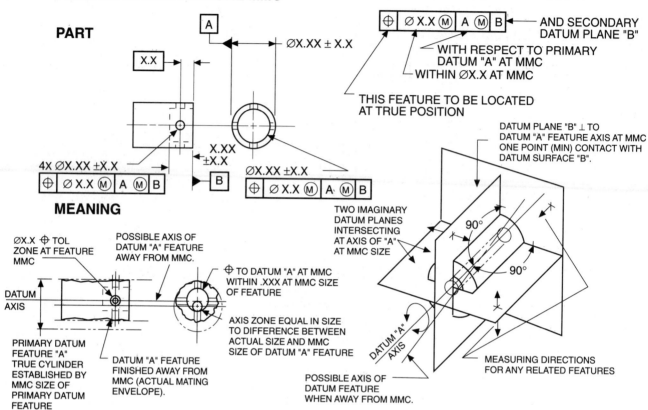

SECONDARY DATUM FEATURE RFS

ØX.XX ±X.X

↗ │ X.X │ A │ B

A

↗ │ X.X │ A │ B ◄── TO DATUM AXIS "B", RFS

WITH RESPECT TO PRIMARY DATUM PLANE "A"

WITHIN X.X FIM

THIS FEATURE TO BE WITHIN CIRCULAR RUNOUT

ØX.XX ± X.X

B

X.XX ± X.X

PART

// │ X.X │ A

PRIMARY DATUM PLANE "A" THREE POINT (MIN) CONTACT WITH DATUM SURFACE "A".

↗ TO DATUM "B" RFS WITHIN X.X FIM, RFS

PRIMARY DATUM PLANE "A"

MEASURING DIRECTIONS FOR ANY RELATED FEATURES

DATUM AXIS

TRUE CYLINDER ⊥ TO "A" *

POSSIBLE AXIS OF SECONDARY DATUM FEATURE

90°

DATUM AXIS (AXIS OF TRUE CYLINDER ⊥ TO PRIMARY DATUM PLANE "A" AND CONTACTING EXTREMITIES OF SECONDARY DATUM FEATURE "B", RFS.)*

90°

DATUM AXIS

TWO IMAGINARY DATUM PLANES INTERSECTING AT DATUM AXIS

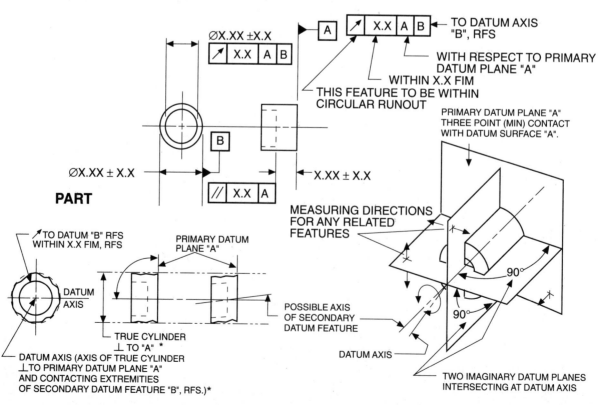

*ACTUAL MATING ENVELOPE

SECONDARY DATUM FEATURE MMC

PART

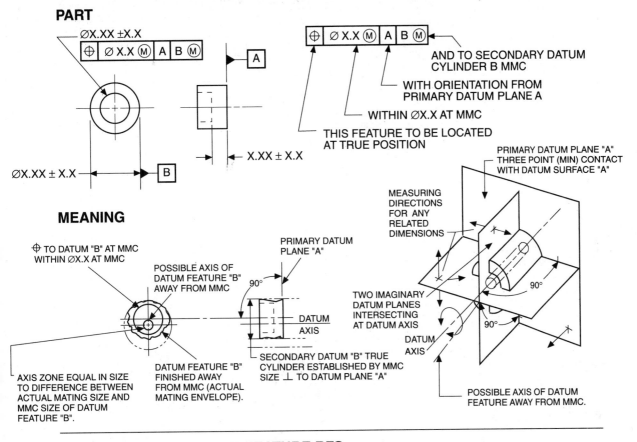

| ⊕ | ∅ X.X Ⓜ | A | B Ⓜ |

AND TO SECONDARY DATUM
CYLINDER B MMC

WITH ORIENTATION FROM
PRIMARY DATUM PLANE A

WITHIN ∅X.X AT MMC

THIS FEATURE TO BE LOCATED
AT TRUE POSITION

∅X.XX ±X.X

| ⊕ | ∅ X.X Ⓜ | A | B Ⓜ |

A

B

∅X.XX ± X.X

X.XX ± X.X

MEANING

⊕ TO DATUM "B" AT MMC
WITHIN ∅X.X AT MMC

POSSIBLE AXIS OF
DATUM FEATURE "B"
AWAY FROM MMC

PRIMARY DATUM
PLANE "A"

90°

DATUM
AXIS

AXIS ZONE EQUAL IN SIZE
TO DIFFERENCE BETWEEN
ACTUAL MATING SIZE AND
MMC SIZE OF DATUM
FEATURE "B".

DATUM FEATURE "B"
FINISHED AWAY
FROM MMC (ACTUAL
MATING ENVELOPE).

SECONDARY DATUM "B" TRUE
CYLINDER ESTABLISHED BY MMC
SIZE ⊥ TO DATUM PLANE "A"

MEASURING
DIRECTIONS
FOR ANY
RELATED
DIMENSIONS

PRIMARY DATUM PLANE "A"
THREE POINT (MIN) CONTACT
WITH DATUM SURFACE "A"

90°

TWO IMAGINARY
DATUM PLANES
INTERSECTING
AT DATUM AXIS

90°

DATUM
AXIS

POSSIBLE AXIS OF DATUM
FEATURE AWAY FROM MMC.

TERTIARY (THIRD) DATUM FEATURE RFS

PART

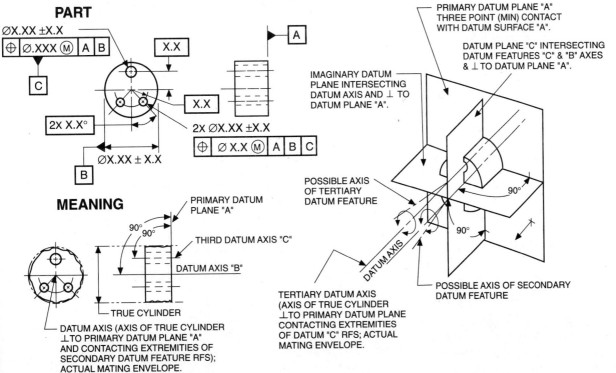

∅X.XX ±X.X

| ⊕ | ∅.XXX Ⓜ | A | B |

C

A

X.X

X.X

2X X.X°

2X ∅X.XX ±X.X

| ⊕ | ∅ X.X Ⓜ | A | B | C |

B

∅X.XX ± X.X

PRIMARY DATUM PLANE "A"
THREE POINT (MIN) CONTACT
WITH DATUM SURFACE "A".

DATUM PLANE "C" INTERSECTING
DATUM FEATURES "C" & "B" AXES
& ⊥ TO DATUM PLANE "A".

IMAGINARY DATUM
PLANE INTERSECTING
DATUM AXIS AND ⊥ TO
DATUM PLANE "A".

POSSIBLE AXIS
OF TERTIARY
DATUM FEATURE

90°

90°

POSSIBLE AXIS OF SECONDARY
DATUM FEATURE

TERTIARY DATUM AXIS
(AXIS OF TRUE CYLINDER
⊥TO PRIMARY DATUM PLANE
CONTACTING EXTREMITIES
OF DATUM "C" RFS; ACTUAL
MATING ENVELOPE.

MEANING

PRIMARY DATUM
PLANE "A"

90° 90°

THIRD DATUM AXIS "C"

DATUM AXIS "B"

TRUE CYLINDER

DATUM AXIS (AXIS OF TRUE CYLINDER
⊥TO PRIMARY DATUM PLANE "A"
AND CONTACTING EXTREMITIES OF
SECONDARY DATUM FEATURE RFS);
ACTUAL MATING ENVELOPE.

DATUM AXIS

ESTABLISHING DATUM CENTERPLANES FROM DATUM FEATURES— DATUM REFERENCE FRAME (THREE PLANE CONCEPT), DATUM PRECEDENCE

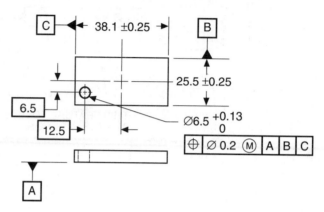

MEANING NO. 1

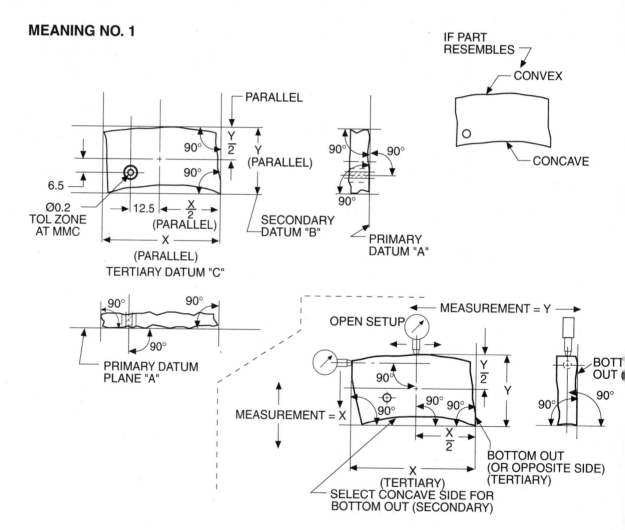

MEANING NO. 2

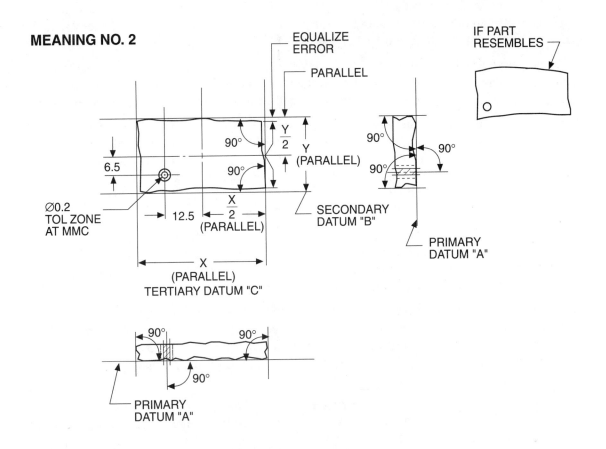

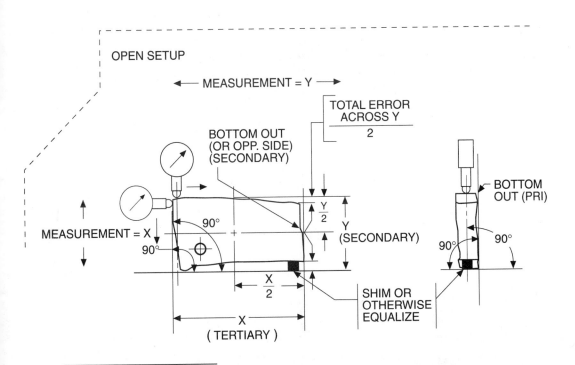

NOTE The "open set-up" interpretations above and on the preceding page, MEANING 1, demonstrate the principles involved. Production gaging methods, or special gages patterned after these techniques, can facilitate more effective inspection, e.g., equalizing mechanisms can be used in lieu of the measuring and shimming method (see MEANING NO. 2).

DATUM APPLICATIONS

The following section presents details of selection, application, and meaning of datums under sample part conditions. Not all conditions are covered, but the examples shown are representative of a variety of typical datum applications.

The proper selection of datum features is a very important aspect of geometric tolerancing. As previously emphasized, datums must be specified. In determining datums, we must consider first of all the part and feature relationships within the framework of the design requirements.

Where the relationships of the features are not of critical importance to part function or design intent, datums may not be required in the drawing specifications. However, usually some form of orientation for part manufacture and inspection is necessary. Therefore, where no specific functional mating part or feature relationship needs to be included in the engineering drawing detail, the designer is depending on the production and inspection departments to provide good workmanship, discretion, and the probabilities that satisfactory parts might result.

Where specific relationships of features, datum precedence, and part orientation are of critical concern to design and manufacture, datums must be specified. Of prime concern is the clarity with which the drawing conveys the design intent. Ambiguity should always be avoided, since it causes costly confusion and leads to misinterpretations.

Datum specification may appear to be a costly approach. Not so—on the contrary, it provides considerable advantages that pay off in economic production. It provides for clearer meaning, protects the design intent, and ensures production and inspection follow-through in keeping with the intent. If properly applied per the standard methodology, uniform meaning of the datum reference frame is universal.

The examples shown highlight the importance of adequate and proper datum specification. The illustrations on the left are not provided with datums, and the feature relationships are in doubt. The addition of datums in the examples on the right clarifies the relationships and thus ensures uniform interpretation.

SELECTION OF DATUMS

1. *Features* which are selected to establish *datums* must be clearly identified and/or easily recognizable. Datums must be specified, and must clearly represent necessary design intent.

2. *Corresponding features* on mating parts should be used in establishing datums to facilitate calculations and ensure proper part assembly.

3. To be useful for measuring, a datum on an actual piece should be accessible during manufacture so that *measurements* from it *can be made* readily.

4. *Avoid ambiguity* of design requirements by specifying datums where necessary for clarity.

5. Consider the *precision* of the features selected as datum features.

SELECTION OF DATUMS

EXAMPLES (WITHOUT DATUMS)   ## EXAMPLES (WITH DATUMS)

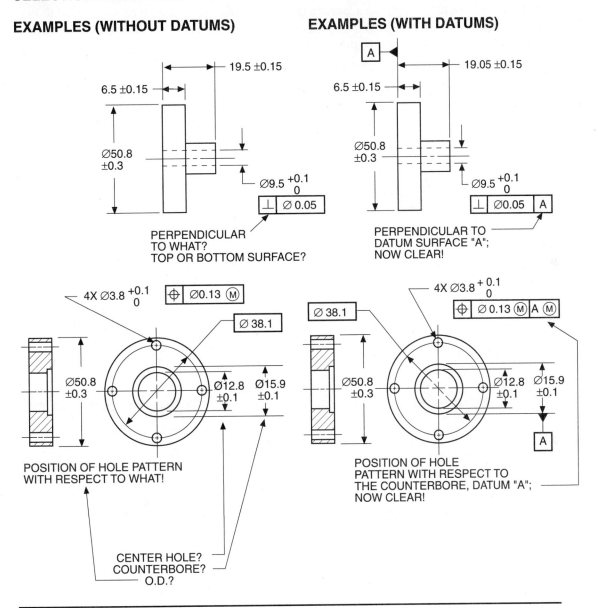

ACCURACY OF A DATUM SURFACE OR FEATURE

The accuracy or quality of surfaces used for datums depends on the design requirement. The desired control, such as the form tolerance of flatness or orientation tolerance for perpendicularity, etc., may be specified where necessary.

Some earlier practices of the past have suggested the need to control datum surfaces to a finer degree of accuracy than that provided by dimensions or relationships taken *from* the datum surfaces. Often design requirements may call for this refinement and the need for added precision of the datum surfaces (or features) should therefore be considered. However, as an overall rule, a request for specific accuracy of the datum surface should be based upon the design requirements and the functional purpose of the surface.

By the definition of datum plane, a datum surface may have inaccuracies which will have no effect on the features located from the datum plane (as established from the extremities of that surface).

The illustration shows the two conditions of datum accuracy for features located from the datum, one where datum surface accuracy is *less* than that for feature location related to the datum (a), and one where the datum surface is held *more* accurate than the feature location related to the datum (b). Either requirement could be valid depending upon the design requirement.

Unless otherwise specified, part surface quality is normally a result of the type of machining process used to achieve size tolerances. Often, workmanship or local manufacturing tolerances will be the determining factors. Thus, where surface accuracy is of no critical concern, it need not be specified, yet that surface can serve as a datum reference from which other relationships can be specified.

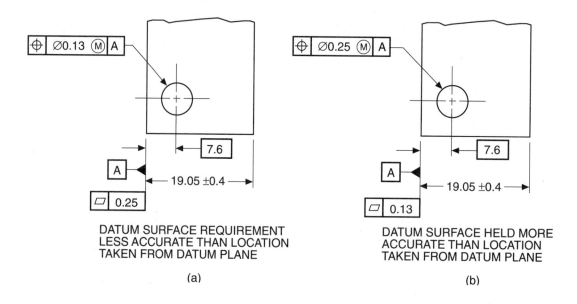

DATUM SURFACE REQUIREMENT
LESS ACCURATE THAN LOCATION
TAKEN FROM DATUM PLANE

(a)

DATUM SURFACE HELD MORE
ACCURATE THAN LOCATION
TAKEN FROM DATUM PLANE

(b)

SPECIFIED DATUM SURFACES USING THE FORM TOLERANCE ⊥ AND POSITION ⊕ DIMENSIONING AND TOLERANCING

When feature interrelationships involving form, orientation, and position tolerances are to be more accurately and directly controlled, it becomes necessary to specify datums on the drawing so that such relationships can be clearly indicated.

In the example on the following page, the top surface of the part is identified as datum A, the left edge as datum B, and the lower edge as datum C. Once these identifications are made, the specific relationships between surfaces and features can be stated.

As previously described, datum planes are established from the extremities of the actual datum surfaces. Each relationship is thus taken from the datum plane. The datum surface inaccuracies, not coincidental with the datum plane, are not involved in the relationship.

The interpretation illustrates the establishment of the datum planes from the specified datum surfaces. The specified datums were used in this example to control orientation or perpendicularity of certain surfaces to precise limits. They were also used to give datum precedence orientation from the part surfaces to the position toleranced pattern.

(Continued on page 246).

EXAMPLE

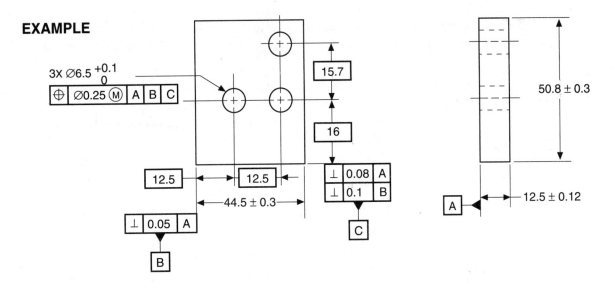

MEANING

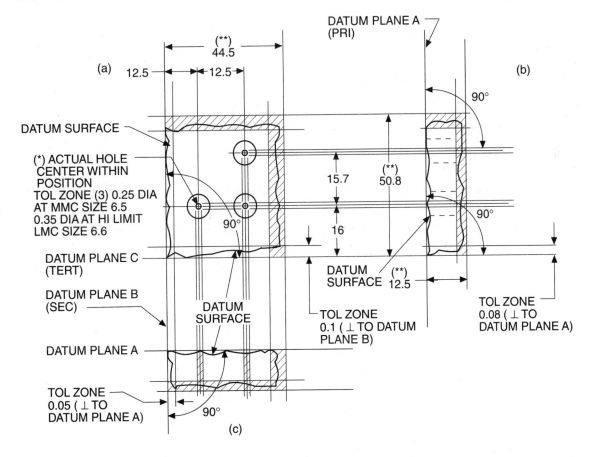

NOTE *Toleranced zones apply as normally measured from the datum planes. If measured by some method where reference on the datum surface is not coincidental with the datum plane, measurement is affected by the datum surface inaccuracy (deviation between datum surface and datum plane) at that point.

**Size measurement of individual feature per Rule 1.

The left datum surface is specified as perpendicular to datum plane A within 0.05. The lower datum surface is specified as perpendicular to datum plane A within 0.08 and to datum plane B within 0.1.

In view (b) in the lower interpretation the 0.08 wide perpendicularity tolerance zone is established between two parallel planes exactly 90° BASIC (or perpendicular) to datum plane A. The actual surface must lie within this 0.08 wide tolerance zone when it is measured relative to datum plane A anywhere along the entire length of its surface.

In (a), the 0.1 wide perpendicularity tolerance zone is established between two parallel planes exactly 90° BASIC (perpendicular) to datum plane B. The actual surface must be within this 0.1 wide tolerance zone when it is measured relative to datum plane B anywhere along the entire length of its surface. In (c), datum surface B is to be held perpendicular to datum A within a 0.05 width tolerance zone 90° BASIC to datum plane A.

Note that the measurements are taken from the datum planes which are theoretically perfect as established by the extremities of the actual datum surfaces. The perpendicularity tolerance zones establish the limit of the perpendicularity or form inaccuracies of these surfaces.

As in the note to view (a) at left, the actual hole centers must fall within the position tolerance zone of $\oslash$ 0.25 when the hole is at MMC size of 6.5. The tolerance zone can increase to $\oslash$0.35 if the holes are produced to the high limit (LMC) size of 6.6. The position tolerance includes both orientation (perpendicularity) and position error control.

The position tolerance zones are basically parallel to each other and the center planes of the pattern are basically perpendicular and parallel to each other. In addition, the position tolerance zones are basically perpendicular to the primary datum plane A, and parallel to the secondary and tertiary datum planes B and C.

The edge surface tolerance zones illustrated relative to the size dimensions 44.5 and 50.8 (**in the preceding figure) are intended to relate the edge extremities to the total part requirements. However, note that the interrelationship (perpendicularity) of these edge surfaces to each other is not controlled by the stated requirements; instead, it is subject to the considerations of Rule 1. The edge tolerance zones shown are representative of the size tolerance relationship of the individual 44.5 ± 0.3 and 50.8 ± 0.3 sizes only. Control of these surfaces in the relationship to one another would require specified orientation tolerances (e.g., perpendicularity), or it may depend on local workmanship or shop tolerances.

SPECIFIED DATUM SURFACES USING FORM TOLERANCES ⊥, ▱, AND POSITION ⊕ DIMENSIONING AND TOLERANCING

In the example on the facing page, the stated requirements and interpretation are identical to those in the preceding one except for the added flatness control of each of the datum surfaces. The accuracy or quality of the datum surfaces is specified.

Note that the flatness control is a refinement of other controls and is contained *within* the other orientation tolerance zones. Where combined form and orientation tolerances are applied, the predominant control (in this case perpendicularity) confines the tolerance latitude of additional controls (such as flatness) as a further refinement.

Datum planes are established from the datum surfaces as previously explained; the orientation to these planes is as shown in the preceding example.

EXAMPLE

MEANING

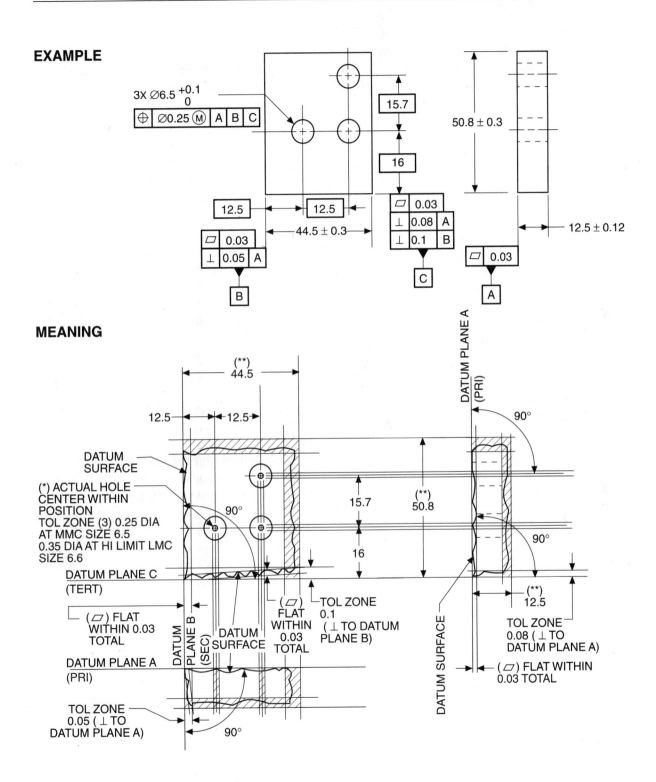

NOTE *Tolerance zones apply as normally measured from the datum planes. If measured by some method where reference on the datum surface is not coincidental with the datum plane, measurement is affected by the datum surface inaccuracy (deviation between datum surface and datum plane) at that point.

**Edge tolerance zone interrelationship to each other, subject to considerations of Rule 1, size measurement of individual feature within boundary of perfect form at MMC.

SPECIFIED DATUM FEATURES (SURFACE, HOLES) USING FORM TOLERANCES ⊥, ▱, //, AND POSITION ⌖ DIMENSIONING AND TOLERANCING

In the example on the following page, (Flange Mount), we use a variety of geometric controls. Note the manner in which datums are applied so thatdirectrelationships may be controlled per the part function as seen in the lower illustration.

The large base surface in Fig. (a) is to be held flat within 0.03 and is also identified as datum plane A. Note that the ∅ 19.05 $^{+0.15}_{0}$ center hole diameter is to be perpendicular to datum plane A within ∅0.03 and is also established as datum B.

Also in Fig. (a), the ∅ 31.8$^{0}_{-0.15}$ diameter is to be held to a positional tolerance of ∅ 0.1 at MMC with respect to datum B at MMC; the M39-6g thread is to be held to a positional tolerance of ∅ 0.1 at MMC with respect to datum B at MMC; the face adjacent to the thread undercut is to be parallel within 0.06 with respect to datum A; and the ∅ 44.5 $^{+0.15}_{0}$ counterbored hole at left is to be held in total runout within 0.08 FIM with respect to datum B and is also established as datum C.

The six holes shown in Fig. (b) are to be located at true position within ∅0.1 at MMC with respect to datum plane A and datum C at MMC. Datum plane A identifies the surface from which the positional pattern is related and orients the hole center tolerance zones as basically perpendicular in orientation to this plane. The theoretical center of datum hole C establishes the center of the 85.7 BASIC diameter and the 60° BASIC angles.

The lower Fig. (c) illustrates the part function (the Flange Mount) as it relates to the mating parts in assembly. From this assembly requirement the critical nature of the perpendicularity, runout and position (mounting holes) can be seen. The interrelationship of all the related features is controlled by the appropriate geometric tolerances and datum references.

EXAMPLE (FLANGE)

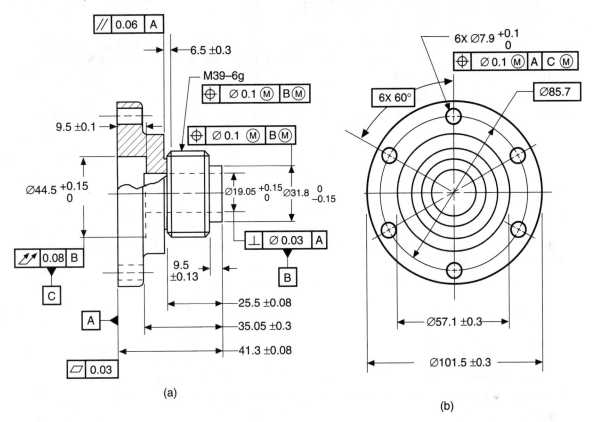

(a)

(b)

ASSEMBLY

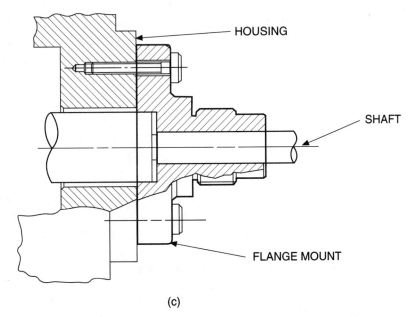

HOUSING

SHAFT

FLANGE MOUNT

(c)

SPECIFIED DATUM POINTS ON SURFACES USING POSITION $\oplus$ DIMENSIONING AND TOLERANCING

This example illustrates a part whose function requires that specific locating or reference points be established on the edge surfaces. These points are *datum* points and are identified by datum target symbols.

Surface datum points are used either to simulate the desired location of contact with the mating parts and thus hold the related dimensions and tolerances with respect to these points, or to assure repeatable location and orientation of the part for successive fabrication and inspection operations on castings, forgings, sheet metal parts or other irregular surfaces.

Note that the datum points are dimensioned with BASIC or untoleranced dimensions. By interpretation, BASIC dimensions are theoretically exact and, in this case, represent the ideal location for which to strive in order to relate the holes with the part edges. In these circumstances and in actual applications, the BASIC dimensions are assumed to represent exactness within tooling, gaging, or shop tolerances.[*] If desired, toleranced locations or sizes may be used with datum target symbols.

Actually, as previously explained, the datum points establish datum planes. Thus, the position pattern is simultaneously related to these points and the planes as established from these points.

The two holes on the left, the $\oslash 76^{+0.1}_{0}$ holes, are to be located at true position within $\oslash 0.15$ at MMC with respect to datum planes A, B, and C. The third hole at the right, the $\oslash 3.3^{+0.05}_{0}$ hole, which also is a part of the hole pattern, is to be located at true position within $\oslash 0.1$ at MMC with respect to datum planes A, B, and C.

The meaning illustrates the relationship of the hole pattern to the datum points and established planes.

Datum points B1 and B2 in the left view establish a datum plane B at 90° BASIC to the top datum plane A. From this plane B, the vertical dimensions are taken. Thus, these dimensions are related *only* to points B1 and B2 and their common plane and *not* to the entire surface.

Datum point C1 establishes a third plane, C, 90° BASIC to both datum planes A and B. Only one point is, of course, necessary on the third plane. The horizontal dimensions are related *only* to point C1 and its common plane, C, and *not* to the entire surface.

Note that the positionally toleranced pattern of holes is oriented with respect to the established datum planes, since, by interpretation, all centerplanes are basically parallel and perpendicular to the edge datum planes.

The actually produced hole centers are permitted to vary within the tolerance zones stated on the drawing. However, as the positional relationships are specified at MMC, the positional tolerance of the two larger holes may possibly increase from $\oslash 0.15$ (at MMC size of $\oslash 7.6$) to $\oslash 0.25$ when the actual size holes are at LMC size, or largest size, of $\oslash 7.7$. The positional tolerance of the third hole may possibly also increase from $\oslash 0.1$ (at MMC size of 3.3) to $\oslash 0.15$ at LMC size, or largest size, or $\oslash 3.35$.

---

[*]Author advisory: Since the usual application of datum targets is for the repeatability between the manufacturing and verification operations via special fixtures (or processes), tolerances for the target pin locations are provided on the design and make of the fixtures. It is rare to see any physical evidence of targeting on the finished part.

EXAMPLE

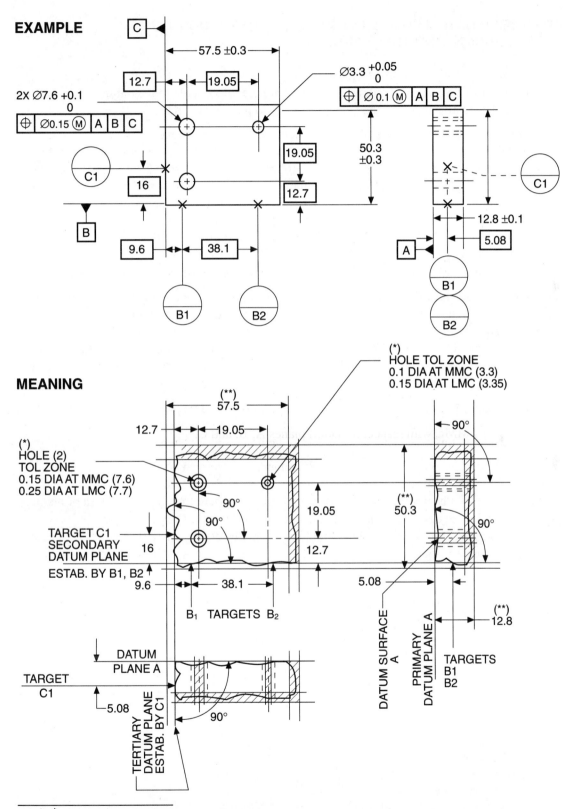

MEANING

NOTE *Tolerance zones apply as measured from the datum planes which have been established by the datum targets.

**Edge tolerance zone interrelationship to each other, subject to considerations of Rule 1, size measurement of individual feature within boundary of perfect form at MMC.

SPECIFIED DATUM POINTS ON SURFACES USING POSITION $\oplus$ DIMENSIONING AND TOLERANCING

Note again that when using this form of position dimensioning and tolerancing with datum points, the position pattern and thus the individual holes are related to the datum points and their established planes only—they are *not* related to the entire surface.

The deviations or imperfections of the remainder of the actual surfaces on which the specified datum points are established are controlled by the overall coordinate dimensions. See note (**) at lower part of illustration. The edge surface tolerance zones illustrated are intended to relate the edge extremities to the total part requirements. However, note that the relationship (perpendicularity) of the edge surfaces to each other is *not* controlled by the stated requirements. This interrelationship is subject to Rule 1. The edge tolerance zones shown are representative of the size tolerance relationship of the individual 50.3 ± 0.3 and the 57.5 ± 0.3 sizes only. Control of these surfaces in relationship to one another would require specified form or orientation tolerances (e.g., perpendicularity), or may depend on local workmanship or shop tolerances.

Typical applications of this form of dimensioning and tolerancing are an irregular surface, such as a casting, which requires specific tooling or gaging pickup for repeatability, or a part whose function requires specified location of points of contact or relationship with a mating part. When we locate these points and identify them, we ensure that specific design or tooling and gaging requirements are clearly stated and that the manufacturing and inspection processes will follow well-defined instructions in agreement with the part requirements.

Usually, in an application such as this, the datum points would be picked up by locating buttons. There is no implication that detents are to be applied to the part surface. If detents are desired, a note to this effect should be added to the drawing.

If positional dimensioning located by BASIC dimensions from an edge surface is employed and *no* datum points nor planes are specified, the high points or extremities of the implied datum surface will be used to establish datum planes and orient the position pattern of features to the edges. In this case, we would use conventional *flat* locating surfaces instead of buttons. Although the resulting control of the relationship of the position pattern of features to specific portions of the part edges would be less stringent, it may be adequate in many applications.

EXAMPLE

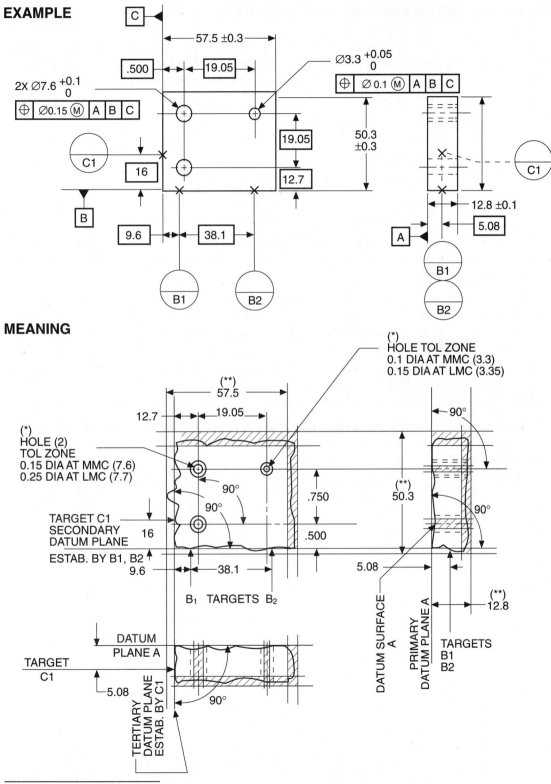

MEANING

NOTE *Tolerance zones apply as measured from the datum planes, which have been established by the datum targets.

**Edge tolerance zone interrelationship to each other, subject to considerations of Rule 1, size measurement of individual feature within boundary of perfect form at MMC.

SPECIFIED DATUM LINES ON SURFACES

In this example, we expand the use of datum points to establish datum lines on surfaces. We see that selecting specific datum or reference lines as the basis for establishing relationships to other features often has the advantage of providing effective and economical methods of achieving design intent or part function.

This illustration shows a rotor. It is not completely dimensioned—only enough to explain how the datums are applied.

Datum lines A1 and A2 specify the location of two specific lines on the rotor shaft which are to be the basis for the runout relationship with the rotor outside diameter.

Datum lines A1 and A2, because they are points on a diameter, actually apply to an infinite number of points around the circumference of the shaft to establish the datum lines. As was previously explained with respect to the BASIC dimensions, exactness within tooling, gaging, or shop tolerances is implied. Where necessary, toleranced locations or sizes may be used with datum target symbols.

In actual application, the gaging pickup of the datum lines could, for example, be accomplished by two vee blocks of very narrow width, or more correctly, a special collet, chuck, or holding device which would close and hold the part at the target location as it is rotated. The datum lines essentially simulate the location of the bearing races and the accuracy required at these lines. This eliminates the need for holding the critical runout requirement on a larger portion of the shaft length than is necessary to ensure proper function of the part. The remainder of the shaft is then controlled by shop tolerances or other specifications.

The runout notation would be interpreted to mean "the rotor diameter is to be within 0.03 FIM total runout with respect to datums A1 and A2 (and their common datum cylinder and axis) simultaneously."

EXAMPLE

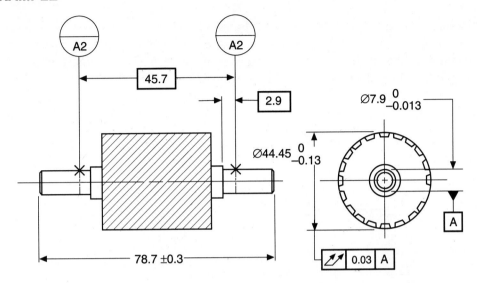

MEANING

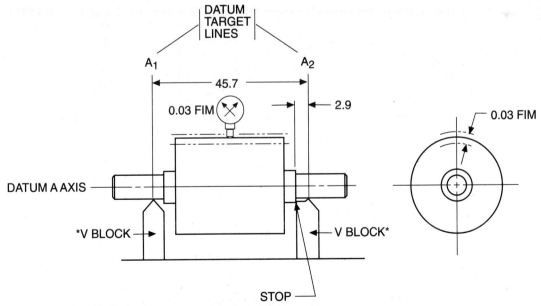

DATUM LINES A_1, & A_2 CREATE DATUM AXIS A.

44.45 DIA TO BE WITHIN TOTAL RUNOUT OF
.001 FIM WHEN MOUNTED ON LINES A_1 & A_2.

* = OR COMPARABLE HOLDING DEVICE

SPECIFIED DATUM POINTS AND LINES ON SURFACES

Figure 1 shows a selection of two datum points, A1 and A2, used to specify a relationship of parallelism between two points on the part surface and the elongated slot having first stabilized the part on datum B as the primary datum and located from datum C. The slot location is non-critical but its parallelism to the points is design functional.

FIGURE 1

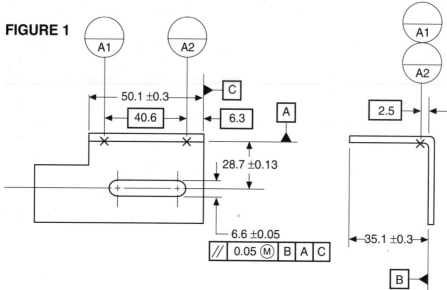

Figure 1 shows that these two datum points, A1 and A2, or more specifically the theoretically exact datum line A extending between these two datum points (and not the entire surface), establish the datum reference from which the parallelism of the slot is to be related.

The surface upon which datum points A1 and A2 are established could have curvature or inaccuracies which would have no effect on the parallelism requirement. The accuracy of the entire surface is controlled by the vertical 28.7 ± 0.13 dimension. The parallelism requirement is specified to be from the common datum line A established by datum points A1 and A2 to the center line of the slot within 0.05 total.

Figure 2 shows datum line A1 (at the left) established on the angular surface of the part. Datum line A1 is used in conjunction with the center of datum feature B to establish an axis and centerplane which determines the reference from which the 30° BASIC angle to the slot is taken.

FIGURE 2

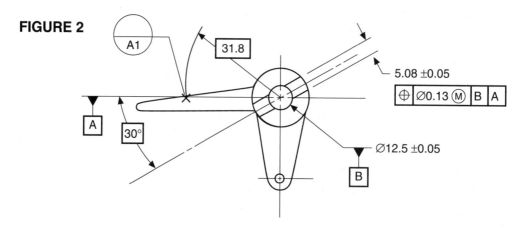

SPECIFIED DATUM AREAS ON SURFACES

FIGURE 3

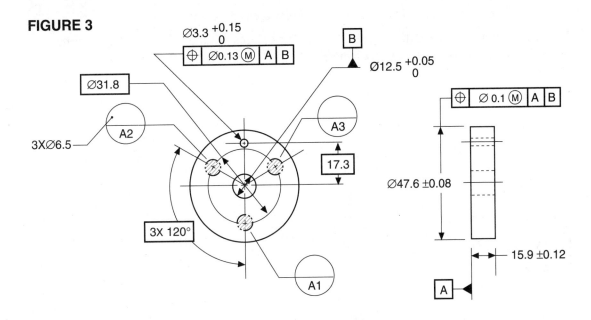

If the datum reference must be taken from given small areas on the primary datum surface (Fig. 3), the desired datum surface area may be shown as a diameter (or other shape), with location and size of the area controlled by BASIC dimensions. The datum surface area limits are enclosed (with phantom-line circles in this case) and the area is identified by diagonal slash lines.

The extremities of the three datum surface areas A1, A2, and A3 combine to form a single, or common, datum plane A.

Three 6.5 flat-ended buttons or locators contacting the surface at the prescribed locations will provide the actual location. The other requirements are then related only to these datum surface areas and their common resulting plane and not to the entire surface.

SPECIFIED DATUM AREAS ON SURFACES

In some applications, it is convenient to select larger areas of surfaces as datums.

If Fig. 1, portions of the shaft diameter identified as datums A1 and A2 are the basis for the runout relationship to the shaft step diameter.

FIGURE 1

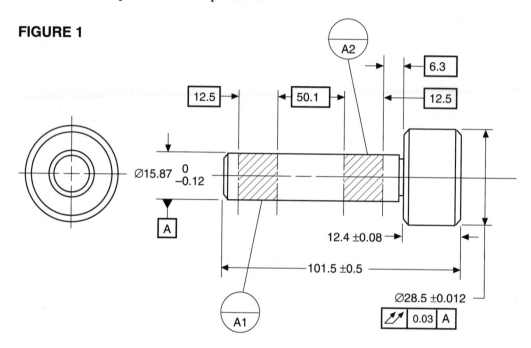

As in the previous example on datum points, the datum reference applies to the circumference around the shaft contained within the limits of the basic dimensions. The extent of the datum surfaces is the 12.5 BASIC wide reference bands or areas on the shaft diameter. These datum surface areas simulate the location of bearing races.

It can be seen that the shaft step diameter of $\oslash$28.5 ± 0.012 is required to be held within total runout of 0.03 with respect to the two datum area diameters A1 and A2 and their established common datum reference axis A. The remainder of the shaft is permitted to deviate within more lenient standard or shop tolerances.

As in the previous examples, exactness within tooling, gaging, or shop tolerances with respect to the basic dimensions is implied.

Figure 2 illustrates a metal blade. The 35.5 ± 0.3 dimension is to be held between the lower flat surface and the contact button. However, in addition, there is a 0.13 total parallelism requirement from that portion of the lower surface identified as datum ∧1 and the button.

The limitations of datum target A and its established datum reference plane are specified by a phantom line and a BASIC dimension, with the area identified by diagonal slash lines drawn across the surface. This clearly shows that the surface so identified determines the datum reference plane.

FIGURE 2

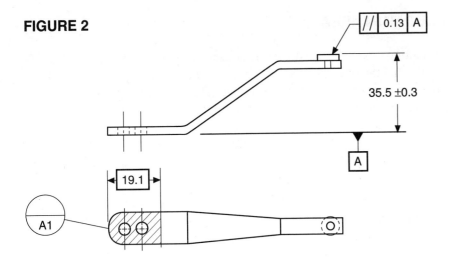

Figure 3 is another example in which a portion or area of a surface serves as a datum reference. This application was based on the part function and the anticipated difficulty of holding accuracy on any more of the formed part flat surface than necessary.

FIGURE 3

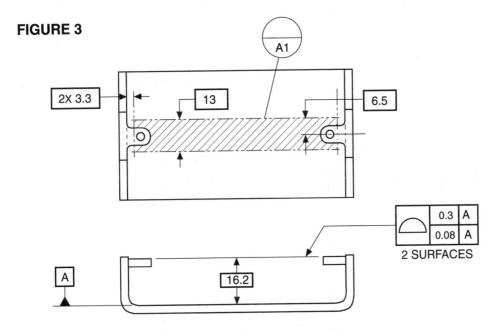

Note the manner in which the basic dimensions have been used to define the limits of the datum area A1 and the two ear relationships. Area A1 establishes the datum reference plane A from which the relationship is taken.

The formed ears of the part are to be held parallel within 0.3 to datum plane A and coplanar within 0.08 relative to each other. A composite profile tolerance has been used.

Again, exactness within tooling, gaging, or shop tolerances is implied by the BASIC dimensions, to the targets.

SPECIFIED MULTIPLE DATUM AREAS

In this example, three mounting bosses of a cast part are selected as datum target areas. The purpose is to clearly specify that the three bosses *only* are to establish datum plane A for the relationship of the parallelism requirement on the top surface. The extremities of the three individual boss surfaces would combine to provide the three-point (minimum) contact, establishing datum plane A. Normally complete features (i.e., the bosses) are *not* used as datum targets. However, in this case, the three boss ends are used to compose *one* datum plane as the design requirement and plane of reference for the parallelism relationship. Usually datum targets are *portions* of surfaces or features.

The top surface of the part is to be parallel within 0.08 to the established datum plane A. This guarantees the assembled part relationship of the parallelism requirement to the mounting bosses.

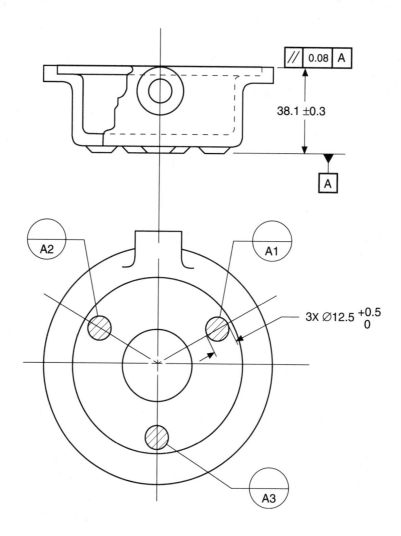

SPECIFIED "STEP" DATUMS

When we establish datum planes of orientation on parts of irregular shape such as a casting, the datum target points (and thus their datum pin locators) may involve surfaces which are at different levels and must therefore be *combined* to establish the desired plane for location or orientation.

Referring to the illustration on the next page, we see that datum points C1 and C2 are situated at different surfaces and at different levels (steps) in order to relate the entire bulk of the part to a common plane of orientation. This plane, identified as the common plane C, may then be related to the primary datum B and the tertiary datum D with the usual implied mutually perpendicular datum plane relationship. The step basic dimension is to be maintained within tooling or gaging tolerance accuracy.

Other features of the part, such as the position toleranced pattern, can then be related to the datum system as established from these "step" datums and the other interrelated datums to ensure compliance with design requirements and repeatability of manufacturing and inspection operations.

SPECIFIED "EQUALIZING" DATUMS

Page 263 illustrates a part on which the orientation is established from specified datum targets (lines and points) as a means of "equalizing" the bulk of the part to centralize the location of the three holes. Through the datum target method as specified, a guarantee of design requirement fulfillment is provided, and tooling and gaging repeatability is assured. The two planes of orientation necessary to this part are provided by the equalizing and centralizing placement of the datum targets. The specified position relationships controlling the hole locations are then established from the datums A, B.

The primary datum plane A is established from points A1, A2, and A3 at the two levels of the part and relative to the step difference. These points and the step difference combine to give the common primary datum plane of orientation.

The secondary datum plane B is established by the equalizing action of the fixed and movable vee's as they contact tangent to the part at an angle of 45° on the left end and at the indicated lines of contact on the right end. Datum center plane B is perpendicular to datum plane A. Some slight "cam-down" or hold-down of the part may be necessary in actual applications.

Note   Where "step" datums or "equalizing" datums are used, it is most likely that no one feature can be identified as *the* datum feature. Therefore, the datum feature symbol cannot be placed conveniently on the drawing on a given feature as normally would be done. In such cases, the datum feature symbol may be placed on one of the common extension lines (as on the "step" datum example) or on the established resulting centerplane (as on the "equalizing" datum example) where target lines on portions of features actually create the datum centerplane. Caution should be exercised that placing the datum feature symbol *on* centerplanes or axes (centerlines on drawings) is done *only* where necessary and convenient and the datum centerplane or axis is established from and by datum target symbols. The datum feature symbol may *be left off* the drawing in such cases with the datum planes established with target symbols (i.e., B1, B2, B3, B4, on the "equalizing" datum example) and referenced in the feature control frame by letter reference as usual.

SPECIFIED "STEP" DATUMS

EXAMPLE

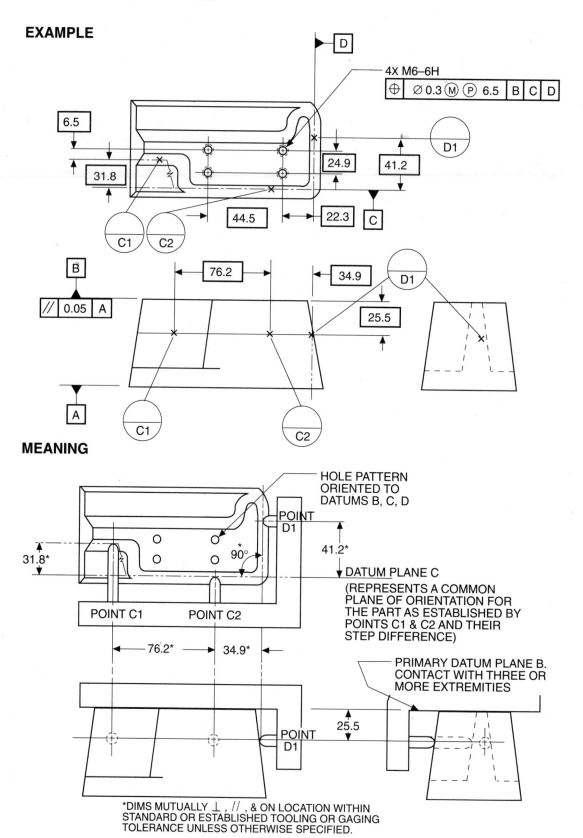

MEANING

*DIMS MUTUALLY ⊥ , ∥ , & ON LOCATION WITHIN
STANDARD OR ESTABLISHED TOOLING OR GAGING
TOLERANCE UNLESS OTHERWISE SPECIFIED.

SPECIFIED "EQUALIZING" DATUMS

EXAMPLE

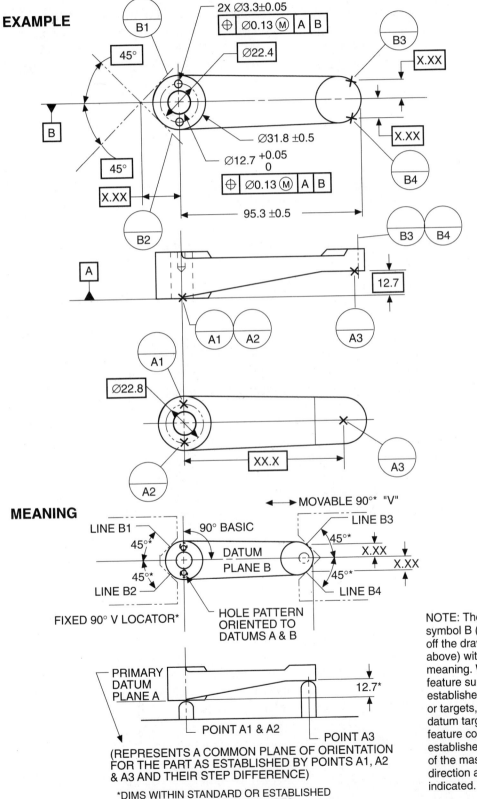

MEANING

NOTE: The datum feature symbol B (⊡) could be left off the drawing (see Example above) with no change in meaning. Where no single feature surface alone establishes the datum target, or targets, reference to the datum target letter in the feature control frame establishes the datum plane of the mass of the part in the direction and precedence indicated.

PRIMARY DATUM PLANE A

POINT A1 & A2   POINT A3

(REPRESENTS A COMMON PLANE OF ORIENTATION FOR THE PART AS ESTABLISHED BY POINTS A1, A2 & A3 AND THEIR STEP DIFFERENCE)

*DIMS WITHIN STANDARD OR ESTABLISHED TOOLING OR GAGING TOLERANCES.

MULTIPLE USE DATUM TARGETS (SWITCHING DATUMS)

Where datum features and their "datum feature symbols" are placed on the drawing, these features create the datums (axes, planes) which may be used in different ways on the same piece part. Other examples in this text illustrate such flexibility of the datum reference frame. As based upon the design requirement, the datum reference frame is established using the conventional 3–2–1 point contact and precedence on the part. However, when the datum references are changed, that is when the primary datum on the part is to be used as a secondary or tertiary datum on another callout on the same drawing, the "switch" is made simply by altering the datum precedence location in the feature control frame. The 3–2–1 datum point precedence sequence is automatically altered since the points of contact on the datum surfaces have not been specifically identified and thus the part has freedom to automatically readjust to the new datum sequence. However, when datum targets are specified to establish specific target points or areas to fix the primary, secondary, or tertiary datums, no such automatic readjustment nor switch of datum precedence, is possible.

To accomplish alteration of the datum target sequence, where the datums are to be used in different ways on the same part, the necessary datum features are *reidentified* and *new* datum targets are specified. The illustration at right clarifies the methods that can be used. Size and basic dimensions have been omitted to accent only the datum target methods which can be used to "switch" them for multiple use.

In the example, datum targets A1, A2, and A3 establish datum plane A and are used in three of the positional tolerance callouts as the primary datum. Datum targets A2 and A3 are reidentified as datum targets H2 and H1 to become the secondary datum plane H in another positional tolerance callout. Datum target C1 establishes the tertiary datum plane for two positional callouts. It is also reidentified as datum target G2 and is associated with added targets G1 and G3 to create a primary datum plane for another positional requirement. Targets G1 and G3 are reidentified as targets F1 and F2 to become the secondary datum plane for the "two coaxial hole" positional callout. Datum feature symbols D and K are conventionally used to establish these datum planes without targets. They are used as primary and secondary datums respectively in one positional specification. Datum target B2 is reidentified as target J1 and establishes a tertiary datum plane; and a new datum target L1 is specified to also establish a tertiary datum plane in other callouts. Datum targets B1 and B2 establish a tertiary datum plane to stabilize the "clocking" (rotation) of the part about datum axis E.

COMPLEX OR CONTOURED FEATURE DATUMS

On complex or contoured surfaces, such as aircraft wing shapes, establishing datum planes would be done by datum targeting methods and would utilize the same reasoning in combination as the "Step" and "Equalizing" examples shown. In such cases it is acceptable to show only the datum target construction with letter identity (i.e., A1, A2, A3, etc.) with no datum feature symbol specified. The target letter identity (i.e., datum plane A) would be referenced in the feature control frame appropriately in an order of precedence.

MULTIPLE USE DATUM TARGETS

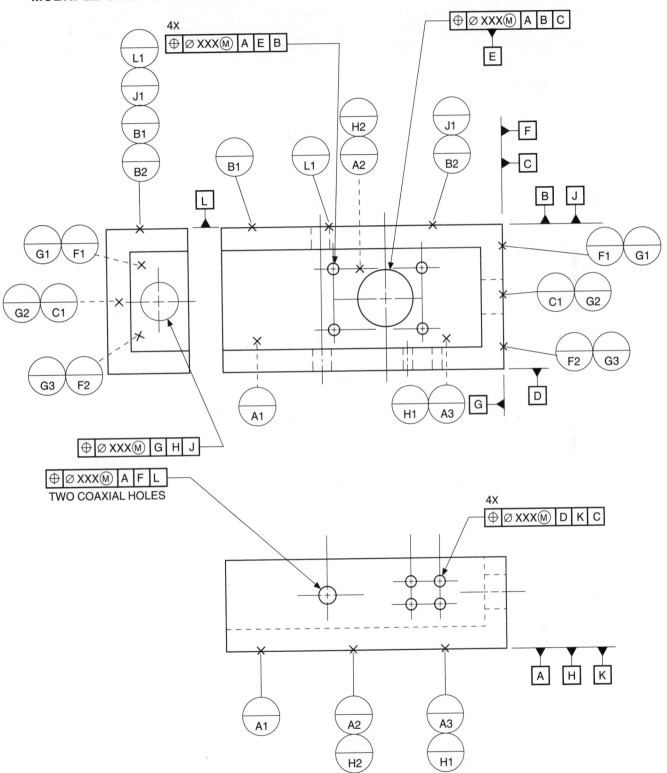

EXTENDED DATUM PRINCIPLES

MULTIPLE DATUM REFERENCE FRAMES WITH NONCOMMON DATUMS

In multiple datum reference frames with noncommon datum references established on the same part to control specific part feature relationships, *no* specific relationship may exist between the datum systems. They are treated separately.

EXAMPLE

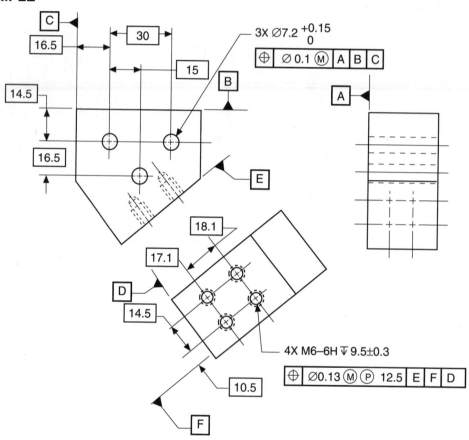

MEANING

FIRST DATUM REFERENCE FRAME

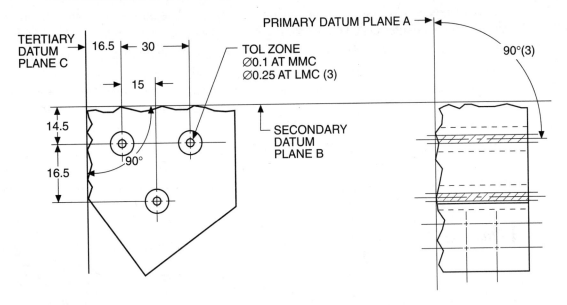

SECOND DATUM REFERENCE FRAME

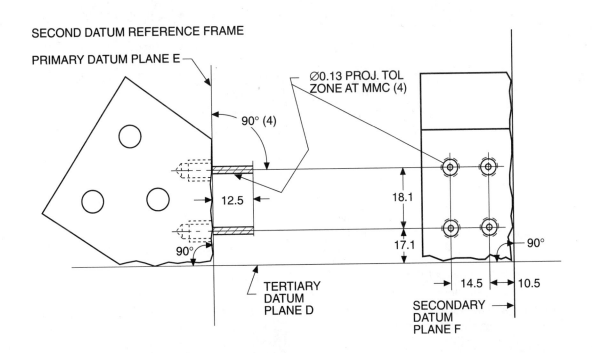

INTERRELATED DATUM REFERENCE FRAMES— WITH COMMON DATUMS

Datum references, either single or double datums, or complete three-plane datum reference frames, which are interrelated on the same part with other datum reference frames, retain a relationship with the other reference frames to the extent of common datum references.

In some cases, one datum reference frame established within or related to another through common datum references involves refinement or extension of the other datum reference frame, and, if necessary for orientation, may retain a relationship with the other datum reference frame as implied by the drawing.

There are two position toleranced hole patterns on this part. The first one involves a single hole which is established relative to the datum reference frame A, B, C. This hole then becomes the secondary datum for the interrelated second datum reference frame A, D, B to which the second position pattern is related.

EXAMPLE

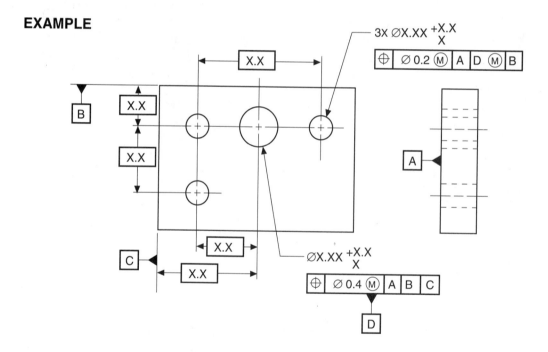

Datum D is directly related to the main datum reference frame (as established by A–B–C) by specification. The three-hole pattern is then related to datums A, D, and B. Datum A is repeated for clarification to ensure correct selection of the primary datum.

Since the establishment of the interrelated datum reference frame A–D–B within the main datum reference frame A–B–C is to achieve a refinement of the relationship between the three small holes and the large hole (datum D), the positional relationship is to be between these features only as specified. Orientation (or rotation) is controlled by datum B as the tertiary datum.

MEANING

FIRST REQUIREMENT (FIRST RELATIONSHIP — TO DATUM REFERENCE FRAME A, B, C)

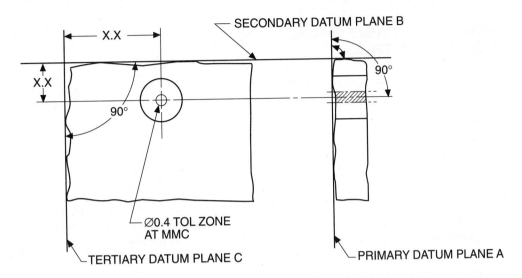

SECOND REQUIREMENT
(SECOND RELATIONSHIP — TO DATUM REFERENCE FRAME A, D, & B)

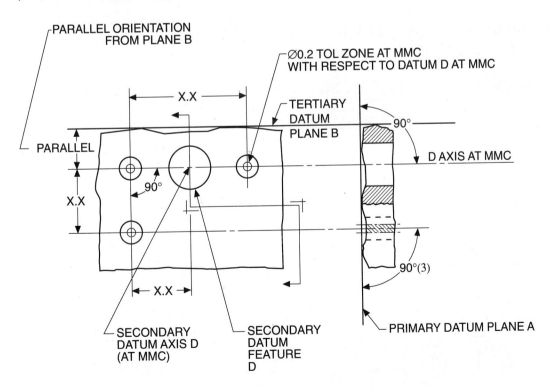

INTERRELATED DATUM REFERENCE FRAMES—WITH COMMON DATUMS—TWO HOLES AS SECONDARY AND TERTIARY DATUMS

USE OF THE TWO DATUM FEATURE (HOLE OR PIN) PRINCIPLES

The use of two holes (or pins) to establish the key interface between mating parts is classic in designing and building hardware. Along with establishment of a primary datum feature (surface) as shown in the example at right and others in this section, two size feature (holes) datums can completely stabilize the part to the three planes (see pages 272–292). Occasionally, due to some unique design requirements, more than two features (or even a pattern) can be used as datums (see below and pages 130,131). Nevertheless, no more than two size features are needed to provide complete orientation to the part. Where more than two features are used as datums, a mutual establishment of the two X and Y planes is derived via a centroid of rotation. Although feasible, and on occasion needed, this approach should be avoided if possible due to complications in interpretation and verification. Only two features are necessary to stabilize the part. Therefore, when a pattern is used as *one* datum, the question as to which two features actually do the locating is left to *any* two. It is seldom discernible (or necessary) to know which two. Yet in verification, unless some functional gage simulation is used, the selection of the appropriate features for orientation will vary by random choice. Where a pattern of features is used as *one* datum (as shown in the following example), the centroid of the pattern establishes a datum axis ("D" in this case) from which all related features take their reference.

COMPOSITE POSITION TOLERANCING—HOLE PATTERN AS SECONDARY DATUM

EXAMPLE

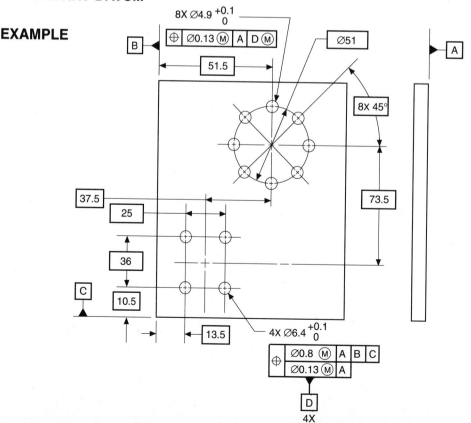

The design requirements determine the manner and precision needed to define the part details. Obviously, its production is also of concern. Therefore, some tempering of the design in consideration of production (with design approval) is normal. However, designing around manufacturing processes, unless pertinent to defining the engineering design requirements, is contrary to recommended design practices and the authoritative standards. The design requirements should predominate.

The *two hole* (or pin) datum application is an extension of the *one hole* (or pin) application covered on pages 134–154 and 298–301. The figure from page 135 is repeated below for information. One hole (or pin) as secondary datum feature establishes a secondary datum axis. However, the x and y planes need to be orientated (rotated) to another feature (datum surface B was selected) as a tertiary datum. This ensures that the hole pattern (9.5 holes) are placed properly relative to the desired datum reference frame.

ONE HOLE AS SECONDARY DATUM

EXAMPLE

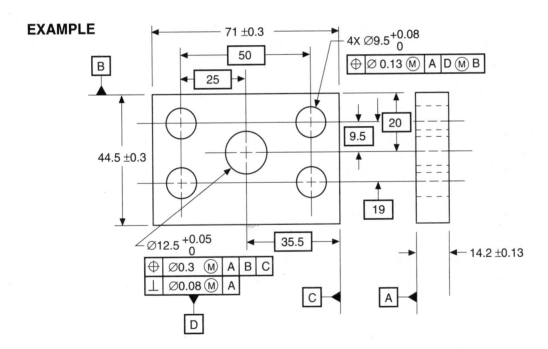

TWO HOLES AS SECONDARY AND TERTIARY DATUMS

Following discussion extends the preceding coverage to two holes used as secondary and tertiary datums. The design requirement is different than the preceding, but the need to stabilize the part to the datum reference frame is the same. In this case, however, it is done with a datum surface and two datum features (holes).

In this section the two feature (hole) application is explored in some detail. It does not exhaust all possibilities; only those most common and typical to two part interface of this variety are covered. The dimensions and tolerances (size and position) selected here are, of course, hypothetical and are only to represent principles and methodology possible in such situations. The magnitude of the values would be established by the real design requirements and the results of calculations and other determinations by the designer. A theme of this section, however, is to address the considerations and "facts" that exist when such requirements are applied. Where a pattern of features must be closely related to "size" datums (i.e., D and E in this section) and must "follow" them, due consideration must be given to the resulting effects of all the features involved (i.e., The Datum Virtual Condition Rule, etc.). The various methods shown progress through different possibilities where the effect of the datum feature size and position is to be considered, and then proceed on to methods where it can be neutralized to some extent. The choice is based upon the best judgment for the design at hand. An example of a typical two hole secondary and tertiary datum application is shown. Further detail on this example and others follow in this section.

EXAMPLE

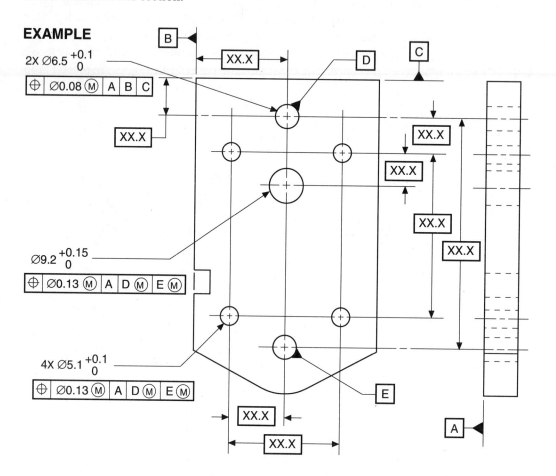

Another aspect of the theme of this section is that some precision is inherent in the subject type application. That is, if in fact a feature, or a pattern of features, is to "follow" other features (datums), a closer relationship must be assured by the tolerances applied. First, some needed precision must be given to the datum features themselves. Then it is seen that improved (greater) precision, or a refinement of some kind, must be specified from the datum features to the pattern or features "following" them. This is needed to ensure the *closer* relationship. The examples shown gravitate toward that which would probably be called rather close precision. This was chosen to represent better the objective of this method. Of course, the values chosen could vary toward a more lenient limit as based upon part function, size, and the relative precision required. The principles remain the same.

FUNCTIONAL GAGING APPLICATION AND ALTERNATIVES

Functional gaging possibilities and options are shown in the examples in this section to explain principles and methodology. A functional gage is in essence the "reconstruction" of the mating part from the requirements indicated on the design. It describes, as well, a representative mating part (or mating situation) which simulates the two parts in assembly. It also represents a "worst case" part which remains an acceptable part. The functional gaging approach is not required on such parts; it is an available option.

A functional gage would never accept a "bad" part but conversely could reject a border line "good" part. This is because in the standard method of allocating gage making tolerances, "some" part tolerance is utilized for the gage.

Functional gaging principles are shown in this section, and elsewhere in this text, for two primary purposes. One, this assists in the understanding of the principles involved where such methods can be used, i.e., when the MMC principle is valid and invoked. Reconstructing the part requirements (e.g., a hole pattern) from the engineering design into a reverse image (a gage) provides a quantified picture of the result. Two, functional gaging is a valid production tool and technique which has numerous technical and economic advantages. See later portions of this section for further details on functional gages.

Functional gaging principles can also be utilized *without* a functional gage. Alternative methods such as graphic analysis, paper gaging, "scatter-grams," etc., using various tools, including the computer, are at our disposal. For example, the results of a coordinately measured part (such as hole patterns shown in this section) using a coordinate measuring machine (CMM) or comparable method can be used to simulate functional gaging. Further, a *mathematical solution* from data derived from a CMM operation can be determined with the assistance of calculators and computer programs.

Therefore, functional gaging principles can be achieved via three different, but complementary, ways: (1) a functional gage, (2) physical graphic analysis (e.g., paper gaging), or (3) mathematically using a calculator or computer programs. Establishment of the computer programs, however, requires a superior knowledge of the technical principles involved and herein described.

See section on Graphic Analysis and Mathematical Analysis of Coordinately Measured Features (pages 293–302) in this text for further discussion of such techniques.

The graphic or mathematical methods may be necessary where the precision of the part may not permit functional gaging (insufficient tolerance can be derived for the gage build), where parts are rejected by a functional gage and are suspected as borderline good parts, where RFS specifications to the features controlled prevent use of a functional gage, where the functional gaging is not justified, etc. Further, it should be noted that the mathematical (calculator/computer) methods may be used to bypass functional gaging and graphic analysis completely, usually with greater accuracy as well as more rapidly.

TWO HOLES AS SECONDARY AND TERTIARY DATUMS AT MMC

There are two positionally toleranced hole patterns on this part. The first pattern is related to a complete datum reference frame established by A, B, C. The two holes established with the first datum reference frame are then used with the common datum A to become the secondary and tertiary datums for the second complete datum reference frame A, D, E.

Note that this is an example illustrating the rather common requirement of pickup of *two* reference holes from which other features are to be related. See also further discussion on following pages using two holes (or pins) as secondary and tertiary datums.

EXAMPLE

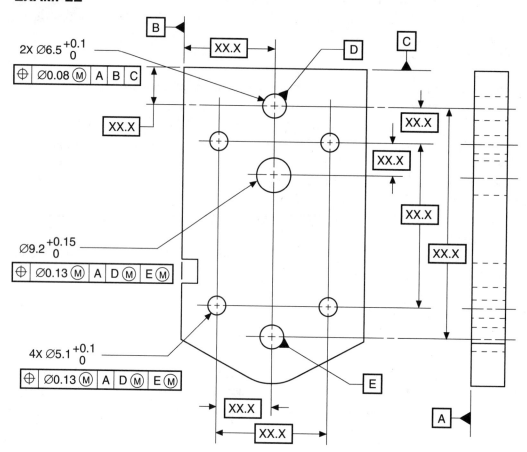

MEANING

FIRST REQUIREMENT (FIRST RELATIONSHIP — TO DATUM REFERENCE FRAME A–B–C)

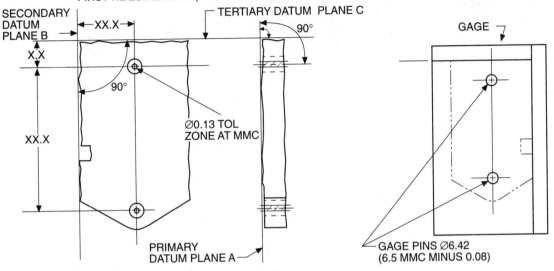

SECONDARY DATUM PLANE B

TERTIARY DATUM PLANE C

90°

XX.X

X.X

90°

∅0.13 TOL ZONE AT MMC

XX.X

PRIMARY DATUM PLANE A

GAGE

GAGE PINS ∅6.42 (6.5 MMC MINUS 0.08)

SECOND REQUIREMENT (SECOND RELATIONSHIP — TO INTERRELATED DATUM REFERENCE FRAME A–D–E)

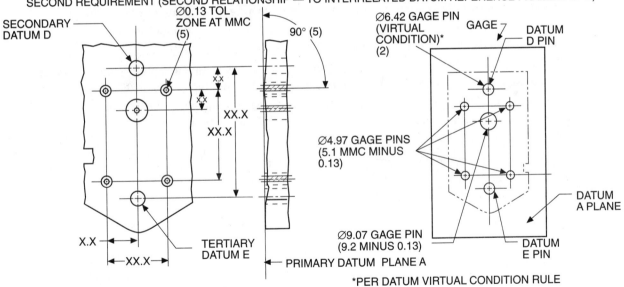

SECONDARY DATUM D

∅0.13 TOL ZONE AT MMC (5)

90° (5)

∅6.42 GAGE PIN (VIRTUAL CONDITION)* (2)

GAGE

DATUM D PIN

x.x

x.x

XX.X

XX.X

∅4.97 GAGE PINS (5.1 MMC MINUS 0.13)

X.X

TERTIARY DATUM E

XX.X

∅9.07 GAGE PIN (9.2 MINUS 0.13)

PRIMARY DATUM PLANE A

DATUM A PLANE

DATUM E PIN

*PER DATUM VIRTUAL CONDITION RULE

The similarity of the above situation to the simpler one shown here on a cylindrical part may serve to clarify this principle of orientation through the secondary and tertiary datums.

Note that the secondary datum cylinder D on this cylindrical part could be compared to the secondary datum hole (cylinder) D on the above flat part and that datum hole E on both parts give orientation (rotation control) as the tertiary datum. A complete datum reference frame is established in both parts to which other features on the parts can be related.

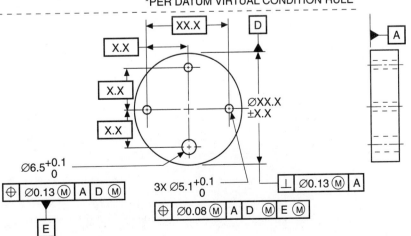

XX.X

X.X

D

X.X

X.X

∅XX.X ±X.X

X.X

∅6.5 +0.1 / 0

3X ∅5.1 +0.1 / 0

⟂ | ∅0.13 Ⓜ | A

⊕ | ∅0.13 Ⓜ | A | D Ⓜ

E

⊕ | ∅0.08 Ⓜ | A | D Ⓜ | E Ⓜ

Where two holes (or pins) are to be used as datum locators for mating parts, numerous possibilities and considerations could be taken into account. This section addresses the major options available and explores in some detail such applications.

First, earlier coverage in this text (page 134, etc.) introduced the use of a "size" feature (hole) as a datum feature. When only *one* hole (or pin) is used as a datum feature, an orientation (rotational) datum is required to stabilize the X, Y (and Z) planes and their relationship to the features (other holes) taken from these datum planes. A surface feature is used in the cited examples as a tertiary datum feature for orientation (rotation), not location. This is to ensure the clarity of the design requirement and for uniformity of interpretation in production and verification. The datum reference frame, in this case, is established by a plane surface (primary datum), a hole (secondary datum), and a plane surface (tertiary datum).

Use of *two* holes as secondary and tertiary datum (see Fig. (a) on page opposite and preceding) provides the location *and* orientation (rotation) for the related features in the X, Y, and Z planes. In this case, the datum reference frame is established by a plane surface (primary datum A), a hole (secondary datum D), and a second hole (tertiary datum E).

The functional interface on this part with its mating part is via the two $\varnothing$ 6.5 holes and the corresponding mating part pins. Thus, selection of these two features as datum features (D and E) for the other related features ($\varnothing$ 9.2 and $\varnothing$ 5.1 holes) seems logical. The related holes, the $\varnothing$ 9.2 and four $\varnothing$ 5.1 holes, are to "follow" the datum locating holes as functional to the interface with corresponding features on the mating part. MMC is used on the datum features (D and E) as desirable to the design requirement.

The positional tolerance ($\varnothing$ 0.08 at MMC) for the two $\varnothing$ 6.5 datums D and E features is determined by the usual calculation relative to the corresponding mating part pins. Note on the representative functional gage (Fig. (b) and preceding page), that the gage pin sizes are $\varnothing$ 6.42 virtual condition of the hole. This gage could verify the part locating holes as in their acceptable location and with respect to datums A, B, and C.

Note that the second functional gage (Fig. (c) and preceding page) to verify the five hole pattern ($\varnothing$ 9.2 and four $\varnothing$ 5.1) also requires that the "pickup" of the datums D and E holes be by *virtual condition* pins, $\varnothing$ 6.42. This is per Datum Virtual Condition Rule.

Since the datum "pickup" pins on *both* gages (Figs. (b) and (c)) are at virtual condition size ($\varnothing$ 6.42), the question can be raised as to how much difference (if any) there is between the use of features D and E as datums or not. Could the entire pattern (all seven holes) have been related as *one* pattern relative to datums A, B, and C and have the same control essentially? Could the two gages have been combined as one and represent the design requirement and part function?

See continued discussion on following pages.

EXAMPLE

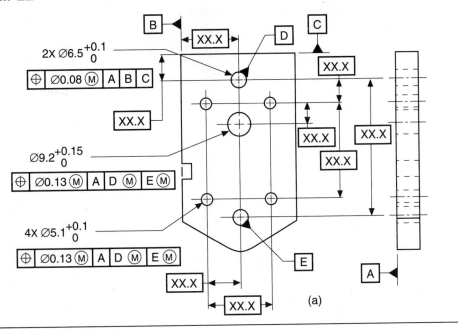

(a)

GAGE

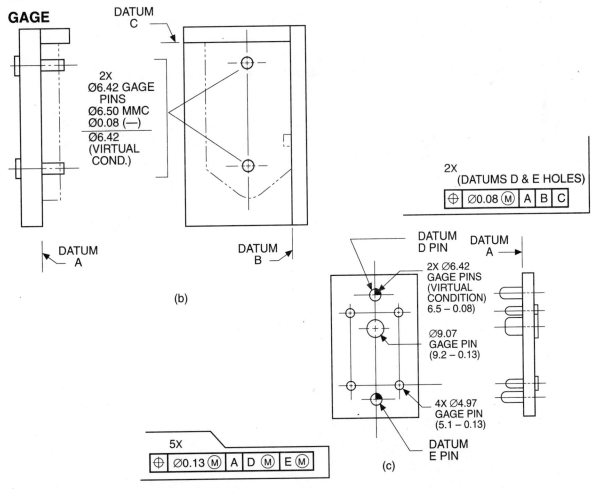

DATUM
C

2X
Ø6.42 GAGE
PINS
Ø6.50 MMC
Ø0.08 (—)
Ø6.42
(VIRTUAL
COND.)

DATUM
A

DATUM
B

(b)

2X
(DATUMS D & E HOLES)

| ⊕ | Ø0.08 Ⓜ | A | B | C |

DATUM
D PIN

DATUM
A

2X Ø6.42
GAGE PINS
(VIRTUAL
CONDITION)
6.5 − 0.08)

Ø9.07
GAGE PIN
(9.2 − 0.13)

4X Ø4.97
GAGE PIN
(5.1 − 0.13)

DATUM
E PIN

5X

| ⊕ | Ø0.13 Ⓜ | A | D Ⓜ | E Ⓜ |

(c)

Figure (a) on the opposite page suggests the option of the preceding page, i.e., all seven holes are related to the *one* datum reference frame A, B, and C (if the B and C datums are inserted instead of the D and E datums). This would be a legitimate requirement, but the desired closer relationship between the ⌀9.2 and ⌀5.1 holes and the two ⌀6.5 holes (as datums) originally intended seems to be no longer specified nor achieved.

From this discussion it can be derived that if the *closer* relationship between the five holes to the datum features D and E is, in fact, desired, further controls may be necessary. Also, from the discussion on the preceding pages, the representative gages reveal the narrow difference between the *with*, or *without*, datums D and E at MMC applications. Where open set-up, coordinate measuring machine (CMM), or other manual methods of verification are implemented, the MMC effect and virtual condition basis could inadvertently be overlooked by common misunderstanding; whereby, the relationship to datums D and E may be incorrectly assumed as on an RFS basis and, thus, would invoke a *closer* relationship to the datum reference frame A, D, E. The meaning of the datums D and E (at MMC) relationship would then be mistaken in its interpretation unless qualified by analysis to invoke MMC. Further, a broad discrepancy could exist in such cases between a functional gage check versus a CMM check, for example.

DATUM FEATURES AT MMC, ZERO PERPENDICULARITY TOLERANCE

When an ensured closer relationship is desired between related features and their "size" datum features, several methods may be used. Applying "zero perpendicularity" at MMC as an orientation refinement of position tolerance, as shown in Fig. (b), is one method. This would relate to datum D at MMC size, ⌀6.5 (MMC size and virtual condition identical) and ensure the closer relationship to datum D as the secondary datum and datum E as the tertiary datum. However, as noted in a representative functional gage for this requirement (Fig. (a), page 281), the datum E pin would need to also compensate for the possible ⌀0.08 *positional* tolerances by use of a diamond pin of ⌀6.5 MMC size by diamond (or flat) width to a value less than 6.34 in the D–E direction as shown.

See further discussion and illustrations on the following pages.

EXAMPLE

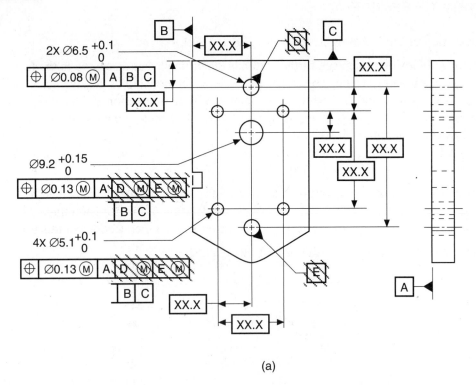

(a)

EXAMPLE

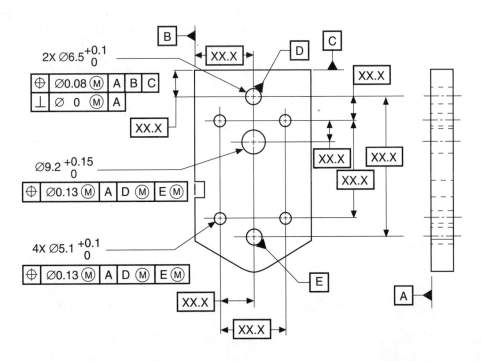

(b)

DATUM FEATURES AT MMC, ZERO
POSITION TOLERANCE

(For discussion of Fig. (a) at right, see pages 278 and 279.)

The use of zero position tolerance further extends the principles as discussed in the foregoing paragraph. This is shown in Fig. (b) at right. The closer relationship of the five holes ($\oslash$9.2 and $\oslash$5.1) are now established with the refinement of both orientation and location. Note in the representative functional gage in Fig. (a) (page 283) that both of the pickup pins of the datums D and E holes are at *MMC size* ($\oslash$6.5). Thus, the orientation question (as a part of position) requiring virtual condition recognition (Datum Virtual Condition Rule) is resolved. The MMC size and virtual condition are identical, $\oslash$6.5. The *closer* relationship is achieved.

Obviously, the use of the zero position tolerance of any kind must be accompanied by good judgment, proper selection of controls, and due consideration of the need and impact on design and production. Zero position tolerancing is also discussed in other sections of this text.

See added discussion and illustrations on following pages.

GAGE

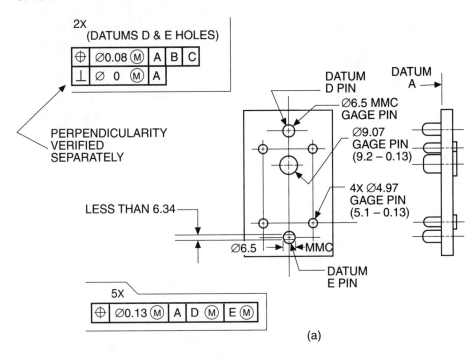

(a)

EXAMPLE

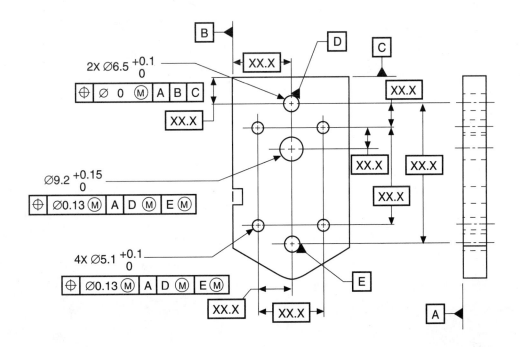

(b)

As discussed on the preceding page, the representative functional gage shown at right (Fig. (a)) would ensure the closer relationship of the five holes to the datums D and E holes. The datums D and E pickup pins are at $\oslash$ 6.5 (MMC size) and, therefore, provide the closer relationship of the five hole pattern to the datums D and E (the datum reference frame A, D, E). That is, instead of $\oslash$ 6.42 pins (D and E) as would be required in the earlier examples (using $\oslash$ 0.08 pos tol at MMC), the movement of the pattern is more restricted. It is again evident that to ensure that the closer relationship results (pattern to datums), the orientation and position tolerance permitted on the datum features must be refined or restricted in some manner.

DATUM FEATURES AT MMC, COMPOSITE ZERO POSITION TOLERANCE

Figure (b) at right utilizes the composite tolerancing method as another option. This would provide the same results as discussed before except, of course, the more lenient location of the datum features D and E from the datum reference frame A, B, C. The reasoning and application of composite positional tolerance is covered in other portions of this text.

The functional gage shown above (Fig. (a)) would also verify this requirement as the feature relating tolerance for the five holes relative to datums D and E are the same as the preceding page example (Fig. (b)).

The upper entry, pattern locating tolerance ($\oslash$ 0.8 at MMC) could be verified with a functional gage similar to that shown earlier in this section where A, B, and C surfaces (simulated datum planes) locate the part. The two gage pin sizes would be $\oslash$ 5.7 (i.e., $\oslash$ 6.5 MMC minus $\oslash$ 0.8) to verify the locating holes (D and E).

See following pages for further possible applications of the two datum hole type of part.

GAGE

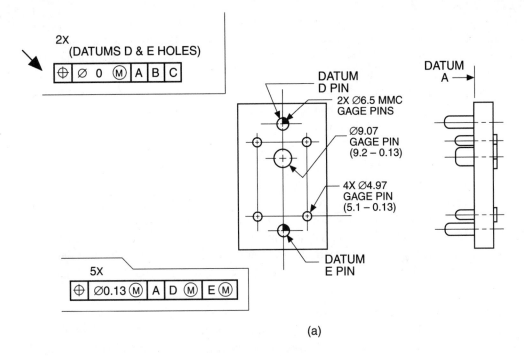

(a)

EXAMPLE

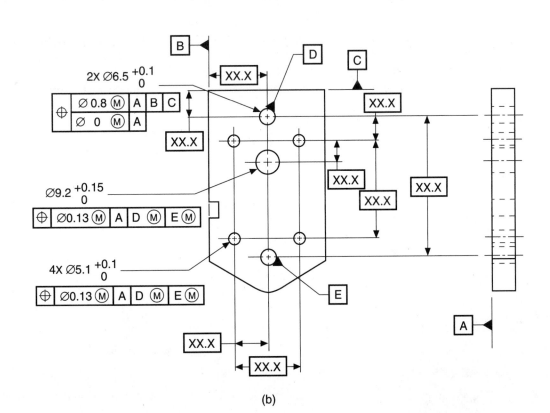

(b)

DATUM FEATURES CONTROLLED AT MMC, FEATURES RELATED TO DATUMS AT RFS

The illustration at right (Fig. (a)) controls the two ⌀6.5 locations (datums D and E) on an MMC basis relative to the datum reference frame A, B, C. A functional gage identical to that shown in the first examples shown in this section could be used. Or, of course, comparable coordinate measuring machine method (CCM) with MMC analysis could be used. This example, as all others in this section, represents a mating part interface control with a mating part (with location pins).

As an addition to the possibilities discussed earlier which could be applied on parts of this kind, this example utilizes RFS on the datum features D and E for the five hole pattern relationship. The closer relationship of the five holes to the datum reference frame A, D, E is then achieved. That is, the effect of departure from MMC size of the datum features (D and E) and the commensurate movement of the five hole pattern relative to that departure is *not* permitted. A shift of the pattern (the five holes) could be detrimental to the two part interface at the datums (D and E) and, thus, the design restricts such movement by specifying RFS to the datums D and E. This will require the pattern to "center-up" on the secondary datum feature D and rotate ("square-up") on the tertiary datum feature E.

Figure (b) at right shows a representative functional gage to verify the A, D, E requirement. Note that a tapered or expanding pin is used to pick up the secondary datum feature hole ("fills" the hole) so that the part cannot shift. The Datum Virtual Condition Rule does not apply here because RFS has been specified on the datum feature D relationship. Rotation of the part is restricted by the RFS (tapered or expanding) pin to the tertiary datum feature E. Due to the potential positional error (⌀0.08 at MMC) of the D and E holes, some "give" must be included in Fig. (b). A diamond pin method could also have been used on the datum E pin if considered to be more practical than the sliding insert.

If a coordinate measuring machine or comparable inspection were used, part stabilization to the primary datum A, set-up on the secondary datum D feature RFS, and rotation against the datum E feature (RFS) would be accomplished identical to the principles of the gage shown at right. Although the RFS method on the secondary and tertiary datums is restrictive and should be carefully evaluated relative to the design, it does have the advantage of ensuring closer compatibility between the alternative methods of open set-up inspection.

See following pages for further discussion on two feature datum parts and other possibilities.

EXAMPLE

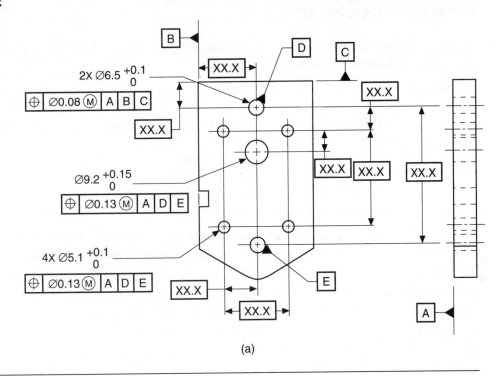

(a)

GAGE

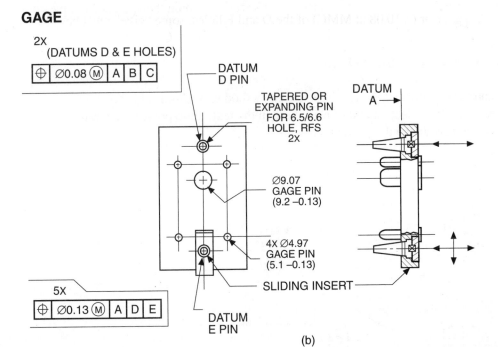

(b)

DATUM FEATURES CONTROLLED AT RFS, FEATURES RELATED TO DATUMS AT MMC

In Fig. (a) at right another method is used to control the hole pattern ($\oslash$9.2 and 4$\oslash$5.1 holes) and the two hole datum (D and E) relationship.

Accuracy of the datum features D and E locations affects the relationship of the pattern of features related to them. In this case RFS has been applied to the $\oslash$6.5 datums D and E holes to ensure greater precision of the datum features themselves. This reduces the error potential of the datum features relative to themselves (i.e., E to D), to the relationship of the five hole pattern, and also to the resulting relationship of *all* holes to the outside surfaces of the part (datums B and C). That is, features D and E movement relative to datums B and C is restricted to $\oslash$0.08, RFS. Thus, the five hole pattern in "following" datums D and E is also more restricted in its ultimate relationship to the outside surfaces.

Figure (b) at right illustrates that the verification of the $\oslash$0.08, RFS requirement would be achieved in some form of open set-up measurement (i.e., coordinate measuring machine, etc.).

Since MMC relationships to datums D and E are specified on the five hole pattern relationship, a gage (and principles) as previously illustrated in this section could be used to verify that requirement. A virtual condition gage pin (to datums D and E) would be used.

Where yet further precision in the relationship is required, the zero perpendicularity or zero position tolerance methods (on an MMC basis, of course) could be specified. Further use of RFS principles could also be stated if necessary. See below for some added possibilities. Gages and comparable principles, as illustrated earlier in this section, would then be added to the considerations as necessary.

SOME ADDITIONAL POSSIBILITIES

Numerous other possibilities of the two hole datum method exist. Some are shown below. From the coverage of this section and of elsewhere in the text, these possibilities can be derived as best represents the design requirements.

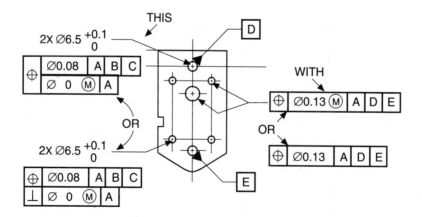

EXAMPLE

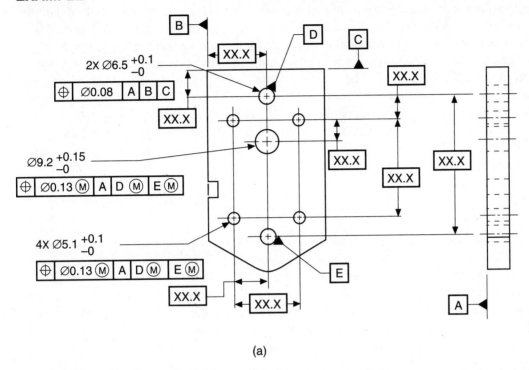

(a)

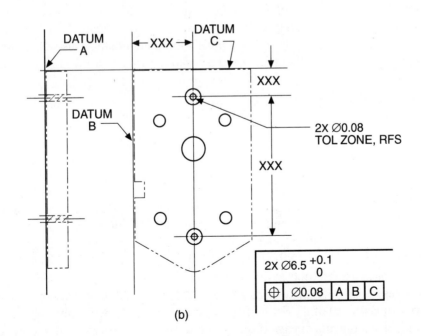

(b)

In Fig. (a) at right, a part (End Plate) is shown which uses a number of specifications and methods explained in earlier portions of this section and elsewhere in the text. Two of the four mounting holes ($\oslash$5.08) are also utilized as the locating holes in the ultimate assembly. These holes will serve both to position and attach the part via the mating part holes and shoulder screws. The two selected datum holes are designated with the darkened quadrants ($\oplus$) and are specified as datums D and E. The four $\oslash$ 5.08 holes are specified with a "pattern locating tolerance" from the datum reference frame A, B, C within $\oslash$ 0.5, at MMC, using the composite tolerancing method. The primary datum plane A is established by datum target areas (A1, A2, A3) on the flat cast surface. This is to establish a plane of orientation which is functionally representative of the part interface (stabilize the related holes) to the mating part features, to minimize effect of the cast surface irregularities and warpage, as well as to clear the six bearing hole ($\oslash$5.59, 4X $\oslash$ 3.55 and $\oslash$ 9.65) support lips. The secondary (B) and tertiary (C) datums are established from the indicated outside part edges. This positions the pattern appropriately on the part.

The two $\oslash$5.08 holes ($\oplus$) used as the datums (D and E) are specified with zero ($\oslash$0) tolerance, at MMC, as a "feature relating tolerance" (hole-to-hole in the pattern). This is to ensure adequate precision (orientation and location) of the locating datum holes and to enable the related six hole pattern to "follow" them with some precision. Earlier discussion in this section noted the reasons for such methods. The other two mounting holes ($\oslash$5.08) are permitted a $\oslash$0.25 positional tolerance, at MMC. This provides sufficient accuracy for these holes at assembly as clearance holes for the attaching screws.

The related six hole pattern features are given positional tolerances of $\oslash$ 0.15, at MMC, relative to datums D and E, at MMC. Via the use of zero ($\oslash$0) positional tolerancing relative to the datum features, the production and verification of these six holes will be to this precision as required by the design. For example, if a functional gage is used (see Fig. (b) at right) to simulate and verify this requirement, relationship ("pickup") with the datums D and E holes would be by MMC (same as virtual condition) sized pins; $\oslash$5.08 nominally. This situation has been discussed earlier in this section. Additionally, gage-maker's tolerances would be required as a further consideration.

If the part were verified by a coordinate measuring machine (CMM) or comparable methods, RFS pick up of datums D and E would probably be done first. Any permitted *pattern* movement relative to the datum feature's size departure from MMC would be determined, if necessary, by the amount of such departure. This would assume that the datum D hole orientation error (out-of-perpendicularity) did not negate the "bonus tolerance" and thus the permitted pattern movement. This would also assume that the datum E hole *size* was at least equal to, or greater than, the datum D hole size, and that it likewise did not negate the "bonus tolerance" and pattern movement via orientation (perpendicularity) *and* position error. The tertiary datum feature must not restrict the pattern relationship to the secondary datum feature. It must, however, provide the rotational orientation to stabilize the true position centerplanes (X, Y, and Z) and establish the basic pattern true positions. Graphic analysis, or computer analysis, can be used in lieu of gages. However, sizes of the holes and the amount of any orientation and/or positional error (quantified) of the datum features may have to be determined to complete such an evaluation. Since a functional gage (Fig. (b)) performs all necessary "judgments" on a "go," "no go" basis (is not quantified), no such exercise is then required. The trade-off negatives of the functional gage approach, however, must obviously also be considered.

The SIM REQT (meaning: Simultaneous Requirement) is added adjacent to the feature relating tolerance feature control frame to ensure that the 4 $\oslash$ 5.08 holes will be related as *one pattern* which is the design requirement in this case. The "ground rule" that "the same datums, in the

EXAMPLE

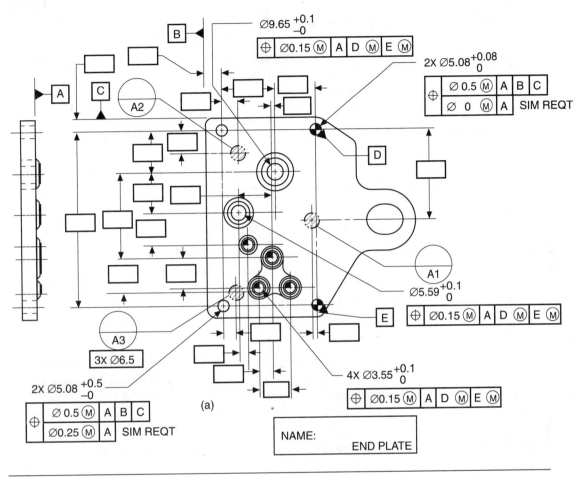

(a)

NAME:

END PLATE

GAGE

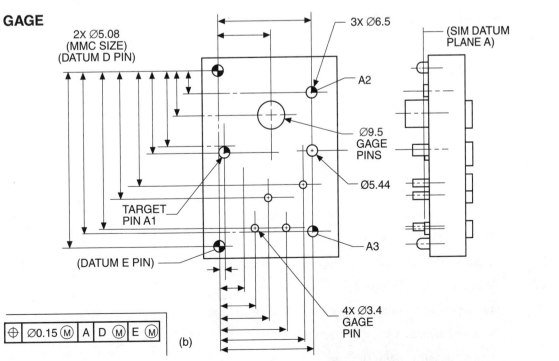

(b)

same order of precedence, under the same conditions are assumed to be one pattern" applicable to the pattern locating tolerance of a composite positional callout, *does not apply* to the feature relating callout automatically. It must be specified as shown.

This form of specification to two hole datum features *does* achieve the "closer" pattern relationship discussed as a theme of this section. Of the numerous distinct options, the method which is most suitable is decided by the design requirements and any further influencing factors, i.e., production quantities, tooling and inspection costs, quality, reliability, time, etc.

Other gaging, or comparable verification (i.e., CMM), of the four $\oslash$ 5.08 hole pattern would be as shown in similar situations discussed in this section. The various methods shown earlier could have been applied to this part. The key determination is the desired design requirement, which will guide selection of the most appropriate method of control and the magnitude of the tolerances.

THE FUNCTIONAL GAGE

Earlier in this section functional gage principles were discussed generally and used to explain positional tolerancing options when relating to two size features (holes) as secondary and tertiary datums. Figure (b), page 289, illustrates a gage for the piece part shown in Fig. (a), page 289. Nominal sizes (i.e., virtual condition) gage pin sizes are given as developed from the part.

To build a functional gage, tolerance for the gage features location must be taken from the piece part feature location tolerance. This is commonly referred to as the 10% rule (or 5% to 10% rule), which means that up to 10% (sometimes slightly more) of the part tolerance limits could be used for gage tolerance. Therefore, a part of borderline (extreme limit of acceptable tolerance) conditions could be rejected by a functional gage if the part were at the fringe edge of the acceptable tolerance range. The gage will not, however, ever accept a "bad" part.

A functional gage has the following advantages:

1. Minimizes time and resources involved to verify parts.

2. Represents functional interface of the concerned features.

3. Recognizes the subtle composite effects of size, orientation, and position as a "go," "no go" result.

4. Provides a "hard" tool which can be utilized by anyone with reasonable technical skill; does not require a highly skilled inspector.

5. Provides alternative methods for verification from surface plate, open set-up, coordinate measuring, etc.

6. Will never accept a "bad" part.

7. Can be implemented via "soft gage" computer programs.

Some disadvantages of a functional gage include:

1. Requires gage-maker's tolerance *taken from* piece part tolerances (up to 10% usually).

2. Could reject borderline good parts.

3. Must be reworked if the part is revised (unless "soft gage" used).

4. Costs for building, storage, and maintenance.

5. Does not quantify results (it's "go" or "no go").

Application of gage design and make tolerances on the gage illustrated in Fig. (b), using the 10% rule, could vary depending upon the gage designer and any local practices. However, the following approach could be considered one option:

Translating to coordinate dimensioning and tolerancing for gage design

Size: Datum pins and all gage pins +0.0025 wear allowance and +0.0025 size tolerance.

Location: Datum pin "E" and all gage pins ± 0.0025 from datum "D" pin in X and Y direction; consider diamond pin shape on datum "E" pin.

Location: Datum target pins A1, A2, A3 ± 0.08 from datum "D" pin in X and Y direction.

Basis for translation to gage pin tolerance within 10%:

⌀ 0.15 position tolerance on holes

10% of ⌀0.15 = ⌀0.015 (Translated to ± coord tol = ± 0.0105) **[SEE NOTES, BELOW]**

.0025 (WEAR) + 0.0025 (SIZE TOL) + .0051 (LOCATION TOL) = 0.0101

Slightly less than 10% of position tolerance ⌀ 0.15 used

The preceding translation to X and Y dimensions, as could be applied to the illustrated gage, depicts the requirements in the language of the toolmaker. That is, the positional requirements of the piece-part, using geometrical tolerances, datums, etc., have been presented as directly relating to the gage-making. The number of gages being considered for construction is normally one, two, or very few. Geometrical tolerance methods, however, can also be used if considered desirable. Any advantages of doing so should be weighed. Someone, somewhere (probably the toolmaker) would be required to translate the "true position" values to coordinates to either make or verify the gage itself. Is this responsibility best placed with the designer, with the toolmaker, or with the gage inspector?

NOTE 1: It may be necessary to avoid this foreshortening of the position tolerance (to equivalent coordinate tolerance, i.e., 70% of ⌀0.015 = 0.0105) to retain sufficient tolerance to make the gage. For example, if 0.015 is retained, the location of the gage pins (holes) could be ± 0.005; (0.0025 + 0.0025 ± 0.01= 0.015). This will, however, permit the possibility of slightly more than 10% of piece-part hole position tolerance (⌀0.15) to accumulate if the produced features of the gage deviate in location from the exact X and Y axes.

NOTE 2: This method uses the fail-safe distribution of position tolerance, permitting no more than 0.005 coordinate error in X or Y about any one exact location. This is because the coordinate system and true position system do not equate exactly to each other (i.e., on which end of the coordinate tolerance of ± 0.0025 is the 0.005 error, or is it equally distributed between, with 0.0025 on each end?) If the calculated risk were to be taken, hoping for or striving for equal distribution of the coordinate tolerance on *each* end, the gage-making pin (hole) location could specify ± 0.005. Using baseline dimensioning as shown (all dimensions from one pin) could encourage such considerations with reduced risk.

If the geometrical tolerance method were used, the following could be considered for the illustrated gage:

Size: Datum pins and all gage pins + 0.0025 wear allowance + 0.0025 size tolerance.

Location: Datum pin "E" $\oslash$ 0.005, RFS, relative to datum "D" pin, RFS (and appropriate datum reference frame). All locating dimensions "basic."

Location: All gage pins positional tolerance $\oslash$ 0.01, RFS, relative to datums "D" and "E" pins, RFS (and appropriate datum reference frame). All locating dimensions "basic."

Location: Datum target pins A1, A2, A3 positional tolerance $\oslash$ 0.1, RFS, relative to datum "D" and "E" pins, RFS (and appropriate datum reference frame).

All locating dimensions "basic."

Basis for translation to gage pin tolerance within 10%:

$\oslash$0.15 position tolerance on holes

10% of $\oslash$ 0.15 = $\oslash$ 0.015

0.0025 (WEAR) + 0.0025 (SIZE TOL) + $\oslash$ 0.01 (LOCATION TOL) = $\oslash$ 0.015

10% of $\oslash$ 0.15 used

The 10% rule and the feasibility of functional gaging is often governed by the tolerances which can be derived from gage-making. For example, the illustrated gage and its tolerances are ranging toward that fine limit which may become prohibitive in gage-making.[*] Therefore, the use of physical functional gages may not always be possible. However, optical or computer techniques may be possible to achieve essentially the same results using comparators, special equipment, coordinate measuring machines, etc. The use of graphic analysis (paper gaging) from coordinate measured results provides a valuable option. See following sections of the text for explanation of such techniques.

Coordinate measuring machine data can be used to derive *mathematical* results which simulate functional gaging principles with great precision. In such cases no "gage tolerance" is used, although inherent machine error is present and may have to be considered in very precise applications.

Mathematical simulation of functional gaging principles to part holes and pins can be achieved via the use of calculators or computer programs. Of course, the knowledge of the principles involved and the computer programs must be present and available to perform such operations. Such techniques can be used with MMC principles involved (i.e., functional gaging principles) or with RFS principles. Results with these methods are derived rapidly, to great precision, at minimum cost, with minimal human error, and are able to be recorded in printouts, graphs, etc., with convenience and efficiency. Refer to section on Graphic Analysis and Mathematical Analysis of Coordinately Measured Features (page 293, etc.) in this text for further discussion of such techniques.

---

[*]See NOTES 1 and 2, page 291.

INTERRELATED DATUM REFERENCE FRAMES WITH COMMON DATUMS

In some cases a relationship between interrelated datums is established only to the extent of the common datum references. Hence separate datum reference frames may be established on the same part which are related only to the extent of their common datum references, with no further orientation implied or necessary; see the part shown below.

MEANING

1.   Several *separate* datum reference frames are established using common datums.

2.   Datum C (slot) is a common tertiary datum for position toleranced hole patterns on each end of the part.

3.   Datum D (hole) is a primary datum for runout of the step diameter, and is a secondary datum for runout of the counterbore and position of the slot.

EXAMPLE

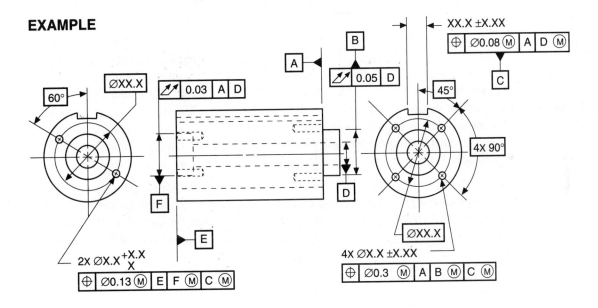

GRAPHIC ANALYSIS AND MATHEMATICAL ANALYSIS OF COORDINATELY MEASURED FEATURES

Graphic analysis or mathematical analysis techniques may be applied as a preferred method or as an option to functional gaging where MMC is applied; or, also, on RFS applications, or coordinately measured part holes or pins.

Other sections of this text have addressed the principles of functional gaging and graphic analysis (paper gaging), e.g., pages 127–129, 136–141. This section expands the concept of graphic analysis and furthers the emphasis placed on mathematical methods, often referred to as soft gaging, as a viable option in part verification.

Suppose the part illustrated at right (Fig. (a)) has been produced and has been measured with a coordinate measuring machine (CMM) oriented and located to its datums A, B, C as shown below. The upper right illustration (Fig. (b)) shows a format which could be used to record the pertinent data obtained from the measurement and the graph (Fig. (c)) for "paper gaging." The circles (tolerance zones) could be a part of the graph or a transparent overlay.

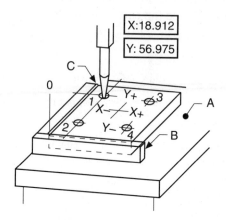

The middle illustration (Fig. (d) at right) shows the four hole locational data recorded, and the dots on the graph are the plotted hole locations with respect to "true position" (at 0 in the center of the circular tolerance zones). The actual hole sizes are also recorded and the available position tolerance (with bonus tolerance), as based upon departure from MMC. From the graphic results, we can see that all holes are acceptable within their respective tolerance zones (using the bonus tolerance on holes 2, 3, 4). Obviously, accuracy of the plot is important, and the scale used is a factor in the precision. Human error plays a role, and some time and care is involved in the plot.

Closer scrutiny of the results, however, reveals holes 2 and 4 are very near their respective "no go" limits at their produced size. There could be a question as to whether the concerned hole is actually good; particularly if the plotted dot is so close to, or on, the limit circle of the tolerance zone. Mathematical verification of the hole's actual location (based upon a diametrical value) can be determined by use of a calculator or computer programs, as based upon the equation $Z = 2\sqrt{X^2 + Y^2}$ (as shown on page 322). The lower results, (Fig. (e)) as calculated, confirm that all holes *are* good mathematically, although holes 2 and 4 need to have been calculated only as necessary.

If the part had been functionally gaged but the gage did not "go," the graphic analysis or mathematical method could provide alternative methods of proof of verification. Or where a functional gage did not "go" and the graphic analysis could not clearly prove acceptability (dots near or on circular lines), the mathematical method could determine a final result. The mathematical method can, in fact, bypass both use of a functional gage *and* graphic analysis with great precision. Obviously, such results are based upon the precision of the coordinate measuring operation. Many CMMs are equipped with standard computer programs to perform such translations. Compact or portable computers or programmed calculators are also readily available to make such computations separately from the CMM computer. This provides portable convenience and less "tie-up" of the larger machines.

Graphic or mathematical analysis is advantageous for many reasons. Such methods provide preferred or optional methods to gaging, provide records which can "see" the results, provide data that can be used in tool maintenance or repair, requires no gaging tolerance, provides precision results, adapts to quality or inspection sampling procedures. The mathematical method provides the highest precision in evaluating the data, and in the shortest possible time.

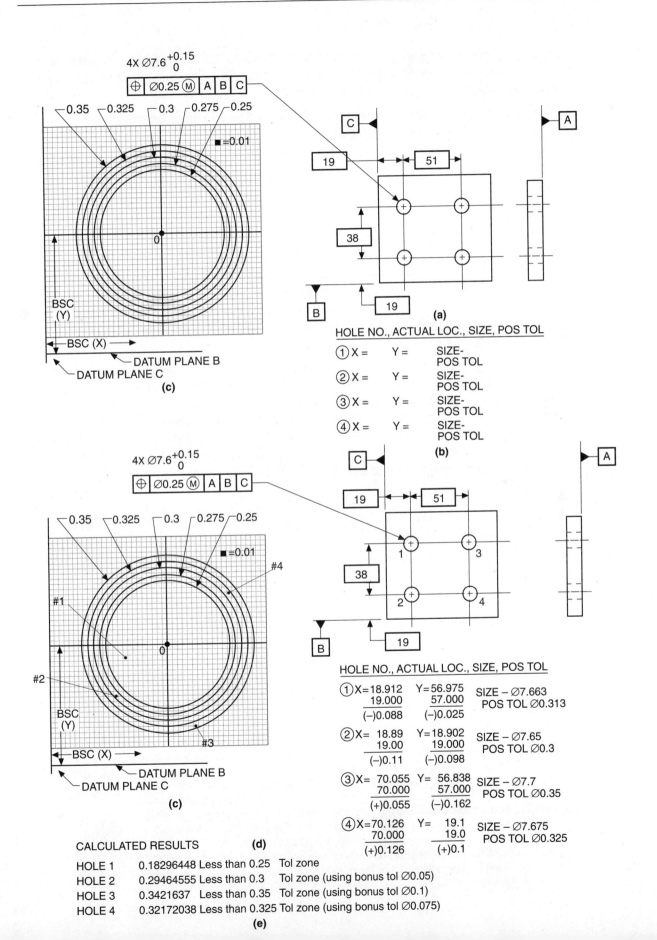

4X ⌀7.6 +0.15 / 0

| ⊕ | ⌀0.25 Ⓜ | A | B | C |

(a)

HOLE NO., ACTUAL LOC., SIZE, POS TOL

① X =   Y =   SIZE–
                 POS TOL

② X =   Y =   SIZE–
                 POS TOL

③ X =   Y =   SIZE–
                 POS TOL

④ X =   Y =   SIZE–
                 POS TOL

(b)

(c)

HOLE NO., ACTUAL LOC., SIZE, POS TOL

① X=18.912   Y=56.975   SIZE – ⌀7.663
     19.000   57.000   POS TOL ⌀0.313
    (–)0.088   (–)0.025

② X= 18.89   Y=18.902   SIZE – ⌀7.65
     19.00   19.000   POS TOL ⌀0.3
    (–)0.11   (–)0.098

③ X= 70.055   Y= 56.838   SIZE – ⌀7.7
     70.000   57.000   POS TOL ⌀0.35
    (+)0.055   (–)0.162

④ X=70.126   Y= 19.1   SIZE – ⌀7.675
     70.000   19.0   POS TOL ⌀0.325
    (+)0.126   (+)0.1

(d)

CALCULATED RESULTS

HOLE 1	0.18296448	Less than 0.25	Tol zone
HOLE 2	0.29464555	Less than 0.3	Tol zone (using bonus tol ⌀0.05)
HOLE 3	0.3421637	Less than 0.35	Tol zone (using bonus tol ⌀0.1)
HOLE 4	0.32172038	Less than 0.325	Tol zone (using bonus tol ⌀0.075)

(e)

Graphic analysis (paper gaging) of the part shown on pages 128 through 129 is repeated at right. Explanation of functional gaging and graphic analysis is also found on the referenced pages and on the preceding pages in this section.

Following are mathematical methods which can be used to determine the part acceptance, or not, in lieu of functional gages or graphic analysis:

STEP 1. Verify Pattern Locating Tolerance ($\oslash$0.8) on three of four $\oslash$6.4 holes. Using a programmed calculator or computer program[*], each hole is *proven acceptable* as the following calculated results:

> Hole 1 $\oslash$0.86023253—Less than $\oslash$0.9 Tol zone (using bonus tol $\oslash$.004)
> Hole 2 $\oslash$0.67082039—Less than $\oslash$0.8 Tol zone
> Hole 3 $\oslash$0.58309519—Less than $\oslash$0.8 Tol zone

STEP 2. Verify Feature Relating (hole-to-hole) Tolerance ($\oslash$0.13) on four $\oslash$6.4 holes. Using a computer program[*] developed with the required mathematical equations, the smallest diameter which will encompass all four produced hole centers while considering size departure from MMC as necessary is determined. The part (its four holes) is found to be within their respective tolerance zones as proven mathematically.

Using the data given at right on holes 1, 2, 3, and 4, and adding the bonus tolerance of $\oslash$0.05 on only hole 4 as shown, all holes are *proven acceptable.* The smallest diameter calculated, which will encompass all four hole centers based on the above conditions and as shown at right, is $\oslash$0.1222532 (less than the $\oslash$0.13 pos Tol). This value has been mathematically "corrected" for the MMC advantage of hole 4. This permits direct comparison to the stated $\oslash$0.13 position tolerance requirement shown on the drawing. The corrected value, $\oslash$0.1222532, is *less than* $\oslash$0.13; therefore, all four holes are acceptable.

Coordinate measuring principles assumed as basis for data:

Step 1. Same as described in preceding example
(pages 294–295).

Step 2. As shown in illustration at right.

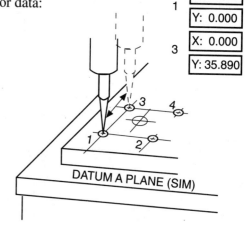

DATUM A PLANE (SIM)

---

[*]The computer programs are based upon the mathematics that is necessary to perform mathematical simulation of the hole location requirements. This simulation is also representative of a functional gage or graphic analysis (paper gaging). The Step 1 mathematical analysis can be performed by a calculator (See Appendix) or computer with the standard equation referenced. The Step 2 mathematical analysis requires a specially developed computer program. This must be based upon knowledge about geometric tolerancing principles, mathematics, and engineering judgment. Such capability now exists in The Lowell W. Foster Geometric Dimensioning and Tolerancing Module, "Geo-Calc," as used with a compact computer (T1 CC40); Step 1 uses Program 1; Step 2 uses Program 2. See "Acknowledgments" section in the front of the book.

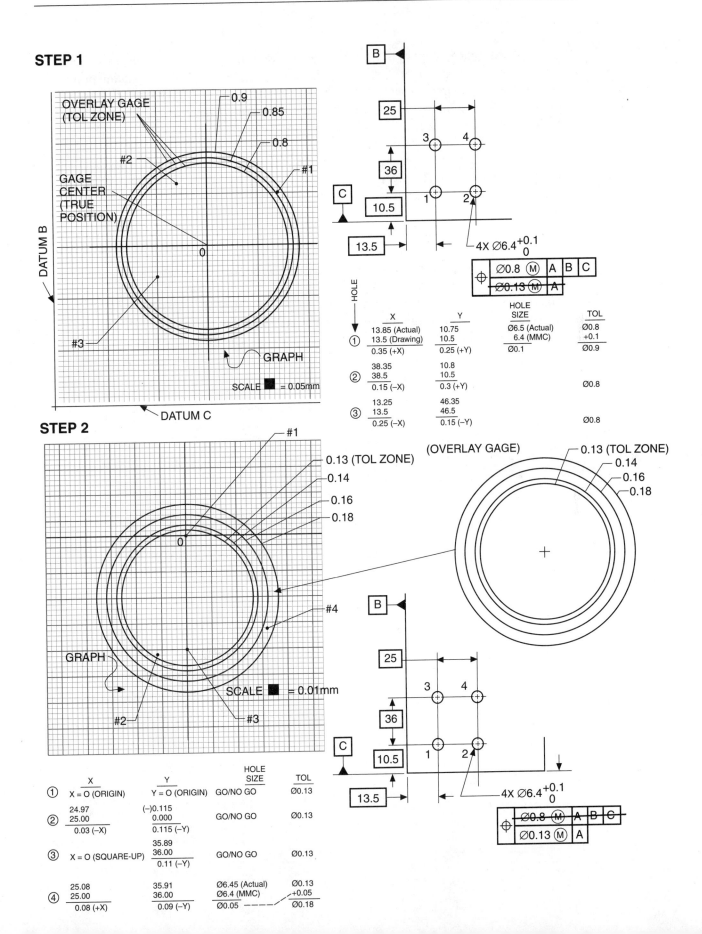

STEP 1

OVERLAY GAGE (TOL ZONE)

0.9
0.85
0.8
#2
#1
GAGE CENTER (TRUE POSITION)
DATUM B
0
#3
GRAPH
SCALE ■ = 0.05mm
DATUM C

B
25
36
3   4
C
10.5
1   2
13.5
HOLE
4X Ø6.4 +0.1/0

⊕ | Ø0.8 Ⓜ | A | B | C
Ø0.13 Ⓜ | A

	X	Y	HOLE SIZE	TOL
①	13.85 (Actual) / 13.5 (Drawing) / 0.35 (+X)	10.75 / 10.5 / 0.25 (+Y)	Ø6.5 (Actual) / 6.4 (MMC) / Ø0.1	0.8 / +0.1 / 0.9
②	38.35 / 38.5 / 0.15 (−X)	10.8 / 10.5 / 0.3 (+Y)		0.8
③	13.25 / 13.5 / 0.25 (−X)	46.35 / 46.5 / 0.15 (−Y)		0.8

STEP 2

#1
0.13 (TOL ZONE)
0.14
0.16
0.18
0
#4
GRAPH
SCALE ■ = 0.01mm
#2
#3

(OVERLAY GAGE)
0.13 (TOL ZONE)
0.14
0.16
0.18
+

B
25
36
3   4
C
10.5
1   2
13.5
4X Ø6.4 +0.1/0

⊕ | Ø0.8 Ⓜ | A | B | C
Ø0.13 Ⓜ | A

	X	Y	HOLE SIZE	TOL
①	X = O (ORIGIN)	Y = O (ORIGIN)	GO/NO GO	Ø0.13
②	24.97 / 25.00 / 0.03 (−X)	(−)0.115 / 0.000 / 0.115 (−Y)	GO/NO GO	Ø0.13
③	X = O (SQUARE-UP)	35.89 / 36.00 / 0.11 (−Y)	GO/NO GO	Ø0.13
④	25.08 / 25.00 / 0.08 (+X)	35.91 / 36.00 / 0.09 (−Y)	Ø6.45 (Actual) / Ø6.4 (MMC) / Ø0.05	0.13 / +0.05 / Ø0.18

Graphic analysis (paper gaging) of the four hole pattern of the part shown on pages 135 and 137, and repeated at upper right, is shown in the lower right illustration. Explanation of the functional gaging and graphic analysis principles is found on pages 127 through 129 and preceding in this section. In this case, however, a datum feature of size ($\oslash$12.5, datum "D") is referenced. In such a case, the four $\oslash$9.5 holes must be located with respect to datum "D" and thus "follow" its location. Datum "A" and "B" surfaces are specified to give the hole pattern orientation and rotation to the mass of the part. The previously discussed principles of graphic analysis remain the same. However, the resulting center of the four hole pattern may "float" within a zone equal to the departure from MMC of datum "D."

GRAPHIC ANALYSIS

From the sample set-up and resulting data shown below, developed from a coordinate measuring operation, the graphic analysis of the four hole pattern (hole-to-hole) and the pattern relationship (center) is illustrated. This represents one of many possible solutions where the center of the pattern could float within the $\oslash$0.038 zone (per the datum hole $\oslash$12.538) and the hole centers also fall within their respective tolerance zones. As seen in the example, both hole-to-hole and pattern relationship to the datum "D" are acceptable within their respective zones.

	in X	*diff.*	*in Y*	*diff.*	*Hole size*	Bonus *Tol*	+	Pos *Tol*	
Hole 1	25.064	(–)0.064	9.553	(–)0.053	$\oslash$ 9.563	$\oslash$ 0.063	+ $\oslash$ 0.13	= $\oslash$ 0.193	
Hole 2	24.916	(+)0.084	9.462	(–)0.038	$\oslash$ 9.54	$\oslash$ 0.04	+ $\oslash$ 0.13	= $\oslash$ 0.17	
Hole 3	24.947	(–)0.053	9.469	(–)0.031	$\oslash$ 9.563	$\oslash$ 0.063	+ $\oslash$ 0.13	= $\oslash$ 0.193	
Hole 4	24.929	(–)0.071	9.518	(–)0.018	$\oslash$ 9.54	$\oslash$ 0.04	+ $\oslash$ 0.13	= $\oslash$ 0.17	
Datum "D"	0	———	0	———	$\oslash$ 12.538	$\oslash$ 0.038	———		

Measurement results

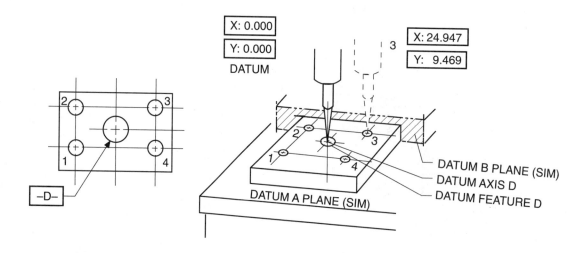

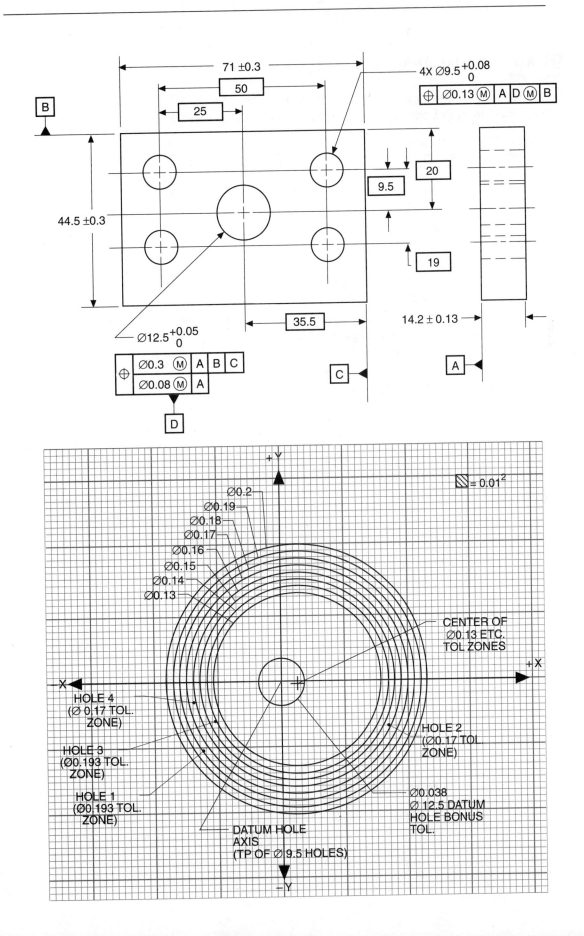

MATHEMATICAL ANALYSIS

Mathematical analysis of this part using the foregoing data can be utilized as a preferred method, or in lieu of graphic analysis (paper gaging) or even a functional gage.

Using a computer program[*] developed with the required mathematical equations, the smallest diameter which will encompass all four produced hole centers is determined. The program must also consider the effect of feature size departure from MMC of each of the four holes and also size departure from MMC of the datum feature and, thus, the permitted pattern float.

HOLE-TO-HOLE RELATIONSHIP

Based upon the data in the table on page 298 (including bonus tolerances as shown), all holes *proved to be acceptable* in their *hole-to-hole* relationship and with respect to their tolerance zones about true position. The smallest diameter calculated which will encompass all four hole centers is ⌀0.1162850—less than ⌀0.13 stated on the drawing. This value has been mathematically "corrected" to take advantage of each hole departure from MMC. This permits a result which can be compared directly to the stated ⌀0.13 position tolerance requirement on the drawing.

HOLE PATTERN RELATIONSHIP

Based on the data in the table on page 298, the "hole pattern" also *proved to be acceptable* in its relationship to the datum "D" axis and its zone permitted by the datum feature "D" size

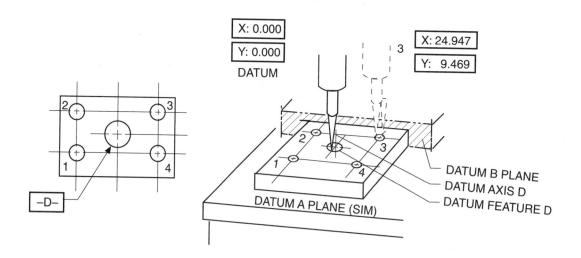

[*]The computer programs are based upon the mathematics necessary to perform mathematical simulation of the hole location requirements. This simulation is also representative of a functional gage or graphic analysis (paper gaging). The programs must be based upon knowledge about geometric tolerancing principles, mathematics, and engineering judgment. Such capability now exists in The Lowell W. Foster Geometric Dimensioning and Tolerancing Module, "Geo-Calc," as used with a compact computer (TI CC40). See "Acknowledgments" section in the front of the book.

departure from MMC. The result of the hole pattern location relationship calculation indicates the minimum bonus tolerance zone diameter required is ⌀0.0143909—less than ⌀0.038 bonus tolerance provided by the datum "D" departure from MMC. See also page 302.

The graphic analysis method of page 299, and repeated below, "pictures" what has mathematically been done.

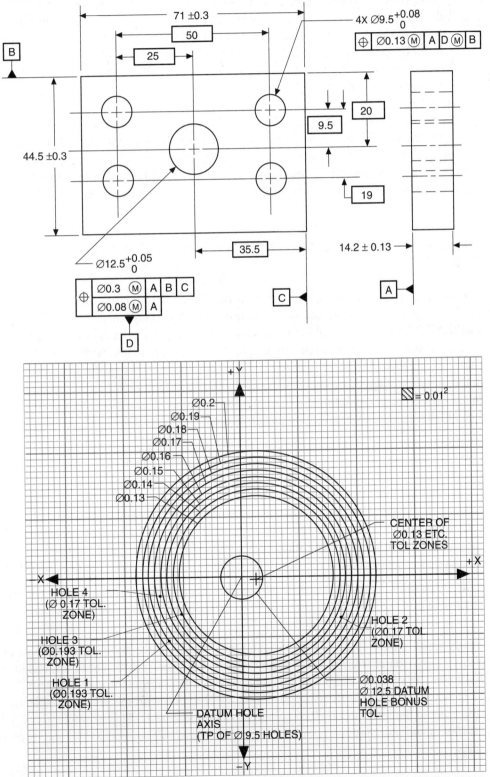

DATUM VIRTUAL CONDITION EFFECT

Additional "float" is available an amount equal to the difference between the $\oslash$12.5 MMC size and its "actual mating envelope size" (the largest inscribed cylinder, perpendicular to datum A which can be brought into contact against the $\oslash$12.5 datum A feature). This recognizes the possible virtual condition size of $\oslash$12.42 as invoked by the Datum Virtual Condition Rule and as found on the functional gage shown on page 137. For example, if the "actual mating envelope size" was $\oslash$12.46, an *additional "float"* of the center of the four hole pattern could be $\oslash$0.04 ($\oslash$12.5 MMC $-\oslash$12.46 $=\oslash$0.04). Thus, the total float ("Bonus tolerance") of the example part described in the data above and illustrated on the graphic analysis on Pages 299 and 301 could be up to $\oslash$0.078; $\oslash$0.038 (departure from MMC) $+\oslash$0.04 (difference between $\oslash$12.5 MMC size and $\oslash$12.46 "actual mating envelope size"); i.e., $\oslash$0.038 $+\oslash$0.04 $=\oslash$0.078. The $\oslash$0.078 value could have been used in the illustrated graphic analysis and in the mathematical analysis (page 298–301) as the "bonus tolerance" (instead of $\oslash$0.038 as shown) on this part as necessary. However, in this example of possible results in a hypothetical situation, the part was found acceptable without this consideration and the added inspection steps (determine "actual mating envelope size") were not needed.

POSITION AND PROFILE ALL AROUND TOLERANCE — DATUM FEATURES AT RFS

Preceding discussion has covered several methods of relating to two holes (or pins) as secondary and tertiary datums. Numerous other options, though varying the same principles in various RFS, MMC, and LMC combinations, are possible. In the preceding examples the two datum features (holes) have been related to *outside* part surface features; with subsequent location from these established to other features in the other patterns. This approach is probably found in the majority of applications. However, whether so-called *outer* features or *inner* features are established as the initial origin (datums) for relationship to other features either method is, of course, valid as based upon design choice.

Some families of parts such as component boards, PC boards, and numerous other possible parts can best be established from *inside* datums; with the outside features (surfaces) then established from them. The upper example at right (Fig. (a)) locates the two holes as a pattern, relative to datum A thus controlling the two hole pattern in location relative to itself and stabilized perpendicular to datum plane A. Then these two holes are identified as the B and C datums and referenced in the smaller hole pattern feature control frame as the secondary (B) and tertiary datums at MMC. Previous examples adequately describe the meaning of such callouts via functional gages or CMM verification relative to the positional tolerances specified.

The *outside* surfaces of the part are then located and controlled in size, form, and orientation by the all around profile of a surface callout. Note that the datums A, B (RFS), and C (RFS) relate the peripheral surface *from* the datum reference frame A, B, C.

The lower example at right (Fig. (b)) shows a similar situation to the above example; i.e., profile of a surface all around relative to the datum reference frame A, B, C. However, this example uses another variation, by establishing one hole perpendicular to datum A with zero tolerance at MMC. This hole is then established as the B datum with the second hole feature control referencing datum A and B (RFS) to create datum feature C. This example gives slightly different and more precise results because of the zero tolerance at MMC and reference to B at RFS.

Both parts can be verified with (hard) functional gages as discussed earlier in this text section and also with CMM or soft-gage techniques. The lower example would adapt more directly to CMM methods with the B datum providing a fixed datum axis and the C datum for orientation (rotation, clocking) control.

Numerous other variations of such applications as found below, using the various other datum options and material conditions, are possible. See other sections of this text for suggestions of these possibilities.

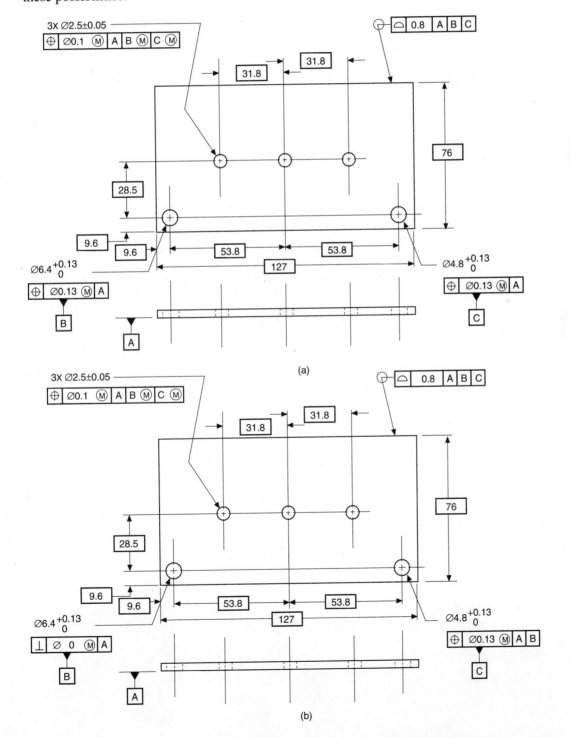

(a)

(b)

CONCENTRICITY

CONCENTRICITY ◎

Definition. Concentricity is that condition where the median points of all diametrically opposed elements of a figure of revolution (or corresponding-located elements of two or more radially disposed features) are congruent with the axis (or center point) of a datum feature.

Concentricity tolerance. A concentricity tolerance is a cylindrical (or spherical) tolerance zone whose axis (or center point) coincides with the axis (or center point) of the datum feature(s). The median points of all correspondingly-located elements of the feature(s) being controlled, regardless of feature size, must be within the cylindrical (or spherical) tolerance zone. The specified tolerance and the datum reference can only apply on an RFS basis.

EXAMPLE

MEANING

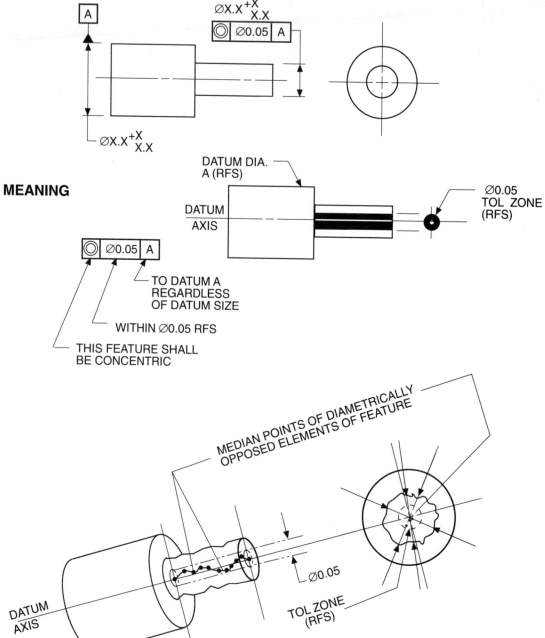

Concentricity is a type of locational tolerancing. It involves coaxial features of size and controls the relationship of the median points of all correspondingly-located elements on the surface of the feature, or features, relative to a datum axis. Concentricity tolerance, due to its unique nature of precision or relationship to the datum axis is always applied on an RFS basis.

Where interrelated coaxial features are concerned, it is well to first consider the possibility of specifying position tolerance (MMC) or runout tolerance as more readily achieved and economical types of control. (See also "coaxial features—selection of proper control," page 312).

Concentricity involves a condition where two or more features such as cylinders, cones, spheres, hexagons, etc., relate to a datum axis. The median points of these features must lie within the cylindrical tolerance zone, the axis of which coincides with the datum axis. Since the median points will be derived from the surfaces of the feature, or features, surface irregularities of form or size of these features will have an effect. For instance, the surface may be bowed, have "flats," be out-of-circularity, etc., in addition to being offset from the datum axis. This could involve a time consuming, complex, and costly procedure to perform the inspection and analysis of such parts.

Concentricity requirements are required less frequently than position (MMC) or runout requirements. However, where concentricity is required, it provides effective control over the unique applications of coaxial relationships. For example, concentricity might be applied to the coaxiality requirements of a tape drive pulley or of a capstan on a computer mechanism or a motor generator rotor. Often where balance is required, the out-of-circularity or lobing effect (and possible other form errors) may be permissible although it may exceed the conventional FIM requirement. Hence, any basically symmetrical form of revolution (hexagons, cones, etc.), or consistently symmetrical variation of such shape, could satisfy a concentricity tolerance where a runout requirement may not.

As stated before, the median points of a feature are determined by the *surface* of that feature. The resulting derived median point locations are compared with the feature's datum axis. The result of this comparison will determine whether all median points are contained within the tolerance zone and whether or not the feature satisfies the concentricity requirement.

A common method of determining the coaxiality of a feature with its datum (such as in runout) uses dial indicator (FIM) readings, with the part mounted and rotated on its datum feature. This method is also often adequate as a means of determining whether concentricity requirements have been satisfied. However, in concentricity, by the very nature of its characteristics, a FIM reading in excess of the stated permissible tolerance may not necessarily mean that the feature median points lie outside the concentricity tolerance zone. That is, if a feature has a surface error such as out-of-circularity, its surface will register a FIM dial reading relative to its datum diameter but its axis may not have actually exceeded the concentricity tolerance zone.

Figures 1 through 5 on page 309 illustrate this comparison with FIM and the principles involved in concentricity tolerancing. Using the sample requirement shown, Figure 1 shows a part that is theoretically perfect; it also shows the location of the cylindrical concentricity tolerance zone about the datum axis.

Figure 2 illustrates the part feature at its maximum displacement (or combination of various form errors). Note that the displacement registered by 0.05 FIM at the *surface* is equivalent to the *diametral* displacement of the median points within the concentricity tolerance zone. Figure 3 illustrates an out-of-circularity condition for which the 0.05 FIM reading is also *within* the specified concentricity zone. Under the conditions illustrated in Figs. 2 and 3, since the FIM is within the stated concentricity tolerance, the parts are acceptable with no further considerations. In this case, the results are essentially identical to a requirement which might have been stated using total runout on the basis of the FIM only. Therefore, to some extent, concentricity and runout can be considered identical except for interpretation, that is, surface variation (FIM direct) vs. median point displacement (as derived from FIM or some other comparable method).

Also, the above provides the uniform interpretation of any drawings for which concentricity (RFS) as previously defined (now generally equivalent to a total runout requirement) is used. However, where the stated concentricity tolerance is exceeded by the FIM, the characteristics of concentricity as now interpreted may possibly authorize acceptance of the part as being concentric within the specified limits.

Figure 4 illustrates the part with a 0.1 FIM reading which appears to exceed the stated concentricity tolerance. Yet closer observation will show that the part is actually *perfectly concentric* (as a hypothetical example) because the out-of-circularity error (possibly including other form errors) is symmetrically uniform. The FIM reading as conventionally applied is no longer conclusive and, in fact, could reject good parts. Closer scrutiny of Fig. 3 also shows that the FIM reading is again inconclusive with respect to the part's actual concentricity. The part is "perfectly concentric" although a FIM reading was registered.

Therefore, where a part is being evaluated for concentricity with conventional FIM techniques and is found to be within the concentricity tolerance stated, the part has met the concentricity tolerance requirement. However, when the part *exceeds* the FIM as conventionally checked or when actual concentricity is to be checked, more extensive inspection techniques are required, i.e., further consideration must be given to the question of whether additional form errors, such as out-of-circularity, are permissible within the concentricity tolerance. The form errors may vary anywhere within the tolerances of size so long as the concentricity tolerance is met.

Figure 5 illustrates the principles involved in checking actual concentricity through establishing median points. A precision collet and indicator set-up method are illustrated for purposes of explanation. However, other methods can be used, such as precision spindle inspection machines which are capable of establishing a datum axis for part rotation.

In the method shown, a measurement (M1) is taken of the part feature at a given point (X). The part is rotated 180°, and another measurement (M2) is taken.

(Continued on page 310).

COMPARISON WITH FIM METHODS

EXAMPLE

MEANING

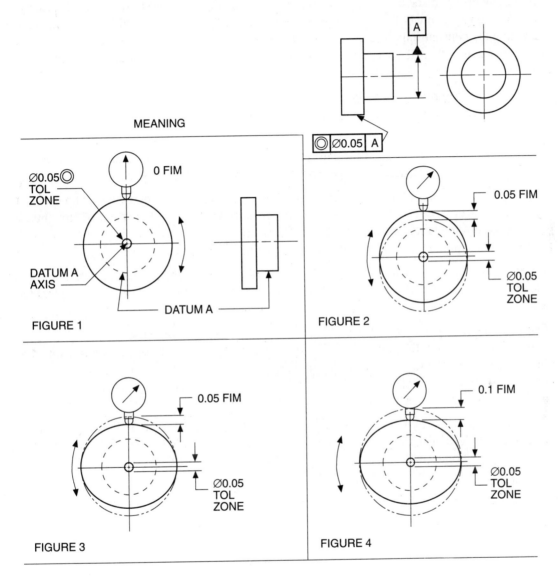

FIGURE 1

FIGURE 2

FIGURE 3

FIGURE 4

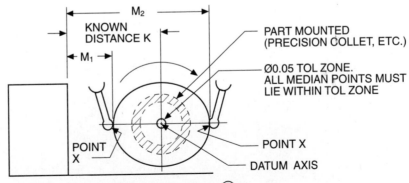

$$(M_2 - K) - (K - M_1) = \text{DEVIATION FROM} \ ◎$$
DIAL INDICATOR READINGS TAKEN ⊥ TO DATUM AXIS

FIGURE 5

The difference between these readings (see formula) is the deviation from concentricity (as a diameter) at that cross-sectional plane which is perpendicular to the datum axis.

Measurements are made at as many diametrical cross sections on the entire surface as necessary to adequately satisfy the requirement. For part acceptance, the derived median points must be within the concentricity tolerance zone. Since the deviation from concentricity determined is expressed as a diameter, it can be compared directly with the stated concentricity tolerance zone.

As previously described, when checking concentricity, all the irregularities of errors of form of a feature surface have an effect when determining its median points, and thus their displacement relative to the datum axis.

Since the dial indicator (or other measuring method) contacts the surface, the type of errors involved are, for all practical purposes, indistinguishable from one another.

To illustrate the various basic types of error involved, the example shown on the next page explains the general effect of feature eccentricity, error of parallelism of axis, out-of-straightness of axis, out-of-circularity, and out-of-cylindricity on concentricity tolerance requirements. Note that the sample part has a concentricity tolerance requirement of 0.03

Eccentricity is seen as the allowable displacement of the feature median points with respect to the datum axis. The 0.015 eccentricity is a displacement to one side but must be considered as a ⌀ 0.03 total tolerance cylinder.

Out-of-straightness and error of parallelism is seen as a tip up, down, sideways, or bow of the feature with respect to the datum axis. The 0.03 cylindrical tolerance zone represents the limits of the allowable error of this kind in determining the median points. Out-of-circularity and out-of-cylindricity errors reflect themselves in offset of the various median points about the datum axis in varying directions normal to the measuring plane.

As is seen, concentricity tolerance is a composite control of all elements of the feature concerned as they establish the median points. It is the resulting median points which are the basis for part evaluation or inspection.

For all practical purposes, when concentricity is the requirement, these errors are viewed as indistinguishable from one another; as a result they are considered in composite when we determine whether the part complies with the concentricity tolerance.

EXAMPLE

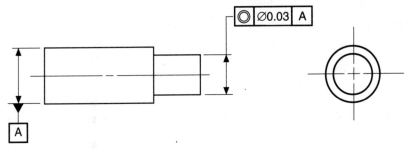

MEANING

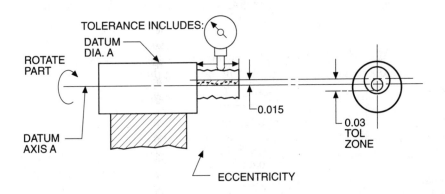

TOLERANCE INCLUDES:

DATUM DIA. A

ROTATE PART

DATUM AXIS A

0.015

0.03 TOL ZONE

ECCENTRICITY

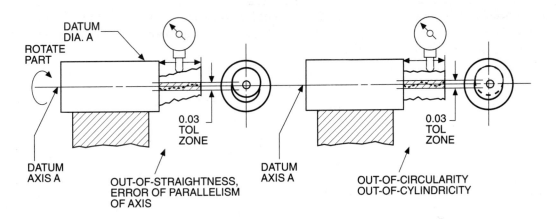

DATUM DIA. A

ROTATE PART

DATUM AXIS A

0.03 TOL ZONE

OUT-OF-STRAIGHTNESS, ERROR OF PARALLELISM OF AXIS

DATUM AXIS A

0.03 TOL ZONE

OUT-OF-CIRCULARITY OUT-OF-CYLINDRICITY

COAXIAL FEATURES—
SELECTION OF PROPER CONTROL

There are four characteristics for controlling interrelated coaxial features:

1.   RUNOUT TOLERANCE (circular or total)   (RFS)

2.   POSITION TOLERANCE   (MMC or RFS)

3.   CONCENTRICITY TOLERANCE   (RFS)

4.   PROFILE OF A SURFACE   (RFS DATUM)

Any of the above methods provides effective control. However, it is important to select the *most appropriate* one to both meet the design requirements and provide the most economical manufacturing conditions. (See also details of preceding and following sections.)

Below are recommendations to assist in selecting the proper control:

If the need is to control only CIRCULAR cross-sectional elements in a composite relationship to the datum axis, RFS, e.g., multidiameters on a shaft, use:

CIRCULAR RUNOUT ↗   **EXAMPLE** | ↗ | 0.1 | A – B |

(This method controls any composite error effect of circularity, concentricity, and circular cross-sectional profile variations.)

If the need is to control the TOTAL cylindrical or profile surface in composite relative to the datum axis RFS, e.g., multi-diameters on a shaft, bearing mounting diameters, etc., use:

TOTAL RUNOUT ↗↗   **EXAMPLE** | ↗↗ | 0.1 | A – B |

(This method controls any composite error effect of circularity, cylindricity, straightness, coaxiality, angularity, and parallelism.)

NOTE   Runout is always implied as an RFS application. It cannot be applied on an MMC basis, since an MMC situation involves functional interchangeability or assemblability (probably of mating parts), in which case POSITION tolerance would be used. See below.

If the need is to control the total cylindrical or profile surface and its actual mating envelope axis relative to the datum axis on an MMC or RFS basis, e.g., on mating parts to assure interchangeability or assemblability, use:

POSITION ⊕   **(IF MMC)**   **EXAMPLE** | ⊕ | Ø 0.1 Ⓜ | A Ⓜ |

(IF RFS)   **EXAMPLE** | ⊕ | Ø 0.1 | A |

OR RFS DATUM | ⊕ | Ø 0.1 Ⓜ | A |

If the need is to control the *axis* of one or more features in composite relative to a *datum axis*, RFS, e.g., to control such as balance of a rotating part, use:

CONCENTRICITY ◎   **EXAMPLE** | ◎ | Ø0.1 | A – B |

NOTE   Concentricity is always implied as an RFS application. Variations in size (departure from MMC size, out-of-circularity, out-of-cylindricity, etc.) do not in themselves conclude *axis* error.

If the need is to control the total cylindrical or profile surface simultaneously with the size dimension(s) (using basic dimensions for both), relative to a datum axis, e.g., precise fit, multi-diameters, etc., use:

PROFILE OF A SURFACE ⌒   **EXAMPLE** | ⌒ | 0.1 | A |

EXAMPLE

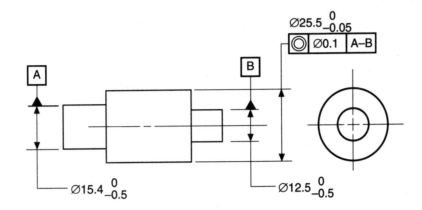

MEANING

MEANING

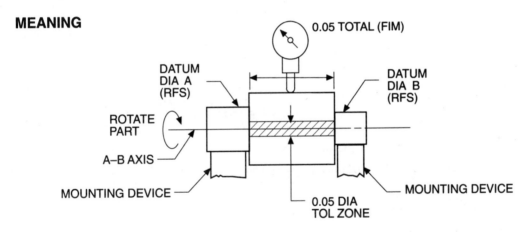

The example above illustrates a concentricity tolerance, using two datums to establish a common datum axis.

The $25.5_{-0.05}^{0}$ outside diameter is to be concentric to the common datum axis established by datum diameters A and B, within $\oslash 0.1$ regardless of the sizes to which any of the diameters are produced. The diameter size tolerances must be evaluated separately.

The lower portion of the illustration shows the part mounted on both datum diameters A and B, so that the concentricity of the 25.5 diameter with respect to both diameters A and B and thus to their common axis can be checked in one operation. The FIM method is not conclusive when the reading exceeds the stated concentricity tolerance. In this case, other methods must be used (see previous pages) to determine whether the feature does exceed the concentricity tolerance in an actual concentricity check.

TANGENT PLANE

EXAMPLE

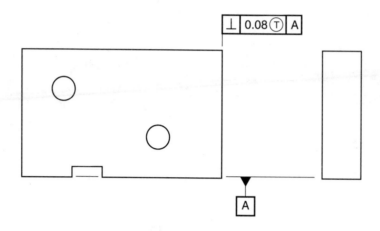

MEANING

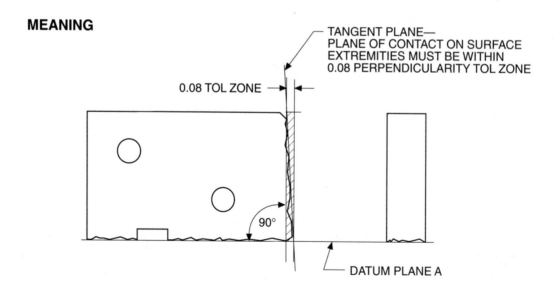

Where an orientation tolerance (e.g., Perpendicularity) is critical *only* to those portions of a flat surface which may be in contact with a mating part as a plane of tangency, the *Tangent Plane* symbol is added in the feature control frame. The resulting tangent plane (e.g., inspection plate or representation) only, must be within the orientation tolerance zone.

FREE–STATE VARIATION: "RESTRAINED" APPLICATION

Free-state variation is the amount a part distorts after removal of external forces applied during manufacture, for instance, parts consisting essentially of shells or tubes with a thin wall thickness in proportion to the diameter. Geometric tolerances (such as circularity, cylindricity, and concentricity) cannot be properly applied without controlling free-state variation on parts of this type.

Variations in the free state can exist in two ways: (1) distortion due to the weight or flexibility of the part, or (2) distortion due to internal stresses set up in fabrication.

Where free-state variation control is necessary, any datums and the features in control may require specification of their allowable free-state variation or the maximum force necessary to restrain each of them to drawing tolerance—so that the desired assembly relationship can be stated and represented for evaluation of the requirement, thus achieving compliance.

EXAMPLE

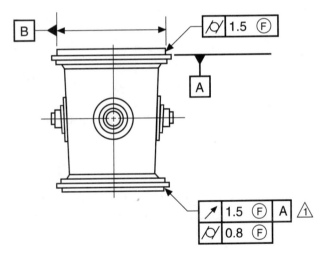

⚠ APPLIES WHEN DATUM ⌀B IS RESTRAINED TO DRAWING TOLERANCE
AND DATUM SURFACE A IS HELD FLAT WITH THE FORCE OF 150 NEWTONS
WITH PART IN HORIZONTAL POSITION.

MEANING

Compliance with the drawing note as stated.

AVERAGE DIAMETER

An average diameter is the mean of several diameters (not less than four) across a circular or spherical part used to determine conformance to *diameter* tolerance only. If practicable, the average diameter may be determined by using a periphery tape. Only when a diameter is allowed a maximum circularity tolerance in the free state should it be specified as AVG $\oslash$.

A part of this kind is normally expected to flex to proper form or shape at assembly. The reason for control in its free state is to facilitate hand or automated assembly and handling while the part is in its free-state configuration, to restrict distortion to within safe elastic limits of the material, etc.

EXAMPLE

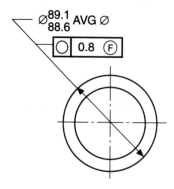

MEANING

1.   The high and low diameter size measurements (four minimum) must differ no more than 1.6 (because tolerance is 0.8 round on R). The calculated *average* (mean) of these diameters must be within the specified size tolerance 88.6–89.1 diameters.

2.   The entire surface at the cross section measured must lie within the 0.8 wide tolerance zone.

If the part in free state is of the general shape shown in the figure below, the 89.9 could be the highest diameter measurement, and 88.3 could be the lowest diameter measurement.

For example, four measurements are:

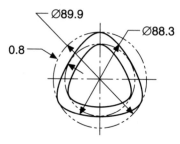

88.9 (high)

89.4

88.8

88.3 (low)

─────────

356.4

89.1 average diameter (within *size* tolerance)

89.9 − 88.3 = 1.6
average dia = 89.1

High and low measurements do not differ by more than 1.6; therefore, part is within freestate *circularity* tolerance.

If the part in free state is of the general shape shown in the figure below, the 89.4 could be the highest diameter measurement, and 87.8 could be the lowest diameter measurement. For example, four measurements taken are:

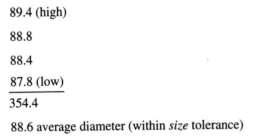

89.4 (high)

88.8

88.4

87.8 (low)
──────
354.4

88.6 average diameter (within *size* tolerance)

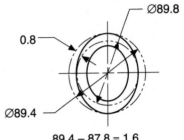

89.4 − 87.8 = 1.6
AVERAGE DIA = 88.6

APPENDIX A

CONVERSION OF ⊕ POSITION (CYLINDRICAL) TOLERANCE ZONES TO/FROM COORDINATE TOLERANCE ZONES

⊕ TO ±

TOTAL ⊕ TOL ZONE x .70711 = TOTAL COORDINATE TOLERANCE ZONE

EXAMPLE: 0.17 ⊕ TOL x .70711 = 0.1202 $\begin{cases}$ .0.12 TOTAL COORDINATE TOL
OR
±0.06 BILATERAL TOL

RULE OF THUMB:
USE .7 (OR 70%) OF TOL ⊕ ZONE TO CONVERT QUICKLY IN NON-CRITICAL APPLICATIONS, e.g. .7X0.17 = 0.119 OR 0.12 (±0.06)

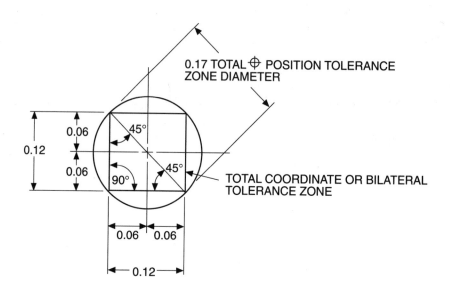

0.17 TOTAL ⊕ POSITION TOLERANCE ZONE DIAMETER

TOTAL COORDINATE OR BILATERAL TOLERANCE ZONE

±TO ⊕

TOTAL COORDINATE TOL ZONE x 1.4142 =TOTAL ⊕ TOL ZONE

EXAMPLE: 0.12 TOTAL COORDINATE TOL
OR
0.06 BILATERAL TOL × 2
$\Big\}$ × 1.4142=0.17 TOTAL ⊕ TOL

RULE OF THUMB:
USE 1.4 TIMES TOTAL COORD TOL ZONE TO CONVERT QUICKLY IN NON-CRITICAL APPLICATIONS, e.g. 1.4 × 0.12 =0.168 OR 0.17 ⊕ TOL

CONVERSION CHART

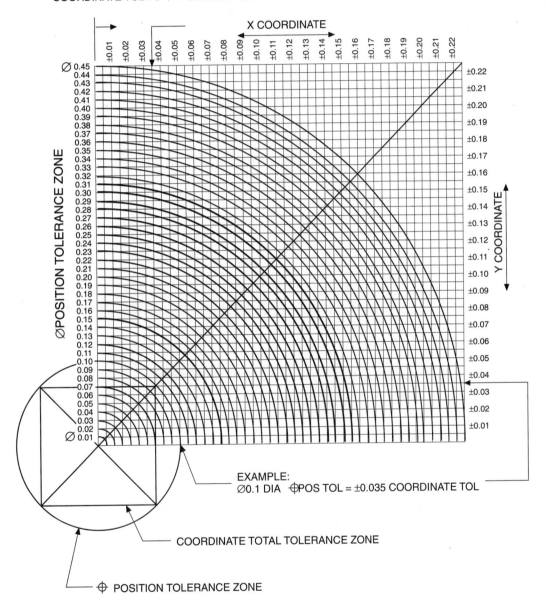

⊕ POSITION TOL TO COORDINATE TOL
COORDINATE TOL TO ⊕ POSITION TOL

X COORDINATE

⌀POSITION TOLERANCE ZONE

Y COORDINATE

EXAMPLE:
⌀0.1 DIA   ⊕POS TOL = ±0.035 COORDINATE TOL

COORDINATE TOTAL TOLERANCE ZONE

⊕ POSITION TOLERANCE ZONE

CONVERSION OF COORDINATE MEASUREMENTS TO ⊕ POSITION LOCATION

Y \ X	0.01	0.02	0.03	0.04	0.05	0.06	0.07	0.08	0.09	0.10	0.11	0.12	0.13	0.14	0.15	0.16	0.17	0.18	0.19	0.20
0.20	0.400	0.402	0.404	0.408	0.412	0.418	0.424	0.431	0.439	0.447	0.456	0.466	0.477	0.488	0.500	0.512	0.525	0.538	0.552	0.566
0.19	0.380	0.382	0.385	0.388	0.393	0.398	0.405	0.412	0.420	0.429	0.439	0.449	0.460	0.472	0.484	0.497	0.510	0.523	0.537	0.552
0.18	0.360	0.362	0.365	0.369	0.374	0.379	0.386	0.394	0.403	0.412	0.422	0.433	0.444	0.456	0.469	0.482	0.495	0.509	0.523	0.538
0.17	0.340	0.342	0.345	0.349	0.354	0.360	0.368	0.376	0.385	0.394	0.405	0.416	0.428	0.440	0.453	0.467	0.481	0.495	0.510	0.525
0.16	0.321	0.322	0.325	0.300	0.335	0.342	0.349	0.358	0.367	0.377	0.388	0.400	0.412	0.425	0.439	0.452	0.467	0.482	0.497	0.512
0.15	0.301	0.303	0.306	0.310	0.316	0.323	0.331	0.340	0.350	0.360	0.372	0.384	0.397	0.410	0.424	0.439	0.453	0.469	0.484	0.500
0.14	0.281	0.283	0.286	0.291	0.297	0.305	0.313	0.322	0.333	0.344	0.356	0.369	0.382	0.396	0.410	0.425	0.440	0.456	0.472	0.488
0.13	0.261	0.263	0.267	0.272	0.278	0.286	0.295	0.305	0.316	0.328	0.340	0.354	0.368	0.382	0.397	0.412	0.428	0.444	0.460	0.477
0.12	0.241	0.243	0.247	0.253	0.260	0.268	0.278	0.288	0.300	0.312	0.325	0.339	0.354	0.369	0.384	0.400	0.416	0.433	0.449	0.466
0.11	0.221	0.224	0.228	0.234	0.242	0.250	0.261	0.272	0.284	0.297	0.311	0.325	0.340	0.356	0.372	0.388	0.405	0.422	0.439	0.456
0.10	0.201	0.204	0.209	0.215	0.224	0.233	0.244	0.256	0.269	0.283	0.297	0.312	0.328	0.344	0.360	0.377	0.394	0.412	0.429	0.447
0.09	0.181	0.184	0.190	0.197	0.206	0.216	0.228	0.241	0.254	0.269	0.284	0.300	0.316	0.333	0.350	0.367	0.385	0.402	0.420	0.439
0.08	0.161	0.165	0.171	0.179	0.189	0.200	0.213	0.226	0.241	0.256	0.272	0.288	0.305	0.322	0.340	0.358	0.376	0.394	0.412	0.431
0.07	0.141	0.146	0.152	0.161	0.172	0.184	0.198	0.213	0.228	0.244	0.261	0.278	0.295	0.313	0.331	0.349	0.368	0.386	0.405	0.424
0.06	0.122	0.126	0.134	0.144	0.156	0.170	0.184	0.200	0.216	0.233	0.250	0.268	0.286	0.305	0.323	0.342	0.360	0.379	0.398	0.418
0.05	0.102	0.108	0.117	0.128	0.141	0.156	0.172	0.189	0.206	0.224	0.242	0.260	0.278	0.297	0.316	0.335	0.354	0.374	0.393	0.412
0.04	0.082	0.089	0.100	0.113	0.128	0.144	0.161	0.179	0.197	0.215	0.234	0.253	0.272	0.291	0.310	0.330	0.349	0.369	0.388	0.408
0.03	0.063	0.072	0.085	0.100	0.117	0.134	0.152	0.171	0.190	0.209	0.228	0.247	0.267	0.286	0.306	0.325	0.345	0.365	0.385	0.404
0.02	0.045	0.056	0.072	0.089	0.108	0.126	0.146	0.165	0.184	0.204	0.224	0.243	0.263	0.283	0.303	0.322	0.342	0.362	0.382	0.402
0.01	0.028	0.045	0.063	0.082	0.102	0.122	0.141	0.161	0.181	0.201	0.221	0.241	0.261	0.281	0.301	0.321	0.340	0.360	0.380	0.400

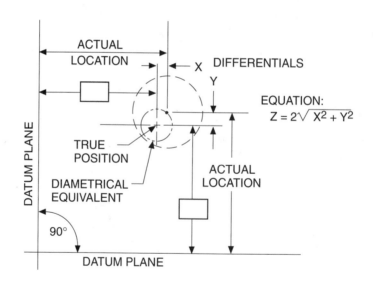

ACTUAL LOCATION

X DIFFERENTIALS
Y

EQUATION:
$$Z = 2\sqrt{X^2 + Y^2}$$

DATUM PLANE

TRUE POSITION

DIAMETRICAL EQUIVALENT

ACTUAL LOCATION

90°

DATUM PLANE

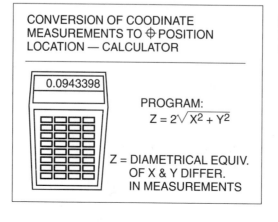

CONVERSION OF COODINATE MEASUREMENTS TO ⊕ POSITION LOCATION — CALCULATOR

0.0943398

PROGRAM:
$$Z = 2\sqrt{X^2 + Y^2}$$

Z = DIAMETRICAL EQUIV. OF X & Y DIFFER. IN MEASUREMENTS

MAY BE DONE BY CALCULATOR OR COMPUTER

EXAMPLE

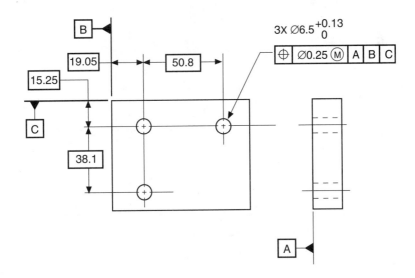

CONVERSION

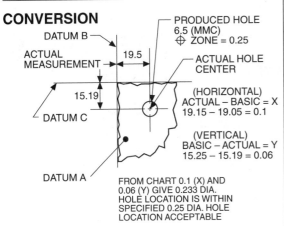

PRODUCED HOLE
6.5 (MMC)
⊕ ZONE = 0.25

ACTUAL HOLE
CENTER

(HORIZONTAL)
ACTUAL − BASIC = X
19.15 − 19.05 = 0.1

(VERTICAL)
BASIC − ACTUAL = Y
15.25 − 15.19 = 0.06

FROM CHART 0.1 (X) AND
0.06 (Y) GIVE 0.233 DIA.
HOLE LOCATION IS WITHIN
SPECIFIED 0.25 DIA. HOLE
LOCATION ACCEPTABLE

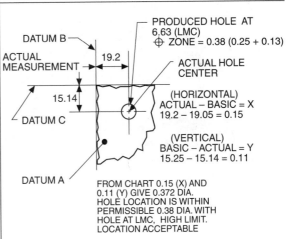

PRODUCED HOLE AT
6.63 (LMC)
⊕ ZONE = 0.38 (0.25 + 0.13)

ACTUAL HOLE
CENTER

(HORIZONTAL)
ACTUAL − BASIC = X
19.2 − 19.05 = 0.15

(VERTICAL)
BASIC − ACTUAL = Y
15.25 − 15.14 = 0.11

FROM CHART 0.15 (X) AND
0.11 (Y) GIVE 0.372 DIA.
HOLE LOCATION IS WITHIN
PERMISSIBLE 0.38 DIA. WITH
HOLE AT LMC, HIGH LIMIT.
LOCATION ACCEPTABLE

COMPARISON BETWEEN ASME Y14.5M AND ISO/1101

CHARACTERISTIC Requirement or Term	ASME – Y14.5M	ISO – 1101
STRAIGHTNESS	—	—
FLATNESS	⬭	⬭
ANGULARITY	∠	∠
PERPENDICULARITY (SQUARENESS)	⊥	⊥
PARALLELISM	//	//
CONCENTRICITY	◎	◎ *
POSITION	⊕	⊕
CIRCULARITY (ROUNDNESS)	○	○
SYMMETRY	≡	≡
PROFILE OF A LINE	⌒	⌒
PROFILE OF A SURFACE	⌓	⌓
RUNOUT (CIRCULAR)	↗	↗
RUNOUT (TOTAL)	↗↗	↗↗
CYLINDRICITY	⌭	⌭
DATUM FEATURE	A ▲	A ▲
MAXIMUM MATERIAL CONDITION (MMC)	Ⓜ	Ⓜ
REGARDLESS OF FEATURE SIZE (RFS)	NONE (ASSUMED UNLESS SPECIFIED MMC OR LMC)	NONE (ASSUMED UNLESS SPECIFIED MMC OR LMC)
LEAST MATERIAL CONDITION (LMC)	Ⓛ	Ⓛ
DATUM TARGET	A 1	A 1
PROJECTED TOLERANCE ZONE	Ⓟ	Ⓟ
BASIC DIMENSIONS (Untoleranced Dim)	50	50
FREE STATE CONDITION	Ⓕ	Ⓕ
TANGENT PLANE	Ⓣ	(PROPOSED)
RECIPROCITY REQUIREMENTS	NONE	Ⓡ (PROPOSED)
ENVELOPE REQUIREMENT	RULE #1	Ⓔ
ALL AROUND	⤷	⤷
SPHERICAL DIAMETER	SØ	SØ

* ALSO USED FOR COAXIALITY

XIV Symbolic methods of indicating common word requirements are optional:

WORDS	SYMBOL	EXAMPLE
C Bore or Spotface	⊔	
Depth or deep	⩒	4X ⌀3.42 ±0.25 ⊔ ⌀6.88 ±0.4 ⩒ 4.32 ±0.13
Quantity	X	
C Sink	∨	3X ⌀3.42 ±0.25 ∨ ⌀6.88 ±0.4 x100° ±5°
Square	□	3X□
All Around	(Symbol)	
Conical Taper		0.2
Flat Taper Slope		0.15 ±0.015:1
Reference Ref	( )	(25)
Radius	R	R7.63 ±0.08
Spherical Radius	S R	SR7.63 ±0.08
Spherical Diameter Spherical Dia	S⌀	S⌀7.63 ±0.08
Arc Length		55

CLARIFICATION OF RULE 1 OF ANSI Y14.5

That Rule 1 applies to "individual features *only* and *not* to an "interrelationship" of individual features and that form variation of the individual feature is contained within the PERFECT FORM AT MMC BOUNDARY will be explained in detail in the following text and illustrations.

Rule 1 applies to "individual" features whether the form variation is through permissible SIZE variation (form tolerance *not* specified) or whether the tolerance is specified.

The illustrations on the following pages show that the individual features involved in the form tolerances of FLATNESS, STRAIGHTNESS, CIRCULARITY, CYLINDRICITY and the orientation tolerance PARALLELISM readily comply with the definition of "individual" feature (see definition in section of "Clarification of Feature") and are to be within the perfect form at MMC boundary. If no form tolerance relationship had been specified, we would nonetheless assume that these features were also confined to the perfect form at MMC boundary.

ANGULARITY is an orientation characteristic involving an *interrelationship* of individual features. Rule 1 does *not* apply to such interrelationship, so angularity control must be specified. However, an extremity of the angular surface which is adjacent to, or part of, a surface which is size-dimensioned must be contained within the perfect form at MMC boundary, such as, for example, the lower left surface extremity near the angle vertex in the figure illustrating angularity.

PERPENDICULARITY is an orientation characteristic and defines an interrelationship of individual features; thus Rule 1 does *not* apply. Where perpendicularity is required, it must be specified. However, the individual surface features must be contained within the perfect form at MMC boundary as established by the size dimensions (such as the X and Y dimensions shown).

A PROFILE OF A SURFACE tolerance is usually applied as a combined form, orientation and size control (i.e., a BASIC overall dimension with a profile of a surface tolerance). In this case, it is *not* subject to Rule 1. When a PROFILE OF A LINE control is used as a refinement of size control, Rule 1 *would* apply.

RUNOUT, being a unique variety of rotating surface tolerance, involves an interrelationship of individual features; thus Rule 1 does *not* apply. However, the individual features must be confined to their size dimensions and perfect form at MMC boundary.

From the preceding we can state in summary that unless otherwise specified or controlled, the "size" limits of an individual feature control the applicable "form" variation within these size limits and the boundary of perfect form when no form tolerance is stated; furthermore, when a form tolerance *is* specified, the form variation must also be contained within these size limits and the boundary of perfect form at MMC (exception: straightness specified on ⌀ basis and related to size dimension). Rule 1 does not control and is not intended to control "interrelationships" of individual features. When an "interrelationship" is required, it must be specified.

PERFECT FORM AT MMC AS APPLIED TO FORM, ORIENTATION, AND PROFILE TOLERANCE

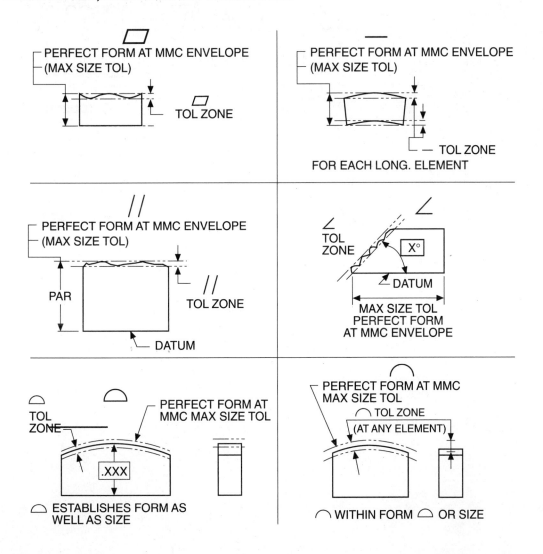

PERFECT FORM AT MMC AS APPLIED TO FORM, ORIENTATION, AND RUNOUT TOLERANCE

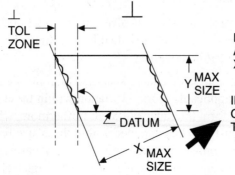

⊥
TOL
ZONE

⊥

Y MAX SIZE

DATUM

X MAX SIZE

PERF. FORM AT MMC BOUNDARY
APPLIES TO
X & Y SEPARATELY (INDIVIDUAL FEATURES)

INTERRELATIONSHIP
CONTROLLED BY ORIENTATION
TOL AS STATED

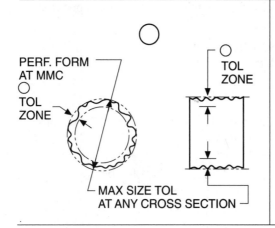

○

PERF. FORM
AT MMC
○
TOL
ZONE

○
TOL
ZONE

MAX SIZE TOL
AT ANY CROSS SECTION

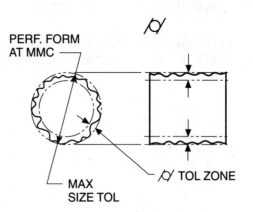

⌭

PERF. FORM
AT MMC

⌭ TOL ZONE

MAX
SIZE TOL

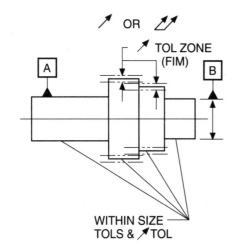

↗ OR ↗↗

TOL ZONE
(FIM)

A

B

WITHIN SIZE
TOLS & ↗TOL

PERFECT FORM AT MMC
ENV. APPLIES TO EACH
FEATURE SEPARATELY

INTERRELATIONSHIP
CONTROLLED BY
RUNOUT TOL

CLARIFICATION OF FEATURE

The meaning of the term "feature" is defined below. The definitions clearly show the flexibility of this term.

Definitions.

Features. A feature is the general term applied to a physical portion of a part and may include one or more surfaces such as holes, pins, screw threads, profiles, faces, or slots. A feature may be "individual" or "related."

Individual Feature. The "individual" feature, as implied by Rule 1, is a feature related to the part size. There are two types of individual features: "single surface" (non-size) and "opposed surface," (size feature).

A "single surface" individual feature, for example, is the flat surface extremity of a size dimension, or a surface given a "flatness" tolerance, etc.

An "opposed surface" individual feature of size, for example, is the cylindrical feature shown under the RULE 1 explanation (pages 26–28), or a specified "circularity" or "cylindricity" tolerance, etc. Specified "parallelism," which involves opposed surfaces, becomes a variety of interrelationship but is considered to retain the continuation of an individual feature and is subject to the confinement of form error within the perfect form at MMC boundary.

Interrelationship (Related) Feature. An "interrelationship" feature involves a relationship of two or more individual features and always requires specified geometric controls and a datum or datum reference frame. "Perpendicularity" is the most predominant example of an interrelationship type feature. One exception to the need for a datum would be the use of "straightness" RFS or MMC applied on a ⌀ basis to the part diameter size dimension. In this instance, an application of straightness (all longitudinal elements assumed collectively to establish a *size* feature consideration) would change from the category of "individual" feature to that of "interrelationship" feature because of the interrelationship of size and form which results in a "virtual condition" size or outer or inner boundary exceeding the MMC perfect form boundary.

Due to the variety of applications in which a "feature" can be used, we cannot clearly define it without relating it to a specific drawing requirement. Just as it would be difficult to describe a form or location tolerance application without a drawing, we need pictorial representation and specification of the involved requirements to clearly "see" the "feature."

A drawing illustrates the "feature" relationship within the specification and clarifies the functional intent of the requirement.

See the drawing that follows as an example.

EXAMPLE

Question How many features are on this part?

Answer At least thirteen. However, they need not and cannot be identified until the drawing specification is completed.

Question What are the features on the part shown below in its completed version?

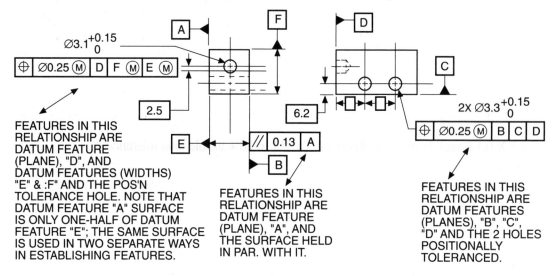

FEATURES IN THIS
RELATIONSHIP ARE
DATUM FEATURE
(PLANE), "D", AND
DATUM FEATURES (WIDTHS)
"E" & :F" AND THE POS'N
TOLERANCE HOLE. NOTE THAT
DATUM FEATURE "A" SURFACE
IS ONLY ONE-HALF OF DATUM
FEATURE "E"; THE SAME SURFACE
IS USED IN TWO SEPARATE WAYS
IN ESTABLISHING FEATURES.

FEATURES IN THIS
RELATIONSHIP ARE
DATUM FEATURE
(PLANE), "A", AND
THE SURFACE HELD
IN PAR. WITH IT.

FEATURES IN THIS
RELATIONSHIP ARE
DATUM FEATURES
(PLANES), "B", "C",
"D" AND THE 2 HOLES
POSITIONALLY
TOLERANCED.

POSITION TOLERANCE CONVERSION TO TOOL TOLERANCE

The question is often raised concerning relationship between a part dimensioned and toleranced with positional tolerance and the tooling necessary to produce the part. Positional tolerancing of cylindrical features normally derives cylindrical tolerance zones, whereas the tooling for the part can be based upon X and Y movement in its design or construction. The two geometrically opposed concepts are at times misunderstood as being incompatible with one another.

This point is often also used as an argument of resistance against the use of positional tolerancing using cylindrical tolerance zones. The uninformed fail to make the important distinction between the necessity (and normal practice) of defining the part *end-product* (reflecting its function) on the engineering drawing as the objective, rather than *methods* of manufacture of the part or its tooling.

Since, however, most hard tooling is produced in a shop provided with machinery based upon conventional X and Y axis movement, the methods of translating requirements between the product drawing and the tooling mode is important.

In the example on the following page, a sample part is shown using position tolerancing of three holes. Below the part representation is a typical box type drill jig design, which for purposes of our explanation, is the method used in producing the holes in this part during its manufacture.

Note that the X and Y nominal coordinates are readily transferable to the jig design from the product design. The nest surface and rails in the jig provide the datum location and origination of the jig coordinate dimensions to the drill bushing holes. The part clamping device in the jig must hold the part down against its primary datum A (for orientation), against the datum B rail (for orientation and location), and in contact with the third datum (for location).

The matter of translating (converting) the product positional tolerance ($\oslash 0.14$) to an X and Y value is illustrated in the jig drawing.

Imagine that in the evolutionary process from part drawing to part build, a tool order is written from the processing engineer to the tool engineer. This tool order requires the tool engineer to design a drill jig capable of producing 5,000 parts. Suppose as a hypothetical situation, that this tool order specifies, within a company standard or a contractual requirement, an "X" class tool to set the life, quality, and thus the percentage of product tolerance which can be used for this part.

The tool engineer simply converts as shown the total product design positional tolerance ($\oslash 0.14$) to the equivalent total coordinate $\pm$ tolerance (0.1) based upon standard conversion factors (.7 X 0.14) and derives 0.1. Reducing the 0.1 to 20% usable total tolerance for the "X" class jig derives 0.02 or ± 0.01 coordinate tooling tolerance for each dimension in X and Y.

The lower illustration recaps the dimension and tolerance allocation in one plane for clarification.

This reasoning of conversion from positional tolerancing on the product drawing to equivalent coordinate $\pm$ tolerancing may be applied to other situations in the process from design to building of parts, e.g., building prototype parts without hard tooling, building temporary tooling without tool design, and inspecting prototype parts.

Note that the transition from positional tolerance to coordinate tolerance is simply done and illustrates the flexibility inherent in the use of this system.

POSITION TOLERANCE CONVERSION TO TOOL TOLERANCE

EXAMPLE

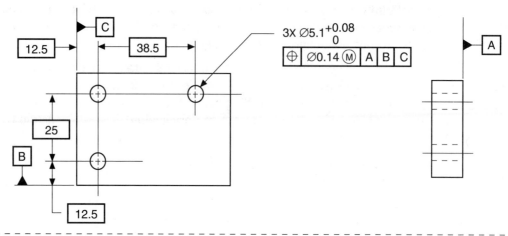

TOOL

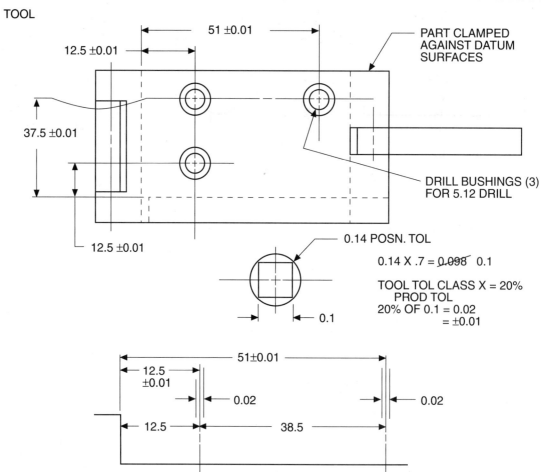

Permitted use of ISO Limits and Fits System methods

USE OF ISO LIMITS AND FITS SYSTEM

Limits and fits may be applied per ISO practices; see ANSI B4.2 for complete detailed information on this system.

Example of use:

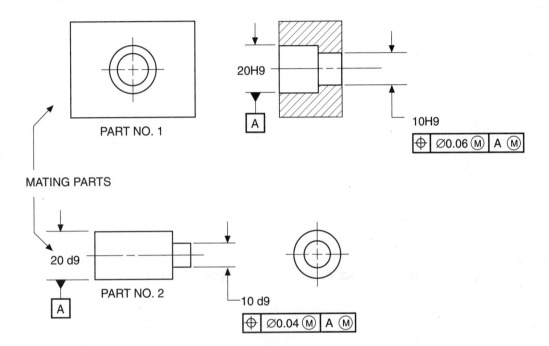

See Pages 113–119 for calculation of positional tolerances. (Also ASME Y14.5M Appendix B)

Depending upon preference, the optional methods of indicating the limits and fits sizes presented below may be substituted in the above examples and in numerous other examples in this text.

Part 1	20.000 20.052	(20 H9)	10.036 10.000	(10 H9)
Part 2	19.935 19.883	(20 d9)	9.960 9.924	(10 d9)
Part 1	20 H9		10 H9	
Part 2	20 d9		10 d9	

APPENDIX B

This Appendix section is to identify key added, revised, expanded or clarification changes found in this text edition as compared to the previous texts, Geo-Metrics II (Customary Inch or Metric) 1986 Editions.

Since these changes are found at varied appropriate places in the text, the following summary can provide the reader with a sense of their general content and identity as general information. Reference to the Glossary, Index, and the various text sections or chapters will enlarge upon the listed "changes" and their details as necessary.

- The Introduction is expanded to include some further information relative to GD&T importance, implementation, harmonizing with ISO standards, USA input to ISO standards, some history and reference to related ISO standards and the future of GD&T.

- New or revised terms and definitions:

  (See Glossary for definitions)

  Actual size

  Actual local size

  Actual mating size

  Actual mating envelope

  Datum feature simulator

  Derived median line

  Derived median plane

  Inner boundary

  Outer boundary

  Resultant condition (size) (MMC or LMC)

  Virtual condition (size) (MMC or LMC)

  Tangent plane

  Center plane

  Inner locus

  Outer locus

  Symmetry

  True geometric counterpart

- Changes under "Limits Of Size" Rule #1:
  Variations in size referred to as "the actual size of an individual feature" is now referred to as "the actual local size of an individual feature" at any cross section.

In numerous places where the term size was used previously, the terms *actual local size, actual mating size,* and *actual mating envelope* are substituted as appropriate for design intent and the expansion in distinguishing between the different uses of the term size. Where the term *actual size* is now used it connotes either of the above terms as appropriate if not fully stated.

- Regarding applicability of RFS and MMC in controlling straightness of an axis or center plane, the tolerance zone must contain the "derived median line" or the "derived median plane" rather than the "derived axis," "center line," or "derived center plane."

- Changes under Rules #2 and #3:
  Former Rules #2 and #3 regarding applicability of RFS, MMC, or LMC are replaced by a new Rule #2 that states that for all applicable geometric tolerances, "regardless of feature size" (RFS) applies with respect to the individual tolerance, datum reference, or both, where no modifying symbol is specified.

  Maximum material condition (MMC) or least material condition (LMC) must be specified on the drawing where it is required.

- Since the "regardless of feature size" condition is implied on all applicable geometric tolerancing for features of size, the RFS symbol is no longer necessary. This harmonizes U.S. practices with universal international (ISO) practices.

  As an alternate interim practice (Rule 2a), RFS may be specified on the drawing as in the previous standard.

- The "Symmetry" characteristic is reactivated and may be applied only on an RFS basis. Likewise, circular runout, total runout, and concentricity are reaffirmed as applicable only at RFS and cannot be modified to MMC or LMC.

- *Virtual Condition* explanation is expanded and described as a constant value and as it relates to *resultant condition* as a worst case value. *Inner locus* and *outer locus* terms are also introduced as an associated method of identifying extreme limits of the concerned feature tolerances.

- *Resultant Condition* is introduced and explained as a worst case inner locus or outer locus condition.

- Added figures to explain *virtual condition boundary* and *resultant condition boundary* as derived from the material condition specified at MMC or LMC.

- Datum features at virtual condition explanation is expanded to include the use of zero tolerance at MMC or LMC where a virtual condition equal to the maximum material condition is desired.

- The universal (ISO) datum feature symbol is adopted and replaces the previous one. Construction and application of the datum feature symbol and its use when establishing datums are added. The datum feature symbol is applied to the concerned feature surface outline, extension line, dimension line, feature control frame, dimension leader line, etc., in keeping with the principles established and the options provided.

- Use of the material condition symbol for RFS is no longer necessary, The "regardless of feature size" condition applies where the symbols for MMC or LMC are not stated on size features.

- The "Tangent Plane" symbol is introduced, explained, and illustrated.

- The "Between" symbol is introduced and explained.

- The "Symmetry" characteristic and symbol are reactivated from earlier texts and explained.

- The "Free State" symbol and explanation are added to the text.

- The "Projected Tolerance Zone" symbol is now placed in the feature control frame following the stated tolerance and any modifier. The dimension indicating the minimum height of the tolerance zone may also be placed in the feature control frame following the "projected tolerance zone" symbol or shown in a sectional view of the feature.

- All illustrations have been revised to show the universal ISO datum feature symbol and remove the RFS material condition symbol.

- Immobilization of the part relative to three mutually perpendicular planes in the datum reference frame is discussed and application relative to the "true geometric counterpart" is expanded.

- A *true geometric counterpart* of a feature is further defined, explained, and examples are provided.

- Circularity Tolerance is redefined for more explicit meaning and better description of this characteristic.

- More explicit terms are provided to describe and explain the datum of a cylindrical feature. The datum of a cylindrical surface is the axis of the *true geometric counterpart* of the datum feature (for example, the *actual mating size* or the *virtual condition boundary*).

- The role of the "simulated datum" is clarified. The term *actual mating envelope* is inserted where appropriate.

- Text on primary, secondary, and tertiary datums for diameter or width features, and under RFS, MMC, or LMC conditions, is expanded and explained using the terms *simulated datum*, *actual mating envelope*, *true geometric counterpart*, *virtual condition*, and *least material condition*.

- Where datum targets or equalizing datums are used on more complex parts and the datum feature symbol cannot be conveniently tied to a specific feature, the datum feature symbol is not required. The datum reference frame will be established by the collective points, lines, areas, or portions of the features involved.

- On equalizing datums, step datums, or contoured surfaces, it is permissible to show the datum feature symbol to identify the equalized or otherwise established theoretical center planes or planes of the datum reference frame established. This is an exception and should be done only when necessary and in conjunction with datum targets.

- For irregular or step datum surfaces, the datum plane should contain at least one of the datum targets.

- The switching datum target methodology is introduced.

- In expansion of the datum nomenclature, all appropriate figures were expanded or revised to include explanation of the relationships between the datum feature; simulated datum feature; simulated datum plane, axis, or center plane; datum feature simulator; true geometric counterpart; and datum plane, axis, or center plane.

- Numerous figures were expanded to provide more information.

- The composite positional tolerancing text is expanded.

  Text and illustrations are added where the composite tolerancing principle is extended to addition of a secondary datum in the lower segment of the feature control frame.

- The boundary positional tolerancing concept is introduced as a requirement on elongated or irregular features of size. The term BOUNDARY is placed beneath the feature control frame.

- The definition of *concentricity* is revised and refined.

- Under "Coaxial Feature Controls" a distinction is made between "Runout" as a control for elements of a surface of revolution (RFS); "Positional Tolerance," either MMC or RFS, to determine the axis of the actual mating envelope; "Concentricity" requiring the establishment and verification of the feature's median points and median line; and "Profile" where surface size and profile are to be controlled simultaneously. Illustrations were either revised or added to explain these principles.

- Under "Noncylindrical Feature Controls," a distinction is made between "Positional Tolerance," either MMC or RFS, to determine the center plane of the actual mating envelope; "Symmetry" requiring establishment and verification of the feature's median points and median line; and "Profile" where size and shape of the feature are to be controlled simultaneously. Illustrations were either revised or added to explain these principles.

- The option, where appropriate, to use profile tolerancing for location of features is added.

- Composite profile tolerance explanation, application, methodology, and illustrations are added.

- Angularity tolerance is expanded to "size feature" application using a total wide or cylindrical tolerance zone.

- Combined profile tolerance and position tolerance to irregular shaped features is introduced and explained.

- Profile tolerance related to "inside the part" datums, such as on PC boards is expanded and explained.

- Position tolerance, least material condition application, is expanded and explained; method of calculating tolerances is introduced.

- Concentricity tolerance is further defined and explained.

- Multiple use of datum target identities is explained, example introduced (i.e. switching datum targets).

- The "Principle of Reciprocity" is introduced and explained for informational purposes (advisory only—not per the standard as yet).

- "Author advisories" and rules-of-thumb are introduced and inserted at various pertinent places in the text.

- This Appendix section is added to identify key changes, additions, expansions, and clarifications found in this text as compared to the preceding printing of this text as an assist to the reader.

GLOSSARY

Actual Size—The general term for the size of a produced feature. This term includes the actual mating size and the actual local sizes.

Actual Local Size—The value of any individual distance at any cross section of a feature.

Actual Mating Size—The dimensional value of the actual mating envelope.

Actual Mating Envelope—This term is defined according to the type of feature, as follows:

(a) For an external feature. A similar perfect feature counterpart of smallest size which can be circumscribed about the feature so that it just contacts the surface at the highest points. For example, a smallest cylinder of perfect form or two parallel planes of perfect form at minimum separation which just contact(s) the highest points of the surface(s). For features controlled by orientation or positional tolerances, the actual mating envelope is oriented relative to the appropriate datum(s); e.g., perpendicular to a primary datum plane.

(b) For an internal feature. A similar perfect feature counterpart of largest size which can be inscribed within the feature so that it just contacts the surface at the highest points. For example, a largest cylinder of perfect form or two parallel planes of perfect form at maximum separation which just contact(s) the highest points of the surface(s). For features controlled by orientation or positional tolerances, the actual mating envelope is oriented relative to the appropriate datum(s); e.g., perpendicular to a primary datum plane.

Angularity—Angularity is the condition of a surface, axis, or center plane at a specified angle (other than 90°) from a datum plane or axis. Symbol: ∠.

Attitude Tolerance—See Orientation Tolerance.

Basic Dimension—A dimension specified on a drawing as BASIC (or abbreviated BSC) is a numerical value used to describe the theoretically exact size, profile, orientation, or location of a feature or datum target. It is the basis from which permissible variations are established by tolerances on other dimensions, in notes, or in feature control frames. A basic dimension is symbolized by boxing it: $\boxed{\text{X.XXX}}$

Basic Size—The basic size is that size from which limits of size are derived by the application of allowances and tolerances using the Limits and Fits system (see ANSI B4.2 standard).

Bilateral Tolerancing—A bilateral tolerance is a tolerance in which variation is permitted in both directions from the specified dimension, e.g., X.XXX ± .XXX.

Boundary, Outer and Inner—See Outer Boundary or Inner Boundary.

Center Plane—Center plane is the middle or median plane of a feature.

Circular Runout—Circular runout is the composite control of circular elements of a surface independently at any circular measuring position as the part is rotated through 360°. Symbol: ↗.

Circularity (Roundness)—Circularity is a condition of a surface where:

(a) For a feature other than a sphere. All points of the surface intersected by any plane perpendicular to an axis are equidistant from that axis.

(b) For a sphere. All points of the surface intersected by any plane passing through a common center are equidistant from that center. Symbol: ○.

Clearance Fit—A clearance fit is one having limits of size so prescribed that a clearance always results when mating parts are assembled.

Coaxiality—Coaxiality of features exists when two or more features have coincident axes, i.e., a feature axis and a datum feature axis.

Concentricity—Concentricity is that condition where the median points of all diametrically opposed elements of a figure of revolution (or correspondingly located elements of two or more radially-disposed features) are congruent with the axis (or center point) of a datum feature. Symbol: ◎ (or, formerly, ⦿).

Contour Tolerancing—See Profile of a Line or Surface.

Cylindricity—Cylindricity is a condition of a surface of revolution in which all points of the surface are equidistant from a common axis. Symbol: ⌭ .

Datum—A theoretically exact point, axis, or plane derived from the true geometric counterpart of a specified datum feature. A datum is the origin from which the location or geometric characteristics of features of a part are established.

Datum Axis—The datum axis is the theoretically exact axis of the true geometric counterpart of the specified datum feature (a center line on the drawing).

Datum Feature—A datum feature is an actual (physical) feature of a part used to establish a datum.

Datum Feature Simulator—A surface of adequately precise form (such as a surface plate, a gage surface, or a mandrel) contacting the datum feature(s) and used to establish the simulated datum(s). NOTE: Simulated datum features are used as the practical embodiment of the datums during manufacture and inspection.

Datum, Simulated—A point, axis, or plane established by processing or inspection equipment such as the following simulators: a surface plate, a gage surface, or a mandrel.

Datum Feature Symbol—The datum feature symbol consists of a capital letter enclosed in a square frame, a leader line extending from the frame to the concerned feature and terminating with a triangle. Symbol: ▢▴.

Datum Line—A datum line is that which has length but no breadth or depth such as the intersection line of two planes, center line or axis of holes or cylinders, reference line for tooling, gaging, or datum target purposes.

Datum Plane—A datum plane is a theoretically exact plane established by the extremities or contacting points of the actual datum feature surface with a reference plane (the simulated datum) as established from the datum feature simulator (surface plate or other checking device) or a plane established by datum targets.

Datum Point—A datum point is that which has position but no extent such as the apex of a pyramid or cone, center point of a sphere, or reference point on a surface for tooling, gaging, or datum target purposes.

Datum Reference—A datum reference is a datum feature and the resulting datum plane or axis.

Datum Reference Frame—A datum reference frame is a set of three mutually perpendicular datum planes or axes established from the simulated datums in contact with datum surfaces or features and used as a basis for dimensions for design, manufacture, and measurement. It provides complete orientation for the features involved.

Datum Surface—A datum surface or feature (hole, slot, diameter, etc.) refers to the actual part, surface, or feature coincidental with, relative to, and/or used to establish a datum plane.

Datum Target—A datum target is a specified point, line, or area on a part (identified on the drawing with a datum target symbol) used to establish datum points, lines, planes, or areas for special function, or manufacturing and inspection repeatability. Symbol: ⊖.

Dimension—A dimension is a numerical value expressed in appropriate units of measure and used to define the size, location, of geometric characteristics of a part or part feature.

Dimension, Basic—A numerical value used to describe the theoretically exact size, profile, orientation, or location of a feature or datum target. It is the basis from which permissible variations are established by tolerances on other dimensions, in notes, or in feature control frames.

Dimension, Reference—A dimension, usually without tolerance, used for information purposes only. A reference dimension is a repeat of a dimension or is derived from other values shown on the drawing or on related drawings. It is considered auxiliary information and does not govern production or inspection operations.

Feature—A feature is the general term applied to a physical portion of a part and may include one or more surfaces such as holes, pins, screw threads, profiles, faces, or slots. A feature may be "individual" or "related."

Feature, Axis of—A straight line which coincides with the axis of the true geometric counterpart of the specified feature.

Feature, Center Plane of—A plane which coincides with the center plane of the true geometric counterpart of the specified feature.

Feature, Derived Median Plane of—An imperfect plane (abstract) which passes through the center points of all line segments bounded by the feature. These line segments are normal to the actual mating envelope.

Feature, Derived Median Line of—An imperfect line (abstract) which passes through the center points of all cross sections of the feature. These cross sections are normal to the axis of the actual mating envelope.

Feature Control Frame—The feature control frame is a rectangular box containing the geometric characteristic symbol and the form, orientation, profile, runout, or location tolerance. If necessary, datum references and modifiers applicable to the feature or the datums are also contained in the box, e.g. ⌖ xxx A .

Feature of Size—A feature of size is one cylindrical or spherical surface, or a set of two opposed elements or opposed parallel surfaces, each of which is associated with a size dimension.

Fit—Fit is the general term used to signify the range of tightness or looseness which may result from the application of a specific combination of allowances and tolerance in the design of mating part features. Fits are of four general types: clearance, interference, transition, and line.

Flatness—Flatness is the condition of a surface having all elements in one plane. Symbol: ⬭.

Form Tolerance—A form tolerance states how far an actual surface or feature is permitted to vary from the desired form implied by the drawing. Expressions of these tolerances refer to flatness, straightness, circularity, and cylindricity, and occasionally profile of a surface and profile of line.

Full Indicator Movement (FIM) (FIR) (TIR)—Full indicator movement is the total movement of an indicator (or comparable measuring device) when appropriately applied to a surface to measure its variation.

Full Indicator Reading (FIR) (TIR) (FIM)—See Full Indicator Movement.

Geometric Characteristics—Geometric characteristics refer to the basic elements or building blocks which form the language of geometric dimensioning and tolerancing. Generally, the term refers to all the symbols used in form, orientation, profile, runout, and location tolerancing.

Geometric Tolerance—Geometric tolerance is the general term applied to the category of tolerances used to control form, profile, orientation, runout, and location.

Geometrics—Geometrics is a colloquial term to describe geometric tolerance (above).

Implied datum—An implied datum is an unspecified datum whose influence on the application is implied by the dimensional arrangement on the drawing, e.g., the primary dimensions are tied to an edge surface; this edge is implied as a datum surface and plane. (Implied datums are no longer permitted in the latest USA standards).

Inner Boundary—A worst case boundary (that is, locus) generated by the smallest feature (MMC for an internal feature and LMC for an external feature) minus the stated geometric tolerance, and any additional geometric tolerance (if applicable), from the feature's departure from its specified material condition.

Inner Locus (Virtual Condition)—An optional term and constant value to describe the "virtual condition size" of an internal size feature in an MMC application, and, the "virtual condition size" of an external feature in LMC application. Sometimes also referred to as inner or outer boundary as appropriate.

Inner Locus (Resultant Condition)—An optional term and variable value (worst case) to describe the "resultant condition size" of an external size feature in an MMC application, and, the "resultant condition size" of an internal size feature in an LMC application. Sometimes also referred to as inner or outer boundary as appropriate.

Interference Fit—An interference fit is one having limits of size so prescribed that an interference always results when mating parts are assembled.

Least Material Condition (LMC)—Least material condition is the condition in which a feature of size contains the least amount of material within the stated limits of size: for example, maximum hole diameter and minimum shaft diameter. It is opposite to Maximum Material Condition (MMC). Symbol: Ⓛ.

Limit Dimensions (Tolerancing)—In limit dimensioning only the maximum and minimum dimensions are specified. When used with dimension lines, the maximum value is placed above the minimum value. When used with leader or note on a single line, the minimum limit is placed first.

Limits of Size—The limits of size are the specified maximum and minimum sizes of a feature.

Line Fit—A line fit is one having limits of size so prescribed that surface contact or clearance may result when mating parts are assembled.

Location Tolerance—A location tolerance states how far an actual feature may vary from the perfect location implied by the drawing as related to datums or other features. Expressions of these tolerances refer to the category of geometric characteristics containing position, concentricity, and symmetry.

Locus, Outer or Inner— See Outer Locus or Inner Locus.

Maximum Dimension—A maximum dimension represents the acceptable upper limit. The lower limit may be considered any value less than the maximum specified.

Maximum Material Condition (MMC)—Maximum material condition is the condition in which a feature of size contains the maximum amount of material within the stated limits of size: for example, minimum hole diameter and maximum shaft diameter. Symbol: Ⓜ.

Median Plane—Median plane is the middle or center plane of a feature.

Minimum Dimension—A minimum dimension represents the acceptable lower limit. The upper limit may be considered any value greater than the minimum specified.

Minimum Material Condition—See Least Material Condition.

Modifier—A modifier is the term sometimes used to describe the application of the "maximum material condition," "regardless of feature size," or "least material condition" principles. The modifiers are maximum material condition (MMC), symbol Ⓜ and least material condition (LMC), symbol Ⓛ. (RFS symbol Ⓢ used in past standards.)

Multiple Datum Reference Frame—A multiple datum reference frame is the condition in which there are one or more related or separate datum frames.

Nominal Size—The nominal size is the stated designation which is used for the purpose of general identification, e.g., 1.400, .060, 12, 30.5, etc.

Normality—See Perpendicularity.

Orientation Tolerance—An orientation tolerance controls the orientation of features to one another. Expression of these tolerances refers to perpendicularity, angularity, and parallelism, and in some instances, profile, as related to datum features. Orientation tolerances are sometimes referred to as attitude tolerances.

Outer Boundary—A worst case boundary (that is, locus) generated by the largest feature (LMC for an internal feature and MMC for an internal feature) plus the geometric tolerance, and any additional geometric tolerance (if applicable), from the feature's departure from its specified material condition.

Outer Locus (Virtual Condition)—An optional term and constant value to describe the "virtual condition size" of an external size feature in an MMC application; and the "virtual condition size" of an internal size feature in an LMC application.

Outer Locus (Resultant Condition)—An optional term and variable value (worst case) to describe the "resultant condition size" of an internal size feature in an MMC application, and, the "resultant condition size" of an external size feature in an LMC application.

Parallelepiped—Shape of tolerance zone. The term is used where TOTAL WIDTH is required and to describe geometrically a square or rectangular prism, or a solid with six faces, each of which is a parallelogram. (Ref ISO standards).

Parallelism—Parallelism is the condition of a surface or center plane equidistant at all points from a datum plane, or an axis, equidistant along its length from one or more datum planes or a datum axis. Symbol: //.

Perpendicularity—Perpendicularity is the condition of a surface, center plane, or axis at a right angle to a datum plane or axis. Symbol: ⊥ .

Position Tolerance—A position tolerance (formerly called true position tolerance) defines a zone within which the axis or center plane of a feature is permitted to vary from true (theoretically exact) position. Symbol: ⊕ .

Profile of a Line—Profile of a line is the condition permitting a uniform amount of profile variation, either unilaterally or bilaterally, along a *line* element of a feature. Symbol: ⌒ .

Profile of a Surface—Profile of a surface is the condition permitting a uniform amount of profile variation, either unilaterally or bilaterally, on a *surface*. Symbol: ⌓ .

Projected Tolerance Zone—A projected tolerance zone is a tolerance zone applied to a hole in which a pin, stud, screw, or bolt, etc., is to be inserted. It controls the perpendicularity of the hole to the extent of the projection from the hole and as it relates to the mating part clearance. The projected tolerance zone extends *above* the surface of the part to the functional length of the pin, screw, etc., relative to its assembly with the mating part. Symbol: Ⓟ.

Regardless of Feature Size (RFS)—Regardless of feature size is the term used to indicate that a geometric tolerance or datum reference applies at any increment of size of the feature within its size tolerance. (Past Symbol: Ⓢ).

Resultant Condition—The variable boundary generated by the collective effects of a size feature's specified MMC or LMC material condition, the geometric tolerance for that material condition, the size tolerance, and the additional geometric tolerance derived from the feature's departure from its specified material condition.

Resultant Condition Size—The actual value of the resultant condition boundary.

Roundness (See Circularity).

Runout—Runout is a composite tolerance used to control the functional relationship of one or more features of a part to a datum axis. Symbols: ↗ or ⌰↗

Runout Tolerance—Runout tolerance states how far an actual surface or feature is permitted to deviate from the desired form implied by the drawing during full rotation of the part on a datum axis. There are two types of runout: circular runout and total runout.

Simulated Datum—A point, axis, or plane established by processing or inspection equipment such as the following simulators: a surface plate, a gage surface, or a mandrel.

Simulated Datum Feature—The practical embodiment of the datum during manufacture and inspection.

Size Feature—A size feature is a three dimensional feature, such as a hole or pin (*not* a single plane surface).

Size Tolerance—A size tolerance states how far individual features may vary from the desired size. Size tolerances are specified with either unilateral, bilateral, or limit tolerancing methods.

Size, Actual—The general term for the size of a produced feature. This term includes the actual mating size and the actual local size.

Size, Actual Local—The value of any individual distance at any cross section of a feature.

Size, Actual Mating—The dimensional value of the actual mating envelope.

Size, Limits of—The specified maximum and minimum sizes.

Size, Nominal—The designation used for purposes of general identification.

Size, Resultant Condition—The actual value of the resultant condition boundary.

Size, Virtual Condition—The actual value of the virtual-condition boundary.

Specified Datum—A specified datum is a surface or feature identified with a datum feature symbol or note.

Squareness—See Perpendicularity.

Straightness—Straightness is a condition where an element of a surface or an axis is a straight line. Symbol: ⎯.

Symmetry—Symmetry is that condition where the median points of all opposed or correspondingly-located elements of two or more feature surfaces are congruent with the axis or center plane of a datum feature.

Tangent Plane—A theoretically exact plane derived from the true geometric counterpart of the specified feature surface.

Tolerance—A tolerance is the total amount by which a specific dimension may vary. The tolerance is the difference between the maximum and minimum limits.

Tolerance, Bilateral—A tolerance in which variation is permitted in both directions from the specified dimension.

Tolerance, Unilateral—A tolerance in which variation is permitted in one direction from the specified dimension.

Total Indicator Reading (TIR) (FIR) (FIM)—See FIM, Full Indicator Movement.

Total Runout—Total runout is the simultaneous composite control of all elements of a surface at all circular and profile measuring positions as the part is rotated through 360°. Symbol: ⟋⟋

Transition Fit—A transition fit is one having limits of size so prescribed that either a clearance or an interference may result when mating parts are assembled.

True Geometric Counterpart—The theoretically perfect boundary (virtual condition or actual mating envelope) or best-fit (tangent) plane of a specified datum feature. The simulated datum feature, established by the datum feature simulator (a gage surface, mandrel, collet, surface plate) is used as the practical embodiment (physical representation) of the datum for manufacture and inspection.

True Position—True position is the theoretically exact location of a point, line, or plane of a feature established by basic dimensions in relationship with a datum reference(s) or other feature(s).

Unilateral Tolerance—A unilateral tolerance is a tolerance in which variation is permitted in one direction from the specified dimension, e.g., $1.400 \, ^{+.000}_{-.005}$ or $20 - ^{0}_{0.1}$.

Virtual Condition—A constant boundary generated by the collective effects of a size feature's specified MMC or LMC material condition and the geometric tolerance for that material condition.

Virtual Condition Size—The actual value of the virtual condition boundary.

INDEX